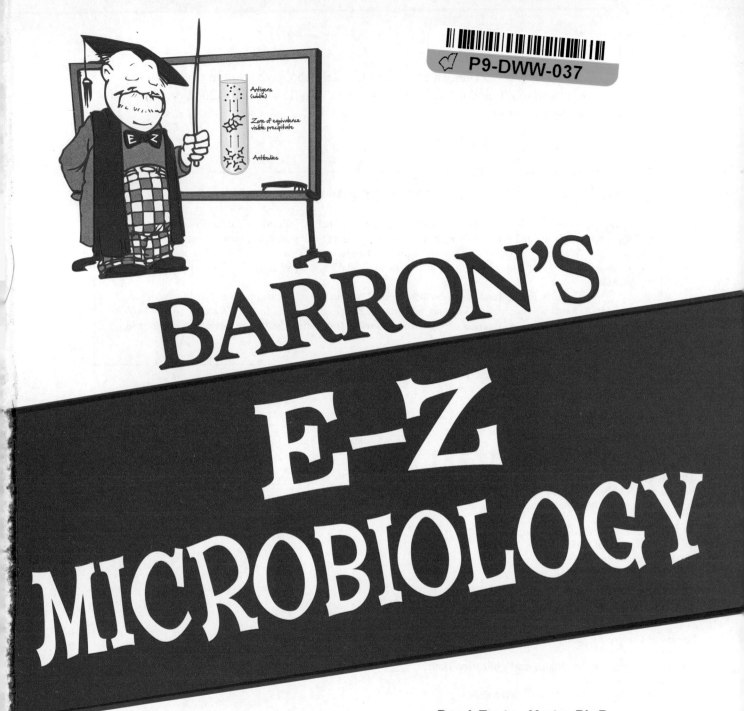

BARRON'S
E-Z
MICROBIOLOGY

René Fester Kratz, Ph.D.
Professor, Department of Biology
Everett Community College

BARRON'S

Dedication

This book is dedicated to Dan, Hueston, and Dashiel, who gave up their time with me so that I could work on it, and to my mother, Annette, who flew in and helped run the household while I was busy.

Better Grades or Your Money Back!

As a leader in educational publishing, Barron's has helped millions of students reach their academic goals. Our E-Z series of books is designed to help students master a variety of subjects. We are so confident that completing all the review material and exercises in this book will help you, that if your grades don't improve within 30 days, we will give you a full refund.

To qualify for a refund, simply return the book within 90 days of purchase and include your store receipt. Refunds will not include sales tax or postage. Offer available only to U.S. residents. Void where prohibited. Send books to **Barron's Educational Series, Inc., Attn: Customer Service** at the address on this page.

All inquiries should be addressed to:
Barron's Educational Series, Inc.
250 Wireless Boulevard
Hauppauge, New York 11788
www.barronseduc.com

ISBN: 978-0-7641-4456-1

Library of Congress Catalog Card No. 2010033288

Library of Congress Cataloging-in-Publication Data
Kratz, René Fester.
  E-Z microbiology / René Fester Kratz.
    p. cm.
  Previous ed: Microbiology the Easy Way, 2005.
  Includes bibliographical references and index.
  ISBN: 978-0-7641-4456-1 (alk. paper)
  ISBN-10: 0-7641-4456-1 (alk. paper)
  1. Microbiology—Outlines, syllabi, etc. 2. Microbiology—Examinations,
questions, etc. I. Title.
  QR62.K73   2011
  579.076—dc22
                                        2010033288

PRINTED IN THE UNITED STATES OF AMERICA
9 8 7 6 5 4 3 2

CONTENTS

1 The Microbial World. 1
Microorganisms and Humans 3
The Science of Microbiology 3

2 Cellular Chemistry 13
Atoms. 14
Chemical Bonds and Molecules 17
Chemical Reactions. 20
Acids, Bases, and pH. 21
The Molecules of Life 22

3 Observing Microbes
Through a Microscope 39
Principles of Microscopy. 41
Types of Microscopy 43
Preparing Specimens for Light Microscopy 46

4 Structure and Function of
Microbial Cells. 55
Prokaryotic vs. Eukaryotic Cells. 56
The Plasma Membrane 58
The Prokaryotic Cell 61
The Eukaryotic Cell. 69

5 Microbial Metabolism. 85
Fundamentals of Metabolism 86
An Overview of Microbial Metabolism 89
Fundamental Metabolic Pathways of
Heterotrophic Metabolism 90
Fundamental Metabolic Pathways of
Autotrophic Metabolism. 101
Anabolic Pathways 106

6 Microbial Growth and
Reproduction 115
Bacterial Growth 116
Conditions That Affect Growth 118
Growth in the Laboratory 123

7 Microbial Genetics. 131
DNA Replication. 133
RNA Synthesis. 137
Protein Synthesis. 139
Gene Regulation 144
Mutation. 150
Gene Transfer 157

8 Molecular Biology and
Recombinant DNA Technology . 173
History of Molecular Biology and Recombinant
DNA Technology 174
Tools of Molecular Biology 175
Applications of Recombinant DNA Technology . 187

9 Classification of Microorganisms . 195
Phylogeny 196
History of Classification Systems 198
Taxonomy 200
Modern Methods Used to Classify and
Identify Bacteria 202

10 The Bacteria 211
Phylum Cyanobacteria. 216
Phylum Proteobacteria. 216
Phylum Firmicutes. 227
Phylum Actinobacteria 230
Phylum Chlamydiae 232
Phylum Spirochaetes. 233

11 The Archaea 237
Phylum Crenarchaeota. 238
Phylum Euryarchaeota. 240

12 Eukaryotic Microorganisms. . . . 245
Protista. 248
Fungi 259

13 Viruses. 267
History of Viruses 268
Properties of Viruses 270
Structure of Viruses. 271
Viral Multiplication Cycles 272
Cultivation of Viruses 278
Identification and Classification of Viruses . . . 278
Other Acellular Infectious Agents 279

14 Environmental Microbiology . . . 285
Microbial Ecosystems 286
Biogeochemical Cycles. 288
Microenvironments and Gradients 293
Microbial Habitats. 295
Microorganisms and Water Quality. 299
Sewage Treatment 301
Bioremediation 304

15 Applied Microbiology 309
History of Fermentation 310
Food Microbiology 311
Microorganisms and Food Production 315
Industrial Microbiology. 318

16 The Study of Disease 325
History of the Study of Disease 327
Pathology. 328
Normal Microbiota 329
Classifying Infectious Disease 330
Development of Disease 331
Spread of Infections. 332
Nosocomial Infections. 334

17 Epidemiology and Emerging Infectious Disease............339
History of Epidemiology..................341
The Science of Epidemiology.............341
Origins of Emerging Infectious Disease......343
The Virulence of Epidemics..............350

18 Innate Immunity.............357
Anatomic and Physiologic Barriers to Infection . 358
Types of White Blood Cells..............359
Phagocytosis.........................360
Inflammation........................362
Fever..............................364
Complement.........................365
Interferon..........................369

19 Adaptive Immunity...........377
Types of Immunity....................378
The Immune System...................379
Antigens............................381
Antigen Presentation and T-cell Activation....383
B-cell Activation......................386
Antibodies..........................388
The Secondary Immune Response.........391
Self-Tolerance........................393

20 Pathogenicity of Microorganisms . 401
Ability to Invade Tissues.................402
Evasion of Host Defenses................405
Damage to the Host....................406

21 Practical Applications of Immunology.................417
History of Vaccination...................418
Type of Immunity Triggered by Vaccines.....419
Types of Vaccines.....................419
The Effect of Immunization on Populations...422
Diagnostic Immunology.................428

22 Control of Microbial Growth...441
History of the Control of Microorganisms....442
Terminology.........................443
Factors Affecting the Success of Microbial
    Control Agents....................444
Physical Methods of Microbial Control......445
Chemical Control of Microorganisms.......448
Resistance to Control Efforts.............452

23 Antimicrobial Drugs..........459
History of the Discovery of
    Antimicrobial Drugs.................460
Important Considerations for Targeting
    Microbial Infections.................461
Mechanism of Action of Antimicrobial Drugs . 463
Determining the Level of
    Antimicrobial Activity...............471
Resistance to Antimicrobial Drugs.........472

24 Infectious Diseases of the Skin and Eyes................481
The Skin............................482
Mucous Membranes...................483
Bacterial Diseases of the Skin.............484
Viral Diseases of the Skin................486
Microbial Diseases of the Eye............487

25 Infectious Diseases of the Respiratory System...........491
The Respiratory System.................492
Bacterial Diseases of the Respiratory System. . 494
Viral Diseases of the Respiratory System.....495

26 Infectious Diseases of the Digestive System.............499
The Digestive System...................500
Bacterial Diseases of the Digestive System ...501
Viral Diseases of the Digestive System......504
Protozoan Diseases of the Digestive System . . 505

27 Infectious Diseases of the Urinary and Reproductive Systems.....509
The Urinary System....................510
The Reproductive System................511
Bacterial Diseases of the Urinary System.....512
Bacterial Diseases of the
    Reproductive System................513
Viral Diseases of the Reproductive System ...514
Protozoan Diseases of the
    Reproductive System................515
Fungal Diseases of the Reproductive System. . 515

28 Infectious Diseases of the Cardiovascular and Lymphatic Systems...........519
The Cardiovascular and Lymphatic Systems . . 520
Bacterial Diseases of the Cardiovascular and
    Lymphatic Systems.................523
Viral Diseases of the Cardiovascular and
    Lymphatic Systems.................526
Protozoan Diseases of the Cardiovascular and
    Lymphatic Systems.................527

29 Infectious Diseases of the Nervous System..............531
The Nervous System...................532
Bacterial Diseases of the Nervous System 534
Viral Diseases of the Nervous System.......535
Protozoan Diseases of the Nervous System ...536

Index........................540

Preface

To study microbiology is to enter another world, the world of the unseen. The microbial world is populated by countless, fascinating creatures that interact with each other to form a complex web of life that is absolutely essential to the survival of all life on Earth. When you study microbiology, your vision of life on Earth will expand in profound ways. You will learn that microbes are not just *germs*, things to get rid of because they make us sick, but that they can actually help keep us well, that they are used to produce food and medicines, and that they are essential recyclers of human wastes. You will get a glimpse of some life forms that seem more bizarre than anything you have seen in a science fiction movie or episode of *Star Trek*. By learning how these microbes live, grow, and reproduce, you will come to a deeper understanding of how humans do all these things and how all life on Earth is connected.

I have tried to write *E-Z Microbiology* in a style that is readable and approachable for both students of microbiology and for anyone who wants to know more about the microbial world. To keep the text readable, I have kept terminology to a minimum and tried, whenever possible, to show the connections between the microbial world and the everyday life of people. For students, I have tried to include the topics that are most frequently covered in college microbiology classes, with a particular emphasis on topics that are relevant to students in the allied health professions, such as nursing and dental hygiene.

If you have comments or questions about anything in this book, you are welcome to contact me by mail at the address below or by e-mail at rkratz@everettcc.edu. I hope you enjoy your voyage into the microbial world in *E-Z Microbiology*.

René Fester Kratz, Ph.D.
Department of Biology
Everett Community College
2000 Tower Street
Everett, WA 98201

How to Use This Book

Microbiology is a challenging subject. Among students in the allied health professions, such as nursing and dental hygiene, it often has the reputation of being the hardest of the biology prerequisites. I think there are a number of reasons that microbiology has earned this reputation. First, microbes are less familiar to people than their own bodies. To study microbes is to study cells and the complex chemical processes that go on in cells. These include subjects that are often very challenging for students, such as metabolism and molecular genetics. Because you often will not be able to relate these subjects to your everyday experience, you will need more time to learn these topics. Second, success in microbiology often requires mastery of process and concept, not just terminology. In other words, you have to learn the "how" and "why" in addition to the "what." Finally, many microbiology instructors emphasize the application of course content to solve problems. This will require you to use your critical thinking skills. Of course, you should figure out the expectations of your particular instructor, but the following are my recommendations for using this book to help you succeed in a traditional microbiology course. My recommendations are based on research about the human brain and learning as well as on my own experiences as an undergraduate and in graduate school.

1. **Set aside enough study time.** The rule of thumb is two hours outside of class for every hour inside of class. However, microbiology may require even more time than this, particularly if you are someone who has to work harder to succeed in science courses. If you had difficulty in a course such as chemistry or cell biology, you should be prepared to put in extra time for microbiology.

2. **Schedule your study time so you have some time available each day.** Studies have shown that our brains retain information better when we take it in smaller chunks, not when we try to cram it in all at once. Longer study sessions on weekends are fine, but you also need some time every day for review. During longer study sessions, take a short break every hour to keep your brain working at its best.

3. **Pre-read chapters before going to class.** Read through the material that will be covered in lecture before you go. Do not worry if you do not understand everything. You are not trying to memorize the material on this first read; you are just trying to familiarize yourself

with it. Pay particular attention to bolded terminology and look at the figures. Note any questions that you have in the margin of the book. As you participate in class, be sure you ask questions about things that confuse you.

4. **Stay alert and take good notes in class.** Sit in the front where you can be seen (that is where most of the A students sit). Because you pre-read the chapter, your brain should be ready to think about the material. Keep your attention focused in class, follow the discussion, and take good notes. Try to write things in your own words. Do not just write down what the instructor writes down. If the instructor tells a story that helps you understand something, jot a note in the margin of your notebook that will remind you of the example. This will help you when you go back over your notes.

5. **As *soon* as you can after class, go back over the material in your notes, using the book to help you review.** Again, studies have shown that going over things again while they are still fresh in your brain is the *best* way to learn them. Each day, you should go back over the notes for the day. If there is something you don't understand, re-read the relevant section in the book and see if that clarifies the material. If it does, make notes about what you read in the book in your notebook so that your notes are now clear to you. If, after re-reading the book, you still do not understand, write down your question and go talk to your instructor.

6. **Review the material.** If you use the suggestions above during the week, you should be staying on top of course material. On the weekends, go back over the material from the week and begin to review and apply it. First, make flash cards for any new terminology. Practice with the cards until you know the terms. When you know a term, remove the card from the stack. Second, use the review questions at the end of the chapter that are relevant to what you did in class and try to answer them. It is best to try to write your answers down. Try to answer the questions without looking at your notes to see what you remember, then go to your notes to fill in things you forgot. Third, use the self-test questions and problems at the back of the chapter to see how well you remember and can apply the material. If your instructor has suggested or assigned other problems, do those too. If you do not understand how to solve a problem, go back to the book or your notes and look for examples. Practice solving the problems until you are confident that you can solve each type. The more you practice and test yourself, the better you will remember and understand the material.

7. **Studying for tests.** Go back over your notes on any especially difficult subjects. Run through your flash cards again, making sure you are solid on the terminology. For any processes that you were asked to learn, practice either saying or writing all the steps of the process in order, including what happens in each step. Review the problems you have solved and try making up new practice problems and solving them. On the night before the test, go to bed on time and get a good night's rest. On the morning of the test, eat a good breakfast and resist the urge to try to do last minute cramming. If you have studied well, you are ready. It is better to be relaxed for the test than to try to squeeze in one more fact.

When you go to take the test, breathe deeply a few times to relax your body. If you are nervous and feel like you cannot remember anything, scan the test until you find a question that you can answer. Answer that one first. This will get your brain going and help you relax.

8. **Studying in groups.** Most of the things I know well, I know because I have taught them to someone else. In my opinion, there is no better way to figure out if you really understand something than to try and explain it to someone else. You may think you "get it" in your head, but unless you can also put it together in words you probably really do not. If it is at all possible, try to hook up with a few people from your class for weekly or pre-test study sessions. Make sure you find a good group of people who are going to focus and not just chat about their social lives. If you cannot form a study group, explain things out loud to your cat, dog, significant other, or just yourself. Although you will not have anyone to check your answers, the exercise of putting things into words is still valuable.

The Microbial World

WHAT YOU WILL LEARN

This chapter introduces the science of microbiology. As you study this chapter, you will:

- discover the importance of microorganisms to humans;
- explore the early history of microbiology;
- learn about the types of things microbiologists study.

SECTIONS IN THIS CHAPTER

- Microorganisms and Humans
- The Science of Microbiology

When you look at the back of your hand, what do you see? Do you see the grooves in your skin, the small hairs? What if you could sharpen your eyesight for a moment, zoom in, and see your skin magnified 100 times, 1000 times, 3000 times? The surface of your skin would look like a landscape, a vast uneven floor made of irregular slabs. The hairs would become giant trees jutting out of the surface of the landscape. And as you zoomed in, you would see them, tucked in the grooves and scattered across the landscape of your skin like small boulders (Figure 1.1). Bacteria are on your skin, in your ears, your nose, and your mouth, and if you looked inward, you would see them again, lining your intestines and genital tract. In fact, for every one of your own body cells you possess, there are about ten bacterial cells living in you or on you.

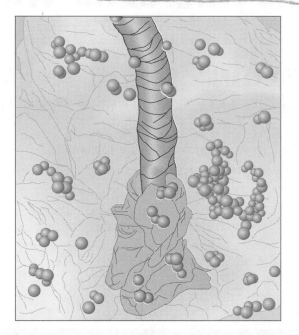

FIGURE 1.1. Staphylococci on the surface of the skin and around a hair follicle.

With an electron microscope, it is possible to see bacteria scattered on the surface of human skin. The bacterial cells in this drawing are round. The skin cells are large and flat. A hair is seen projecting upward out of the skin.

Before you run screaming for the soap or think you have stumbled into a low-budget horror flick, consider this: Most of the bacteria that live on your body are beneficial to your health. They are part of your **normal microbiota**, the normal residents of your body. They help stop other, potentially harmful bacteria from taking up residence. Some of the bacteria that live in your intestines even make necessary vitamins for you that you cannot make yourself. If you have ever had intestinal upsets or a yeast infection after taking antibiotics, you have seen firsthand the negative effect of harming your normal microbiota. You need these bacteria and benefit from their presence. In fact, you might be surprised to consider all the reasons people need **microorganisms** like bacteria, and all the ways we have found to use them in our daily lives.

Microorganisms and Humans

Microorganisms, or **microbes**, are any organisms that are too small to see with the naked eye. They affect human life in almost every way you could imagine. No wine-tasting party could be complete without wine and cheese, both of which are made with microbes. And how could anyone celebrate Oktoberfest without beer and sauerkraut? Any food you can think of that is pickled was made with microbes, as were foods like bread and yogurt. If you spill some food on your clothes and throw them in the washing machine, take a good look at your laundry detergent—it just might contain microbial enzymes to help degrade the greasy compounds on your dirty laundry. We also use microbes to help clean up our sewage and even our oil spills.

In the environment, microbes serve a vital role as **decomposers**, breaking down dead matter so that it can be recycled. Microbes perform over half of the world's photosynthesis, using sunlight energy to convert carbon dioxide and water to glucose and oxygen. These photosynthetic microbes then serve as food for other organisms, forming the basis for food webs. Some microorganisms can take gaseous nitrogen from the atmosphere and convert it to forms usable by plants and other organisms. This process, called **nitrogen fixation**, is essential to life on Earth. Without nitrogen fixation by microbes, other kinds of life like plants and animals would not have enough nitrogen to build proteins or molecules like DNA. In fact, microbes are so important to the recycling of nutrients and energy on planet Earth, no other life could exist without them.

> **REMEMBER**
> Microbes are far more beneficial than they are harmful to humans and life on Earth.

And then, of course, there is disease. Everyone knows that some microbes make you sick. But did you know that microbes can also help you get well? The antibiotics we use to fight infection are made by microbes to fight other microbes. We have simply taken these compounds and used them for our own purposes. We have even modified some bacteria that would not normally produce useful drugs by adding human genes to them so that they will make things like insulin for people with diabetes. The role of microbes in disease is such common knowledge today, it is hard to imagine that, as little as 350 years ago, people had no idea that microbes even existed, much less knew of their important role in human disease.

The Science of Microbiology

THE DISCOVERY OF MICROBES

For the microbial world to be discovered, of course, the microscope had to be invented. People have used curved lenses to magnify things since the days of the Ancient Romans, but it was not until the 1600s that lenses powerful enough to reveal microbes were invented. Using a microscope, an Englishman named Robert Hooke examined thin slices of cork, which is a plant tissue taken from the bark of trees. When he looked at cork under the microscope, he saw lots of tiny chambers (Figure 1.2). The chambers reminded him of the small rooms that monks lived in, so he named the chambers **cells**. The cells were visible even though the tissue was dead because of the strong wall that remains behind even after plant cells have died.

FIGURE 1.2. Cells in cork as drawn by Hooke.

Hooke's discoveries inspired a Dutch cloth merchant named Anton van Leeuwenhoek. Van Leeuwenhoek, who used lenses to inspect cloth, turned his talents to microscope construction. Through means that are not fully known today, van Leeuwenhoek made a finer quality lens than anyone else at the time, achieving magnifications of about 270 times. As a hobby, he turned his microscope upon all aspects of his daily life, looking at samples from lake water, teeth scrapings, minerals, and even blood. To his initial surprise, almost everywhere he looked he found life, tiny microscopic life that no one had ever seen before. He wrote long letters about his discoveries to the Royal Academy of London, which was the premiere scientific organization of the time. His letters were accompanied by drawings of his *animalcules*, drawings that can be recognized today as bacteria and other microbes (Figure 1.3).

> **REMEMBER**
> Anton van Leeuwenhoek was the first person to see living cells.

Van Leeuwenhoek's discoveries shook the human view of the natural world. People traveled from all over Europe to peer through van Leeuwenhoek's microscopes and see microbes firsthand.

Before the discovery of microbes, people thought that disease was caused by sin or by an imbalance in the *humors* of the body (fire, water, earth, and air). The new awareness of van Leeuwenhoek's animalcules laid the foundation for future work that connected microbes with disease (see Chapter 16) and set the stage for the development of treatments that targeted real causes of diseases (see Chapter 23).

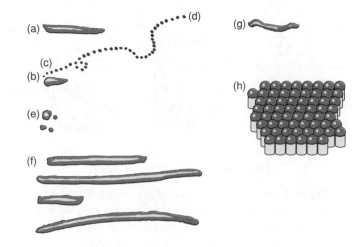

FIGURE 1.3. van Leeuwenhoek's "animalcules."

Drawings in a letter from Anton van Leeuwenhoek to the Royal Academy of London dated September 17, 1683, that described his microscopic observations of the plaque of his own teeth. (a) The largest type of microbe seen by van Leeuwenhoek in the plaque, which he said were "strong and nimble." (b)–(d) A microbe that moved, sometimes spinning, sometimes taking a course as shown by the dotted lines from (b) to (d). (e) Very small microbes that van Leeuwenhoek said moved very quickly and were either round or oval. (f) Unmoving "streaks or threads" seen by van Leeuwenhoek. (g) Another type of the microbe shown at (b–d). (h) Clusters of round microbes seen by van Leewenhoek.

THE DEBATE ABOUT SPONTANEOUS GENERATION

The discovery of microbes also tied into a scientific debate about the origin of life. At the time that microbes were discovered, there was a common belief that living things could arise from nonliving matter. For example, people thought that frogs could come to life from mud, or that maggots grew from decaying meat. There was even the idea that if you mixed wheat with soiled underwear and left it alone for a while, it would turn into mice! The idea that living things could be born from nonliving things is called **spontaneous generation**.

In 1668, an Italian physician named Francesco Redi did some experiments to test the idea of spontaneous generation, specifically to see if maggots could spontaneously arise from decaying meat. He placed meat into different containers, some that were sealed, some that were completely open, and some that were covered with netting (Figure 1.4). He let the meat sit in the containers and found that only the meat in the open container developed maggots. From this he concluded that the meat alone was not enough to generate maggots and that maggots actually came from flies that laid eggs in the meat. His experiment argued against the idea of spontaneous generation and was convincing to many people at the time. Some people, however, agreed that perhaps spontaneous generation did not occur for larger creatures such as maggots, but that it did occur for microscopic organisms like those seen by van Leeuwenhoek.

FIGURE 1.4. Redi's experiment.

Meat was placed in three types of containers: containers that were open to the air, containers that were sealed, and containers that were covered with netting. The meat was allowed to rot and observed for the presence of maggots. Maggots appeared only in the meat in the open container.

This led to a series of experiments by different people, some who seemed to support the idea of spontaneous generation of microbes and others who argued against it. One of the people who supported the idea was an English clergyman named John Needham. In 1745, he did an experiment in which he boiled broth, poured it into flasks, and then sealed the flasks. After a while, the broth grew cloudy as microorganisms grew in it. Although Needham concluded from this that microorganisms had spontaneously arisen in the broth, an Italian scientist named Lazzaro Spallanzani thought Needham's experiment was flawed and that the microorganisms had actually entered the broth after it was boiled but before the flasks were sealed. In 1765, he did an experiment in which he placed broth in flasks, sealed them, and then boiled the broth. This broth did not produce microbes, which Spallanzani said demonstrated that spontaneous generation did not occur. However, one of the other ideas popular around this time was that air contained some sort of "vital force" necessary to life. Because Spallanzani had excluded air by sealing his flasks, some people argued that he had kept the necessary vital force away from the broth.

> **REMEMBER**
> Spontaneous generation, the idea that living things could arise from nonliving things, was disproved by Pasteur's experiment with swan-necked flasks.

The person who finally settled the question of spontaneous generation was the French scientist Louis Pasteur. In 1861, Pasteur repeated the boiled broth experiments but added a clever improvement to settle the question of the role of air. He heated the neck of some flasks to make the glass flexible, then stretched the necks out into an S-curve (Figure 1.5). He then put broth into the flasks, boiled the broth and allowed it to cool. No microbes developed, even though the very tip of the S-necked flask was open to the air. The twist in the neck of the flask prevented air from flowing into the flask and carrying microbes in with it. Only if he tipped the flask so that broth flowed into the neck and back again was he able to generate any microbes. Thus he showed that broth alone did not spontaneously give rise to microbes, even if air was present.

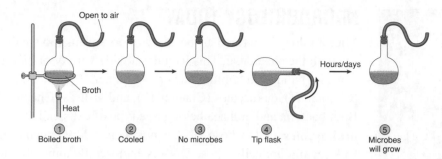

FIGURE 1.5. Pasteur's swan-necked flask experiment.

Pasteur heated the neck of a glass flask and then extended the neck into an S-shape. Although the neck was still open to the air, its long curves prevented air from moving into the main part of the flask. Pasteur heated broth in the flask to kill all the microbes and then allowed the flask to cool. No microbes grew in the broth. If he tipped the flask and allowed broth to run into the neck and then straightened the flask, microbes grew in the broth.

MAJOR IDEAS IN MICROBIOLOGY

Experiments like these during the early years of microbiology resulted in many contributions to the understanding of human health and disease, food production, and control of microbial growth. Some of the major accomplishments are listed in Table 1.1. Many of these will be discussed further in subsequent chapters.

TABLE 1.1. Major Ideas in the History of Microbiology

Date	Name	Accomplishment
1665	Robert Hooke	Saw first cells in plant material
1673	Anton van Leeuwenhoek	Saw first living cells
1798	Edward Jenner	Used scrapings from cowpox blisters to vaccinate a boy against smallpox
1840	Ignaz Semmelweis	Proposed that handwashing could prevent the spread of infection (childbirth fever)
1861	Louis Pasteur	Disproved spontaneous generation
1864	Louis Pasteur	Developed pasteurization
1867	Joseph Lister	Used aseptic techniques during surgery
1876	Robert Koch	Supported the germ theory of disease, the idea that disease is caused by microbes, by his work with anthrax. Established Koch's postulates, the steps necessary to prove a particular organism causes a particular disease.
1882	Ilya Ilich Metchnikoff	Described phagocytosis
1891	Paul Ehrlich	Proposed that antibodies are responsible for immunity
1912	Paul Ehrlich	Discovered salvarsan, the first chemotherapeutic agent for a bacterial disease (syphilis)
1928	Alexander Fleming	Discovered penicillin, the first antibiotic

REMEMBER
The heart of microbiology is the study of microbes–organisms that are too small to see with the naked eye–but microbiology includes the study of many related topics such as the immune system and the importance of microbes to agriculture and industry.

MICROBIOLOGY TODAY

Microbiology is defined as the study of organisms too small to see with the naked eye. This includes a wide variety of organisms, including the familiar bacteria (Chapter 10), the archaea (Chapter 11), eukaryotes (Chapter 12), and viruses (Chapter 13). Both bacteria and archaea belong to a type of cells called **prokaryotes**, cells whose DNA is not enclosed by a membrane. **Eukaryotes** are cells whose DNA is enclosed by a membrane; this includes organisms such as plants, animals, and fungi. **Viruses** are not cells at all and can be as simple as DNA surrounded by protein. Examples and brief descriptions of some common types of microbes are listed in Table 1.2. Although some animals, such as small crustaceans and insect larvae, are microscopic, animals are not traditionally considered microbes, so they are not included in the table.

TABLE 1.2. Common Types of Microbes

Type of Microbe	Name of Group	Distinctive Characteristics	Examples
Prokaryote	Bacteria	Cell wall of peptidoglycan	*Escherichia coli*, *Streptococcus*
Prokaryote	Archaea	No cell wall or cell wall not of peptidoglycan; often found in extreme environments such as hot springs, salt lakes, and thermal vents	*Thermoplasma*, *Halobacterium*
Eukaryote	Protozoa	Usually single celled, non-photosynthetic, often motile	Diatoms, ciliates, dinoflagellates (cause red tide)
Eukaryote	Algae	Photosynthetic	Green algae
Eukaryote	Fungi	Grow as fine threads of cells (filamentous) or as single cells (yeast); not photosynthetic	Bread mold, mushrooms, yeast
Viruses		Made of DNA or RNA surrounded by protein	Flu virus, polio virus

The science of microbiology spans not just the study of microbes themselves but also their impact on humans. This includes topics ranging from food safety to the development of useful products to the treatment of disease. The major disciplines of microbiology are listed in Table 1.3.

TABLE 1.3. Areas of Study Within Microbiology

Area	What Is Studied
Bacteriology	Bacteria
Protozoology	Protozoa
Phycology (Algology)	Algae
Mycology	Fungi
Parasitology	Parasites
Immunology	Immune system
Medical microbiology	Microorganisms of medical significance
Agricultural microbiology	Microorganisms of agricultural significance
Industrial microbiology	Microorganisms involved in production of useful products or cleanup of wastes
Microbial ecology	Interactions of microorganisms with each other and with their environment

REVIEW EXERCISES FOR CHAPTER 1

1. The biologist Carl Woese said, "Microbial life on this planet would remain largely unchanged were all plant and animal life eliminated, but the elimination of microbial life itself would lead in very short order to a completely sterile planet." Explain this statement using what you know about the importance of microbes.

2. Describe five ways microbes are beneficial to humans.

3. Describe the contributions of Hooke and van Leeuwenhoek to the science of microbiology.

4. State the theory of spontaneous generation.

5. Describe one experiment that supported the theory of spontaneous generation and one that did not support it.

6. Describe the contribution of Pasteur to the debate about spontaneous generation.

7. Describe the contributions of the following people to the science of microbiology: Jenner, Semmelweis, Pasteur, Lister, Koch, Metchnikoff, Ehrlich, and Fleming.

8. What types of organisms are considered to be microbes?

9. What are the unique characteristics of each group of microbes?

10. What are the major areas of study in microbiology and what does each of them cover?

SELF-TEST

1. All of the following are beneficial effects of microbes for humans **except**:

   A. Clean up pollution
   B. Are involved in food production
   C. Make useful enzymes
   D. Cause disease
   E. Recycle nutrients

2. True or False. Anton van Leeuwenhoek was the first person to see cells.

3. John Needham boiled broth, poured it into flasks, and then sealed the flasks. After a while, the broth grew cloudy as microorganisms grew in it. The microorganisms grew because

   A. microorganisms can spontaneously arise from broth.
   B. boiling does not kill microorganisms.
   C. the flasks and the air had microorganisms in them.
   D. the theory of spontaneous generation is correct.
   E. all of the above are correct.

4. Match each person with their contribution to the science of microbiology.

Person		Contribution
1. Edward Jenner	A.	Developed vaccination against smallpox
2. Louis Pasteur	B.	Saw the first cells
3. Robert Hooke	C.	Developed pasteurization
4. Joseph Lister	D.	Discovered first chemotherapeutic agent for a bacterial disease
5. Ignaz Semmelweis	E.	Discovered penicillin
6. Paul Ehrlich	F.	Described phagocytosis
7. Ilya Ilich Metchnikoff	G.	Created a protocol for proving a particular disease is caused by a particular organism
8. Robert Koch	H.	Saw the first living cells
9. Alexander Fleming	I.	Encouraged handwashing
10. Anton van Leeuwenhoek	J.	Developed aseptic techniques for surgery

5. True or False. Bacteria are prokaryotes; archaea are eukaryotes.

6. True or False. Viruses are a type of bacteria.

7. Which of the following are **not** correctly matched?

   A. Algae—photosynthetic
   B. Fungi—grow as threads
   C. Bacteria—no nucleus
   D. Protozoa—photosynthetic
   E. Virus—may contain DNA

8. Which of the following is the study of fungi?

   A. Phycology
   B. Mycology
   C. Protozoology
   D. Bacteriology
   E. Parasitology

Answers

Review Exercises

1. Microbes perform over half the photosynthesis on the planet, transferring energy from the sun to chemical energy that is usable by other organisms. They are also essential to the process of recycling nutrients, which makes nutrients available to other life forms. For example, only microbes are capable of capturing nitrogen from the atmosphere and turning it into usable forms. Without plants and animals, microbes would continue to thrive because they can perform all the essential functions for life. However, without microbes, plants and animals would die because they would not be able to obtain essential elements from the environment.

2. Microbes are used for sewage treatment, in food production, as enzymes used for laundry detergents, to help keep us healthy, to recycle nutrients in the environment, and for many other uses.

3. Robert Hooke made thin slices of cork and looked at them under the microscope. He saw and drew the remains of the plant cells that had made the cork. He was the first person to see evidence of cells. Anton van Leeuwenhoek was the first person to see living cells. He made the whole world aware that microbes existed.

4. The theory of spontaneous generation was the idea that living things could arise from nonliving things. For example, frogs could grow from mud; maggots, from decaying meat; or microbes, from broth.

5. The experiment of John Needham appeared to support the theory of spontaneous generation. He boiled broth, placed it into flasks, and then sealed the flasks. After a while, the broth became cloudy as microbes grew. This appeared as if microbes had spontaneously arisen from the broth. The experiment of Francesco Redi argued against spontaneous generation. He placed meat into three different types of containers. One was open to the air, another was completely sealed, and a third was covered with netting. In the open container, maggots appeared in the meat. In

the sealed and netting-covered containers, however, no maggots appeared. The maggots could not just arise from the meat. The meat needed to be exposed to the air so something (flies) could get to it and bring the maggots (as fly eggs).

6. The experiment of Louis Pasteur with the swan-necked flasks disproved this idea, however. Pasteur heated the neck of flasks and drew them out into an S-shape, leaving the end open to the air. He then placed broth into flasks and boiled them. No microorganisms grew in the flasks because air currents carrying microbes could not get through the S-curve. This disproved the idea of spontaneous generation. Because the tip of the S-neck was open to air, it also answered the criticism that *vital force* needed to be able to reach the broth in order for spontaneous generation to occur.

7. Edward Jenner used scraping from cowpox blisters to vaccinate a boy against smallpox. Ignaz Semmelweis showed that handwashing could prevent the spread of disease. Louis Pasteur did many things, including his work to disprove spontaneous generation and develop pasteurization. Joseph Lister showed that patient survival could be increased by the use of aseptic techniques during surgery. Robert Koch demonstrated how to prove a particular organism causes a particular disease. His method is called Koch's postulates. Ilya Ilich Metchnikoff described the process of phagocytosis. Paul Ehrlich proposed the role of antibodies in immunity and also discovered salvarsan, the first chemotherapeutic agent used against a bacterial disease. Alexander Fleming discovered penicillin, the first antibiotic.

8. Microbes are organisms that are too small to be seen with the naked eye. They include bacteria, archaea, viruses, protists (protozoa and algae), fungi, and even some animals.

9. Bacteria and archaea are both prokaryotic, but bacteria have a peptidoglycan cell wall whereas archaea do not. Protists, fungi, and animals are all eukaryotic. Protists are usually single-celled organisms. If they are photosynthetic, they are commonly called algae; if they are not photosynthetic, they are commonly called protozoa. Most fungi grow as filaments. They are not photosynthetic and commonly grow by breaking down dead and decaying matter. Animals are eukaryotic, motile, and not-photosynthetic.

10. The science of bacteriology studies bacteria. Protozoology is the study of protozoa. Phycology is the study of algae. Mycology is the study of fungi. Parasitology is the study of parasites. Immunology is the study of the immune system. Medical microbiology studies microbes of medical importance. Agricultural microbiology studies microbes of agricultural importance. Industrial microbiology studies microbes that are involved in industrial processes or in waste management. Microbial ecology is the study of microbes and their interactions with each other and their environment.

Self-Test

1. D	4. 1, A; 2, C; 3, B;	5. F	7. D
2. F	4, J; 5, I; 6, D; 7, F;	6. F	8. B
3. C	8, G; 9, E; 10, H		

Cellular Chemistry

WHAT YOU WILL LEARN

This chapter presents the fundamental cellular chemistry that is essential for an understanding of the structure and function of microbial cells. As you study this chapter, you will:

- review the structure of atoms;
- learn how atoms bond together to form molecules;
- compare the four types of macromolecules found in living things.

SECTIONS IN THIS CHAPTER

- Chemical Bonds and Molecules
- Chemical Reactions
- Acids, Bases, and pH
- The Molecules of Life

Cells, whether they are yours or those of *Escherichia coli*, are like tiny little factories. They make things, break things down, send and receive signals, and use energy to do work. Normal function of the human body depends on normal function of the cells that make up the body's tissues and organs. When a microbe enters the body, the products and actions of the microbial cells can disrupt or assist the normal physiology of the body. Thus, to understand health and disease, you must understand cells. And because everything cells are and do depends on their chemistry, to understand cells, you must understand their basic chemistry.

Atoms

Cells are made of molecules, which are themselves made of **atoms**. Atoms are the smallest units of matter, the tiny building blocks of everything we see, smell, touch, and taste. Atoms consist of three types of particles: **protons**, **neutrons**, and **electrons** (Table 2.1). Protons and neutrons are grouped in the center of the atom, which is called the **nucleus**, whereas electrons are in motion around the nucleus (Figure 2.1). Protons and electrons both have electrical charge: Protons are positively charged, and electrons are negatively charged. Because the number of electrons in an atom is always equal to the number of protons, they balance each other out, and atoms have no net charge.

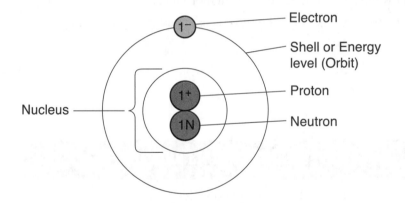

FIGURE 2.1. Structure of the atom.

TABLE 2.1. Atomic Particles and Their Properties

Property	Atomic Particle		
	Proton	**Neutron**	**Electron**
Electrical charge	Positive (+1)	Neutral	Negative (−1)
Mass	Yes (1 Dalton)	Yes (1 Dalton)	Insignificant
Location	Nucleus	Nucleus	Orbiting around nucleus

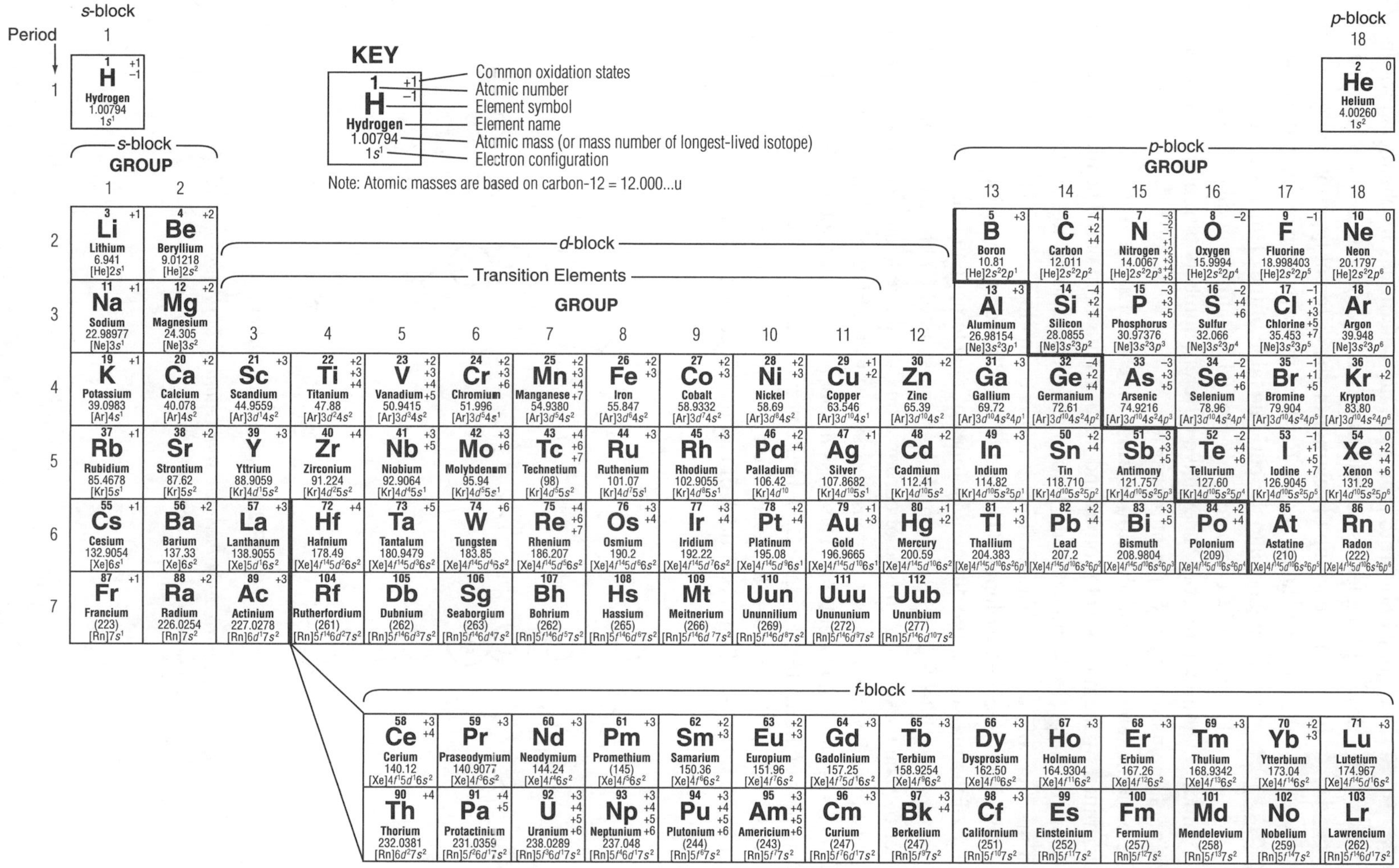

p-block

18	0
2	
He	
Helium	
4.00260	
$1s^2$	

KEY

```
                  Common oxidation states
          +1 ───── Atomic number
   1      -1 ───── Element symbol
   H ──────────── Element name
Hydrogen ──────── Atomic mass (or mass number of longest-lived isotope)
1.00794 ───────── Electron configuration
   1s¹
```

Note: Atomic masses are based on carbon-12 = 12.000...u

s-block
GROUP

p-block
GROUP

d-block
Transition Elements
GROUP

f-block

FIGURE 2.2. The periodic table of the elements.

The number of protons an atom has determines its chemical properties and is called its **atomic number**. Atoms with different numbers of protons make up the **elements**, the different types of matter. All of the elements identified on Earth so far have been organized by their atomic numbers into the periodic table of the elements (Figure 2.2).

For example, compare iron (symbol Fe, atomic number 26) and oxygen (symbol O, atomic number 8). They have different numbers of protons and so are different elements with very different properties.

The periodic table also lists the **atomic mass** for each element. The mass of an atom is determined by the number of protons and neutrons the atom has. The **mass number** is the sum of the protons and neutrons. Within an element, some atoms may have different numbers of neutrons. These different forms of an element, which have different mass numbers, are called **isotopes**. The atomic mass listed in the periodic table is the average of the masses for the different isotopes of the element and takes into account how common each isotope is. For example, carbon (symbol C) has six protons (atomic number 6). The most common type of carbon atom has six neutrons. The mass number of this isotope would be 12 (six protons plus six neutrons, Figure 2.3). However, some carbon atoms have eight neutrons. The mass number for this isotope would be 14 (six protons plus eight neutrons, Figure 2.3). In the periodic table, the atomic mass for carbon is listed as 12.011 because this is an average value for all the carbon atoms on the planet based on the existing proportions of carbon-12, carbon-14, and other isotopes of carbon.

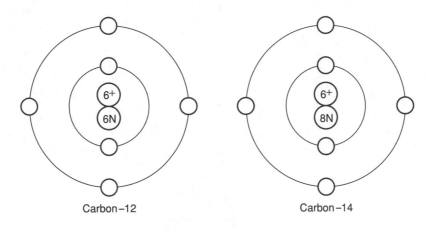

FIGURE 2.3. Isotopes of carbon.

Carbon-12 and carbon-14 have the same number of protons and electrons, but carbon-14 has two more neutrons than carbon-12.

Chemical Bonds and Molecules

In an atom, the number of protons is balanced by the number of electrons. The number of electrons determines how that atom will react toward other atoms. The electrons in an atom are arranged in energy levels, called **electron shells**, around the nucleus. Each shell can hold a specific number of electrons. The shell closest to the nucleus can hold two electrons. The next two shells can each hold eight electrons. Atoms will react with other atoms in a way that allows them to fill their outermost shell. This may involve sharing or taking electrons from other atoms. These types of connections between atoms are called **bonds**.

> **REMEMBER**
> Atoms react with other atoms in order to fill their outermost shell of electrons.

To figure out how an atom will react with other atoms, you must first figure out how many electrons it has in its outermost shell. For example, carbon has six protons. Therefore, it has six electrons. Two electrons would fill the innermost shell, leaving four in the next shell (Figure 2.3), which is filled when it has eight electrons. That means a single carbon atom is seeking four more electrons to fill its outermost shell. Hydrogen atoms have one proton and one electron. The electron is in the innermost shell, which is filled when it has two electrons, so hydrogen atoms are seeking one more electron to fill their outermost shells (Figure 2.4). Carbon and hydrogen can react with each and share electrons to fill their outermost shells. If one carbon atom shares electrons with four hydrogen atoms, all the atoms will have full shells (Figure 2.5). When atoms are joined together by bonds, they form **molecules**. When one carbon atom shared bonds with four hydrogen atoms, a molecule of methane is formed (Figure 2.5). When atoms share electrons, it is called a **covalent bond**. If one pair of electrons is shared, it forms a single bond. If two pairs of electrons are shared, they form a double bond. The four bonds in methane are all single bonds.

> **REMEMBER**
> Atoms form covalent bonds when they share electrons.

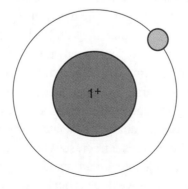

FIGURE 2.4. Structure of the hydrogen atom.

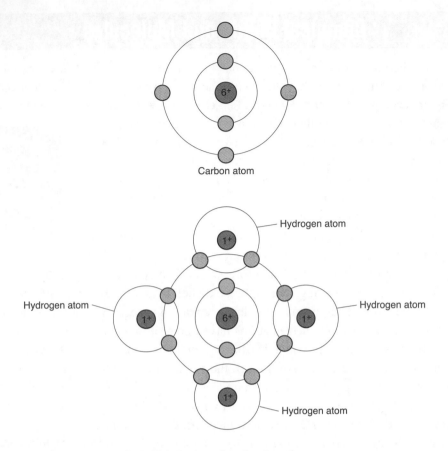

FIGURE 2.5. A molecule of methane gas.

When covalent bonds form between carbon and hydrogen, the electrons are shared equally between the nuclei of the two types of atoms. In some molecules, however, one atom can have stronger pull for electrons than the other. For example, when oxygen and hydrogen share electrons to form water (Figure 2.6), the oxygen has a stronger pull for electrons than does the hydrogen. The electrons spend more time orbiting the oxygen nucleus than they do the hydrogen nucleus. This unequal sharing is called a **polar covalent bond**. Because the electrons spend more time in one part of the molecule than the other, the electrical charge on the molecule is not distributed equally. There is a slight negative charge associated with the oxygen atom and a slight positive charge associated with the hydrogen atoms. Opposite charges attract, so when water molecules are together, the positive ends of one water molecule are attracted to the negative ends of the other water molecules (Figure 2.6). These weak electrical attractions between the water molecules are called **hydrogen bonds**. Hydrogen bonds between water molecules contribute to the unique properties of water like the surface tension that allows bugs to

> **REMEMBER**
> Hydrogen bonds are weak electrical attractions that form between polar groups.

walk on it. Hydrogen bonds are also important in holding large molecules like DNA together. Even though each individual hydrogen bond is weak, if you have a lot of them within a molecule, they can act like molecular Velcro and hold things together.

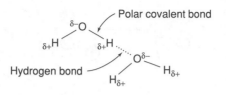
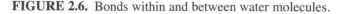

FIGURE 2.6. Bonds within and between water molecules.

The hydrogen and oxygen atoms in a water molecule share electrons unequally. This creates a slight negative charge (δ–) on the oxygen atom and a slight positive charge (δ+) on the hydrogen atoms. Neighboring water molecules are attracted to each other because of the slight opposite charges.

   If an atom that has a very strong pull for electrons interacts with an atom that has a weak pull for electrons, the atom with the strong pull might just take an electron away from the atom with the weak pull. The classic example of this is what happens when sodium and chlorine interact. Sodium is atomic number 11, so it has eleven protons and eleven electrons. Two electrons would fill the innermost shell, eight would fill the next shell, and the final electron would be in the outermost shell (Figure 2.7). This means that sodium either needs to gain seven electrons to fill its outermost shell, or lose one. Chlorine is atomic number 17, so it has seventeen protons and seventeen electrons. Two electrons would fill the innermost shell, eight would go in the next shell, and seven would go in the outermost shell (Figure 2.7). This means that chlorine needs one electron to fill its outmost shell. Chlorine has a much stronger pull for electrons than does sodium, so when they interact, chlorine can pull an electron away from the sodium nucleus. Sodium loses one electron from its outermost shell and becomes a sodium **ion**. An ion is a charged atom. The sodium ion has one more proton than electron and so has a net positive charge. Chlorine has gained an electron, becoming a chloride ion. The chloride ion has one more electron than it does proton and so has a net negative charge. The positive charge of the sodium ion and the negative charge of the chloride ion attract each other. This electrical attraction is called an **ionic bond**.

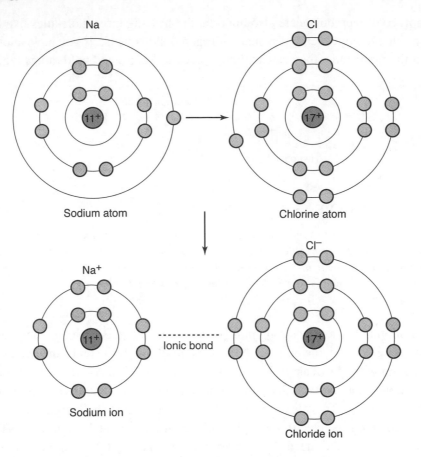

FIGURE 2.7. The ionic bonding of sodium and chlorine.

The sodium and chlorine atoms are electrically neutral. After the sodium atom gives up an electron to chlorine, it becomes the sodium ion and is positively charged. After it accepts an electron, the chlorine atom becomes the chloride ion and is negatively charged. Because they have opposite charges, the sodium ion and chloride ion are attracted to each other and form an ionic bond.

Chemical Reactions

Molecules can interact with each other to form new molecules. These exchanges are called chemical reactions, and they are written as equations. For example, in the following reaction, glucose and oxygen interact to form carbon dioxide and water. The molecules that interact are called the **reactants**, and they are shown on the left side of the arrow. The molecules that are formed are called the **products**, and they are shown on the right side of the arrow. The arrow represents the reaction.

$$C_6H_{12}O_6 \quad + \quad 6O_2 \quad \rightarrow \quad 6CO_2 \quad + \quad 6H_2O$$

glucose $\qquad$ oxygen $\qquad$ carbon dioxide $\qquad$ water

During a chemical reaction, atoms are not created or destroyed; they just exchange bonding partners. For example, in the preceding reaction, six carbon atoms enter the reaction as part of the glucose molecule, and six leave the reaction as part of the six carbon dioxide molecules. As an analogy, think of the reaction as a party: Everyone who comes to the party leaves; they may just leave with different people. When chemical reactions occur, energy is either released or used by the reaction. For example, the preceding reaction releases energy that can be used by cells.

Acids, Bases, and pH

The pH scale measures the concentration of hydrogen ions (H+) in a solution (Figure 2.8). Most cells require a neutral pH of about 7 in order to function. If the pH is higher or lower than 7, it can kill cells. A low pH (<7) indicates a solution that has a high concentration of H+ and is **acidic**. A high pH (>7) indicates a solution that has a low concentration of H+ and is **basic**. Molecules that release H+ into a solution are called **acids**. Molecules that pick up H+ from a solution are called **bases**. **Buffers** are compounds that help keep pH from changing very quickly.

> **REMEMBER**
> Acids have low pH values; bases have high pH values.

Scale of pH Values

pH	Example
0	1 Molar nitric acid
1	Gastric juice
2	Lemon juice
3	Vinegar
4	Tomato juice
5	Sour milk
6	Saliva
7	Pure water
8	Seawater
9	Baking soda
10	Soap
11	Ammonia (11.5)
12	Bleach (12.6)
13	Oven cleaner
14	1 Molar sodium hydroxide

FIGURE 2.8. The pH scale.

Because cells need to maintain a fairly constant pH in order to survive, cells contain buffers that help protect them from pH changes. For example, many cells use a phosphate buffer system to keep their pH from changing. The cells contain two chemicals, dihydrogen phosphate ions ($H_2PO_4^-$) and hydrogen phosphate ions (HPO_4^{2-}). If hydrogen ions (H^+) are released into the cellular fluid, they are picked up by the hydrogen phosphate ions. This prevents the cellular fluid from becoming acidic. If the cellular fluid starts to become basic, the dihydrogen phosphate ions release hydrogen ions to counteract the change. When the hydrogen phosphate ions pick up hydrogen ions, they are converted to dihydrogen phosphate ions. Likewise, when dihydrogen phosphate ions release hydrogen ions, they are converted to hydrogen phosphate ions. Thus, the relative concentration of the two ions is constantly balanced, forming a chemical **equilibrium** as shown in the following reaction:

$$H_2PO_4^- \leftrightarrow H+ + HPO_4^{2-}$$

By releasing or picking up hydrogen ions as necessary, the phosphate buffer system helps keep the cellular fluid at a relatively constant pH.

The Molecules of Life

You have probably heard of DNA, one of the most important molecules to living things. If you've ever looked at a nutrition label, you've also heard of other types of molecules: carbohydrates, fats, and proteins. Molecules are an essential part of the life of cells: Cells are made of molecules, cells break molecules down for energy, cells use molecules for signals and to store information. There are four categories of molecules that are most important to cells: **proteins**, **carbohydrates**, **lipids** (fats and related molecules), and **nucleic acids** (DNA and related molecules).

> **REMEMBER**
> All cells are made of proteins, carbohydrates, lipids, and nucleic acids.

These four types of molecules are called **macromolecules** because they are large and complex (*macro* = big). All of them have a carbon backbone. What makes them different in structure and function is the groups of atoms that are attached to the carbon backbone. These groups of atoms are called **functional groups** (Figure 2.9). Recognizing the key functional groups will help you differentiate between the different types of molecules.

Three of the four groups of macromolecules (proteins, carbohydrates, and nucleic acids) are also considered **polymers**. Polymers are long molecules made up of repeating building blocks called **monomers**. Monomers are joined together by **dehydration synthesis** (condensation) in which a bond is formed between the monomers as a water molecule is removed (Figure 2.10). Polymers can be broken apart by **hydrolysis**, in which the bond between monomers is broken as a water molecule is inserted (Figure 2.10). Lipids are not polymers, but they do have smaller components that are joined together by dehydration synthesis. The polymers and their monomers are listed in Table 2.2.

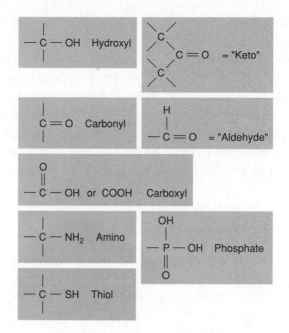

FIGURE 2.9. Functional groups.

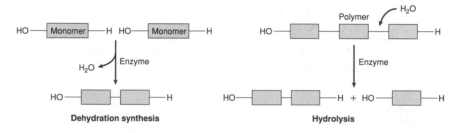

FIGURE 2.10. Dehydration synthesis and hydrolysis.

In dehydration synthesis, two molecules are combined by an enzyme, and a water molecule is removed. In hydrolysis, an enzyme catalyzes the breaking of a bond. Water is used.

TABLE 2.2. Polymers and Their Monomers

Macromolecule	Polymer	Monomer
Protein	Polypeptide	Amino acid
Carbohydrate	Polysaccharide	Monosaccharide
Nucleic acid	Polynucleotide	Nucleotide

Proteins are very important molecules to cells because they perform so many different functions. They can be hormones, defensive compounds (antibodies), helpers in chemical reactions (enzymes), structural components (cytoskeletal proteins), transporters (membrane proteins), movement proteins (cytoskeletal proteins), and signalers (receptor proteins). Proteins are made of chains of **amino acids**. Twenty amino acids make up the proteins of the living organisms on Earth. Each amino acid has the same backbone; what makes them different is the structure of the functional group attached to the backbone. The functional group of an amino acid can be represented by X (Figure 2.11). Some examples of different amino acids are shown in Figure 2.11.

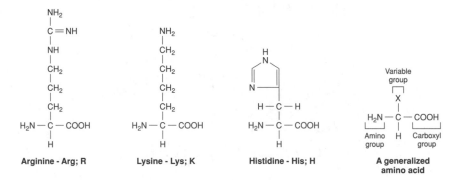

FIGURE 2.11. Examples of amino acids.

Notice that all amino acids have a nitrogen-containing amino group and a carboxyl group. The common structure of all twenty amino acids found in nature is shown in the generalized amino acid. The twenty amino acids differ from each other by what is present at X.

The shape of a protein is very important to its function. For example, a protein might have pockets that are just the right shape to bind to another molecule. In order to create the protein shape, the polypeptide chain folds up. First, amino acids are joined

> **REMEMBER**
> Protein function depends on protein structure.

by dehydration synthesis into **polypeptide** chains, which are called the *primary structure* of the protein (Figure 2.12). Some proteins have small areas that are folded into distinctive shapes, like helices (Figure 2.12). These folded shapes are called the *secondary structure* of the protein. The entire chain folds up into a three-dimensional shape (Figure 2.12), which is called the *tertiary structure*. Many proteins are made of more than one folded polypeptide chain. These proteins have *quarternary structure* (Figure 2.12). If a protein is exposed to a harsh environment, it may unfold or **denature** and no longer be able to function.

Carbohydrates are probably familiar to you as a source of energy, and that is one of their primary functions. They are also an important part of the structure of some types of cells. For example, they are part of the cell walls of plants and bacteria. **Monosaccharides**, or *simple sugars*, have a carbon backbone of three to seven carbon atoms. Hydroxyl groups (–OH) are attached to all the carbons but one, and that carbon is double-bonded to an oxygen atom (Figure 2.13). In the water-filled environment of cells, most monosaccharides adopt a ring form (Figure 2.13).

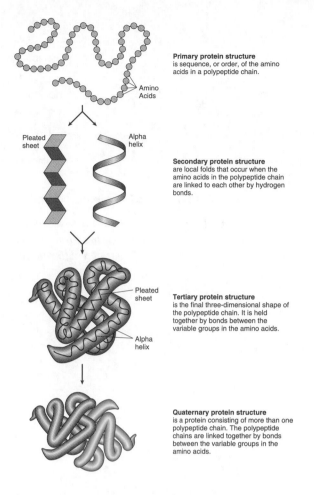

Primary protein structure
is sequence, or order, of the amino
acids in a polypeptide chain.

Amino
Acids

Pleated
sheet

Alpha
helix

Secondary protein structure
are local folds that occur when the
amino acids in the polypeptide chain
are linked to each other by hydrogen
bonds.

Pleated
sheet

Tertiary protein structure
is the final three-dimensional shape of
the polypeptide chain. It is held
together by bonds between the
variable groups in the amino acids.

Alpha
helix

Quaternary protein structure
is a protein consisting of more than one
polypeptide chain. The polypeptide
chains are linked together by bonds
between the variable groups in the
amino acids.

FIGURE 2.12. Protein structure.

The primary structure *of a protein is the sequence of the amino acids in the polypeptide chain.
The* secondary structure *is small folded areas that form structures such as alpha helices or
pleated sheets. The* tertiary structure *is the three-dimensional structure of a folded polypeptide
chain. Many proteins also have* quarternary structure, *where more than one folded polypeptide
chain comes together to form the completed protein.*

Monosaccharides are joined by dehydration synthesis into **disaccharides** (two monomers) and **polysaccharides** (many monomers).
Polysaccharides are commonly referred to as *complex carbohydrates*.
One familiar polysaccharide is starch (Figure 2.14), which is a long
polymer of repeating glucose monomers. Cellulose, a major component of plant cell walls, is also a polymer of glucose; the difference
between starch and cellulose is in the type of bond formed between the glucose
monomers. This small difference has big consequences for the human diet: We have
the enzymes necessary to hydrolyze (break down) starch, and so can get energy from
it, but we do not have the enzymes to hydrolyze cellulose. Thus cellulose, which is
commonly called *fiber*, passes through our systems undigested.

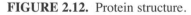

> **REMEMBER**
> Carbohydrates are an
> important source of
> energy for cells.

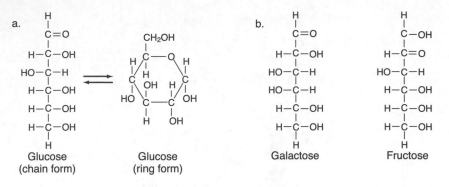

FIGURE 2.13. Monosaccharides.

Glucose, galactose, and fructose are very similar monosaccharides, or simple sugars. a. In the watery environment of cells, glucose adopts a ring form. b. Galactose and fructose have the same number of carbon atoms as glucose, but they differ in the arrangement of their atoms.

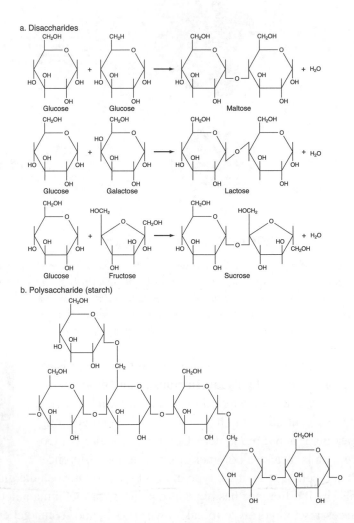

FIGURE 2.14. Disaccharides and the polysaccharide starch.

A. Maltose, lactose, and sucrose are all disaccharides, which are formed by dehydration synthesis of two monosaccharides. B. Starch is a polysaccharide that is formed by the dehydration synthesis of many molecules of glucose.

Deoxyribonucleic Acid (DNA) is the most familiar of the **nucleic acids**. The function of DNA, and the related molecule ribonucleic acid (RNA), is to store and transfer information for the cell. DNA contains the genetic information of the cell, a large part of which is the instructions for building all of the proteins a cell needs to function. Because the genetic information determines how the proteins are built and the proteins do most of the work of the cell, the genetic information ultimately determines how the cell functions. This in turn determines the traits of the organism. For example, people who have the genetic disease hemophilia have the trait of not being able to clot their blood properly. They have this trait because a mistake in their DNA leads to improper construction of a blood-clotting protein.

The structure of nucleic acids is fairly complex. They are polymers of monomers called **nucleotides**. In addition, nucleotides themselves are made up of three components: a five-carbon sugar, a phosphate group, and a nitrogenous base (Figure 2.15). In DNA, the five-carbon sugar is deoxyribose; in RNA, it is ribose. There are four nitrogenous bases found in nucleotides that are part of DNA. Their names are adenine (A), guanine (G), cytosine (C), and thymine (T). In RNA, thymine is replaced by uracil (U). To form DNA or RNA, nucleotides are linked together by dehydration synthesis. In

> **REMEMBER**
> DNA and RNA are both nucleic acids, which are made of nucleotides.

cells, RNA molecules are single polynucleotide strands (Figure 2.16). To complete the structure of DNA, however, a second strand must partner with the first. The nitrogenous bases of the two strands are attracted to each other by hydrogen bonds. Adenine bonds with thymine, cytosine with guanine. The two strands twist slightly, forming the *double helix* of DNA (Figure 2.16).

Nucleotide structure

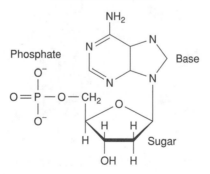

FIGURE 2.15. A nucleotide.

The two strands of the double helix of DNA can only hydrogen bond with each other if their nitrogenous bases are lined up in just the right way. In order for the correct chemical groups to come into contact with each other, the two nucleic acid strands must be lined up antiparallel, or in opposite directions, to each other. Nucleotide strands are polar, that is, one end is different from the other end. Scientists mark these differences by referring to the position of the carbons in the sugar of the nucleotide. The carbon atom attached to the nitrogenous base is designated as the 1′ ("one-prime") carbon. The carbons are numbered 1′ to 5′ going in the direction away from the oxygen in the ring (Figure 2.17). The two ends of a nucleotide strand are called the 5′ end and the 3′ end, because the 5′ carbon connects with the phosphate group in the same nucleotide, and the 3′ end connects with the phosphate group in the next nucleotide. So, when the two strands of the double helix line up and hydrogen bond together, one strand is oriented in the 5′ to 3′ direction, while the other is oriented in the 3′ to 5′ direction (Figure 2.17). The requirement for nucleotide strands to be antiparallel doesn't just apply to the two strands of DNA—it holds true any time two nucleotide strands pair together, whether it is DNA to DNA, DNA to RNA, or RNA to RNA. (The last two examples happen during protein synthesis, which is discussed in Chapter 7.)

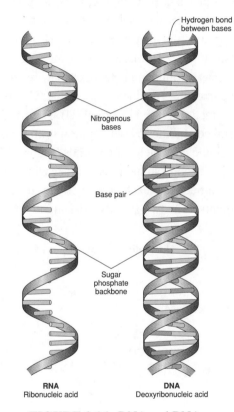

FIGURE 2.16. DNA and RNA.

RNA contains the nitrogenous bases adenine, cytosine, guanine, and uracil.
DNA contains the nitrogenous bases adenine, cytosine, guanine, and thymine.

The last group of macromolecules is the **lipids**. A lipid is any molecule that does not mix with water. There are several different kinds of lipids. Lipids are not polymers, but they are formed by dehydration synthesis. Three important groups of lipids are the fats and oils (triglycerides), the phospholipids, and the sterols. Fats and oils store energy and provide insulation. Phospholipids are important components of the outer boundaries of cells, which are called plasma membranes. Some sterols can also be found in plasma membranes, others are vitamins and hormones.

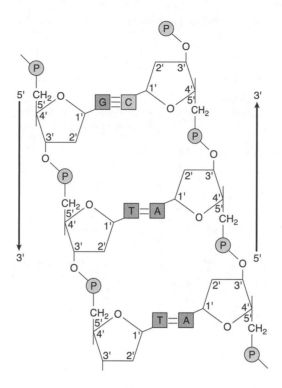

FIGURE 2.17.

Antiparallel bonding between nucleotide strands. In order for two nucleotide strands to hydrogen bond to each other, they must be oriented in opposite, or antiparallel, directions toward each other. Scientists label the orientation of each strand according to the carbons in the sugars of the nucleotides, which are numbered from 1′ to 5′.

Fat molecules consist of a three-carbon backbone to which long hydrocarbon chains, called **fatty acids**, have been attached (Figure 2.18). The three-carbon backbone is made from a molecule of **glycerol**. If the fatty acid tails contain only single bonds between the carbons, they are **saturated**. If they have double bonds between the carbons, they are **unsaturated**. Notice how saturated fatty acid tails are straight, but unsaturated tails are bent (Figure 2.19). Saturated fat molecules pack tightly together and so are more solid at room temperature. Unsaturated fat molecules don't pack together very well because of the kinks in their tails, so these fats are liquid at room temperature. The term "saturated" refers to how many bonds with hydrogen the carbon atoms have. The carbon atoms in saturated fatty acids can bond with two hydrogens each. A carbon atom around a double bond, however, can only form a bond with one hydrogen atom. (Carbon atoms form up to four single bonds each.) So, the chains with single bonds between the carbons are saturated because they are bonded to as many hydrogen atoms as possible.

> **REMEMBER**
> Lipids don't mix well with water because they contain many nonpolar bonds (like C–H and C–C bonds).

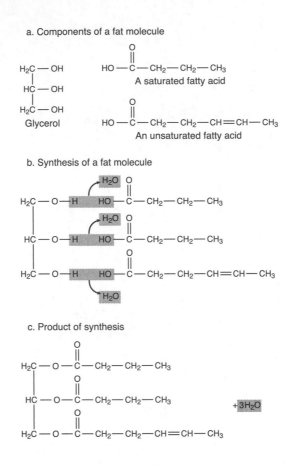

FIGURE 2.18. Fats.

*A. The components of a fat molecule. B. Dehydration synthesis
between the components of a fat molecule. C. Product of synthesis.
One more fatty acid would be added to make a fat.*

The structure of **phospholipids** is very similar to that of fats. The backbone is again made from a molecule of glycerol. In a phospholipid, however, there are only two fatty acid chains attached. In place of the third fatty acid, a phosphate containing a head group is attached (Figure 2.19). This head group is charged and mixes well with water, whereas the fatty acid tails do not mix well with water. The head of a phospholipid is thus said to be **hydrophilic**, or water-loving, whereas the tails are said to be **hydrophobic**, or water-fearing. When placed in water, phospholipids form a double layer called a **bilayer**, turning their hydrophilic heads toward the water and tucking their hydrophobic tails away from the water (Figure 2.19).

The **sterols** are made up of four carbon-containing rings (Figure 2.20). Cholesterol is a component of eukaryotic plasma membranes. Testosterone and estrogen are examples of steroid hormones. Some vitamins, like Vitamin D, are also sterols.

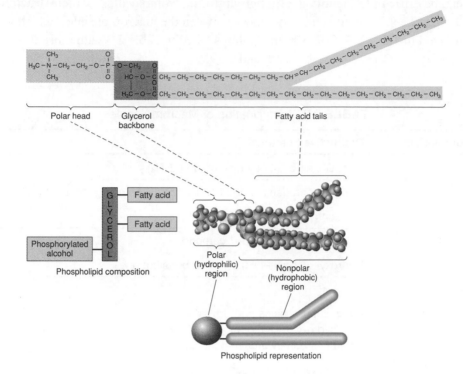

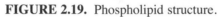

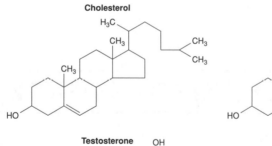

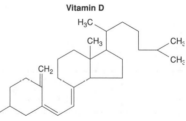

FIGURE 2.19. Phospholipid structure.

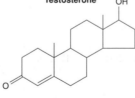

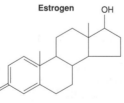

FIGURE 2.20. Sterols.

It can be difficult to identify the four groups of macromolecules by their structures. One thing to keep in mind is the differences between the macromolecules, which are summarized in Table 2.3. Compare the features listed in Table 2.3 with those illustrated in Figures 2.11, 2.13, 2.15, 2.18, and 2.19.

TABLE 2.3. Identification of Macromolecules

Macromolecule	Distinctive Features
Protein	Contains nitrogen in amino group ($-NH_3$)
Nucleic acids	Contains nitrogen in ringed structures called nitrogenous bases
	Contains phosphate
Carbohydrates	Short carbon chains (3–7 carbons)
	Hydroxyl groups (–OH) on all carbons but one
Lipids	
Fatty acids	Long carbon chains One hydroxyl (–OH) group Lots of carbon–hydrogen bonds
Phospholipids	Phosphate in head group No ringed structures

REVIEW EXERCISES FOR CHAPTER 2

1. Describe the structure of an atom.

2. What does the atomic number of an atom tell you?

3. What is the mass number of an atom? How is this different from atomic mass?

4. What is an isotope?

5. How many electrons does each of the first three electron shells hold?

6. What is a covalent bond? How is it different from a polar covalent bond?

7. What is a hydrogen bond?

8. What is an ion? How are ions formed?

9. What is an ionic bond?

10. What happens during a chemical reaction?

11. What is an acid? A base? What are the values on the pH scale and what do they mean?

12. Draw a table with the headings below and fill in the missing information.

Macromolecule	Monomer or Building Block	Structure of monomer or building block	Function of macromolecule
Protein			
Carbohydrate			
Nucleic acid			
Lipid			

SELF-TEST

1. A helium atom has an atomic number of 2. Which of the following are true of the helium atom?

   A. It has two protons.
   B. It has two electrons.
   C. It has a mass number of 2.
   D. Both A and B are correct.
   E. All of the above are correct.

2. Hydrogen is atomic number 1. If a hydrogen atom loses an electron, which of the following are correct?

   A. It will have a net positive charge.
   B. It will have a net negative charge.
   C. Its mass will change significantly.
   D. It will no longer be hydrogen.
   E. Its atomic number will change.

3. Nitrogen is atomic number 7. Hydrogen is atomic number 1. If hydrogen and nitrogen atoms react with each other to form a molecule, which of the following molecules will be formed? (*Hint:* To answer this question, figure out how many electrons each atom has in its outermost shell. Then figure out how to combine the atoms so that each atom has a full outermost shell.)

   A. NH_4
   B. NH_3
   C. NH_2
   D. N_3H
   E. N_2H

4. True or False. The bond between a hydrogen atom and an oxygen atom within a water molecule is called a hydrogen bond.

5. True or False. A solution that has a pH of 4 is acidic.

6. Which orientation shown in Figure 2.21 is most likely for two adjacent water molecules?

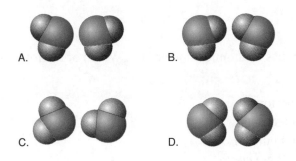

FIGURE 2.21. Hypothetical arrangements of water molecules.

7. To which group of macromolecules does molecule A in Figure 2.22 belong?

   A. Protein
   B. Carbohydrate
   C. Lipid
   D. Nucleic acid

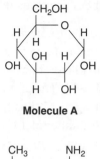

Molecule A

Molecule B

FIGURE 2.22. Test macromolecules.

8. To which group of macromolecules does molecule B in Figure 2.22 belong?

   A. Protein
   B. Carbohydrate
   C. Lipid
   D. Nucleic acid

9. Which of the following is **not** a function of lipids?

   A. Storage of information
   B. Structure of plasma membranes
   C. Storage of energy
   D. Vitamins
   E. Hormones

10. Radioactive nitrogen is fed to some cells. As the cells grow, the radioactive nitrogen will be incorporated into the cells' molecules. In which molecules would you expect to find the radioactive nitrogen?

    A. Amino acids
    B. Nucleic acids
    C. Carbohydrates
    D. Both A and B are correct.
    E. All of the above are correct.

Answers

Review Exercises

1. Atoms are made up of three types of particles: protons, neutrons, and electrons. The protons and neutrons are grouped together in the center of the atom, which is called the atomic nucleus. The electrons orbit the nucleus.

2. The atomic number is the number of protons in the atom.

3. The mass number is the number of protons and neutrons in an atom. It is different from the atomic mass because the atomic mass takes into account the mass of all the atoms of that element on the planet. For example, the most common form of nitrogen on the planet is nitrogen-14. The atomic number of nitrogen is 7, which means it has seven protons. Nitrogen-14 thus has seven neutrons ($14 - 7 = 7$). An atom of nitrogen-14 has a mass number of 14. However, the atomic mass for nitrogen in the periodic table is slightly higher than 14. This is because there are also atoms of nitrogen-15 on the planet. Nitrogen-15 is still nitrogen, so it has an atomic number of 7, which means it has seven protons. If nitrogen-15 has seven protons, it must have eight neutrons ($15 - 7 = 8$). An atom of nitrogen-15 has a mass number of 15. The atomic number listed for nitrogen in the periodic table represents an average weight for all the nitrogen atoms on the planet which are mostly nitrogen-14 (mass 14) but also some nitrogen-15 (mass 15). Thus the atomic mass is slightly higher than 14.

4. An isotope is a form of an element that has a different number of neutrons from another form of that element. For example, carbon-12 and carbon-14 are isotopes of each other. Both are carbon and so have six protons, but carbon-12 has six neutrons whereas carbon-14 has eight neutrons.

5. The first shell holds two electrons; the second and third shells each hold eight electrons.

6. A covalent bond is when two atoms share electrons. A polar covalent bond is also formed when two atoms share electrons, but in this case the two atoms are not sharing equally. One atom is more electronegative, which means it has a stronger attraction for electrons. The electrons spend more time orbiting one nucleus than they do the other. Thus, the molecule has an unequal distribution of charge, or is polar. It is slightly more negative in the area where the electrons spend more time and slightly more positive in the area where the electrons spend less time.

7. A hydrogen bond is a weak electrical attraction that is formed between polar molecules or between a polar group within molecules.

8. An ion is a charged atom. They are formed when an atom gains or loses an electron. For example, a hydrogen atom has one electron and one proton. The charges on the electron and proton are equal and opposite, so a hydrogen atom has no net charge. If the atom loses an electron, it becomes a hydrogen ion. It now has a proton, but no electron, resulting in a net positive charge.

9. An ionic bond is an electrical attraction that forms between ions of opposite charge.

10. During chemical reactions, molecules interact and bonds are broken and reformed. Energy changes also occur.

11. An acid is something that releases hydrogen ions to a solution, thus making it more acidic. A base is something that takes hydrogen ions out of a solution, thus making the solution more basic (alkaline). The pH scale is from 0 to 14. On the scale, 7 is considered neutral. If something has a pH value below 7, it is acidic. If something has a pH value above 7, it is basic (alkaline).

Macromolecule	Monomer or Building Block	Structure of monomer or building block	Function of macromolecule
Protein	Amino acid	Contains nitrogen in an amino group (NH_3) Has a carboxyl group Central atom has a variable group called the R group attached to it	Enzymes, receptors, movement, signaling molecules, defensive molecules, structural molecules
Carbohydrate	Monosaccharide	Three to seven carbons Hydroxyl group on all carbons but one Double-bonded oxygen on one carbon	Energy storage Structure of plant cell walls
Nucleic acid	Nucleotide	Nitrogen in rings called nitrogenous rings Five-carbon sugar Phosphate group	Information storage
Lipid	Glycerol and fatty acids in fats and phospholipids	Glycerol is a three-carbon molecule Fatty acids are long hydrocarbon chains (carbons and hydrogens bonded together) They may have some double bonds between the carbons (unsaturated) oronly single bonds between the carbons (saturated)	Energy storage (fats), structure of plasma membranes (phospholipids, sterols)

Self-Test

1. D	5. T	9. A
2. A	6. C	10. D
3. B	7. B	
4. F	8. A	

Observing Microbes Through a Microscope

WHAT YOU WILL LEARN

This chapter presents the fundamentals of microscopy and the ways that microbiologists use to prepare specimens for the microscope. As you study this chapter, you will:

- compare different types of microscopy;
- review some metric units of measurement;
- distinguish between magnification and resolution;
- learn about the staining protocols microbiologists use to prepare specimens for the microscope.

SECTIONS IN THIS CHAPTER

- Principles of Microscopy
- Types of Microscopy
- Preparing Specimens for Light Microscopy

Looking through a microscope at a drop of pond water is like looking into another world, a world filled with strange creatures of many shapes and sizes that busily swim and wiggle around (Figure 3.1). Invisible to people for thousands of years, this world became visible through the invention of the microscope in the 1600s. The earliest microscopes were small, handheld devices that contained a single lens. Because they contained a single lens, they were called **simple microscopes**. In one type, the specimen was placed on the tip of a screw that could be adjusted in front of the lens to bring the object into focus. Because of poor lens construction, most of these single lens microscopes did not yield very good images. Microscopes were improved when two lenses were combined in a single tube, creating the forerunner of the modern **compound microscope**. Because of the combination of lenses, compound microscopes allow greater magnification of specimens.

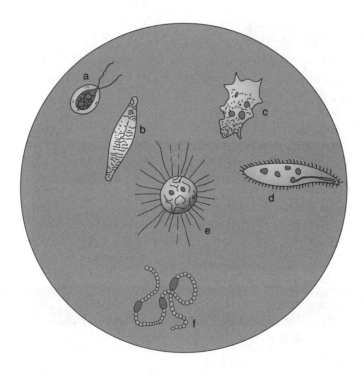

FIGURE 3.1. Organisms from a drop of pond water.

Organisms a–e are single-celled eukaryotes; organism f is a chain of prokaryotic cells. A. Chlamydomonas, *a small green alga. B. A diatom. C. An ameba. D.* Paramecium, *a ciliate. E. A radiolarian. F.* Anabaena, *a blue-green bacterium.*

Principles of Microscopy

METRIC UNITS OF MEASUREMENT

The metric system is the universal system of measurement in science and in most countries of the world. The metric system is based on units of ten. The basic unit of measurement of **length** is the **meter**. A meter is 39.37 inches, or slightly longer than a yard. The abbreviation for meter is **m**. Most rulers today have metric markings along one edge that show **centimeters** and **millimeters**. A centimeter (cm) is 1/100 of a meter. In other words, if a meter is divided up into 100 equal pieces, each one is a centimeter. This can also be represented as 0.01 m or 10^{-2} m. A millimeter (mm) is 1/1000 of a meter, or 0.001, or 10^{-3} m. There are 10 millimeters in 1 centimeter.

> **REMEMBER**
> There are 1000 micrometers (μm) in a millimeter (mm).

Cells, including those of microbes, are much smaller than a meter, and most are even smaller than a millimeter. They are usually measured in micrometers (μm), or 1/1,000,000 of a meter. (There are 1000 micrometers in a millimeter.) A micrometer can also be represented as 0.000001 m or 10^{-6} m. The parts of cells are so small they are often measured in nanometers (nm). A nanometer is 1/1,000,000,000 of a meter, or 0.000000001 m, or 10^{-9} m. Figure 3.2 compares the sizes of some objects and shows the units with which they would be measured.

MAGNIFICATION AND RESOLUTION

The lenses in a microscope **magnify** images or make them larger. Increasing **magnification** of an image, however, does not necessarily reveal more detail. For example, think about blowing up a photograph until the image looks very grainy and unfocused. When an image is captured, whether by an eye, a camera, or a microscope, the amount of detail that is captured is limited by the properties of the system that is capturing the image. For example, the quality of the image captured by a digital camera is determined by the number of pixels it can capture. You can magnify a digital image as much as you want, but the detail you'll see is ultimately limited by the number of pixels you originally captured. The amount of detail you can see in an image is called the **resolution**. Resolution is defined as the ability to see two points as separate. The ability to distinguish detail is called the **resolving power**. Just as the resolving power of a digital camera is determined by its construction, the resolving power of a microscope is determined by the lenses and the type of microscopy (Figure 3.2).

> **REMEMBER**
> Increasing the magnification of a microscope won't necessarily make the image more clear. The clarity, or detail, of an image depends on the resolving power of the microscope and its lenses.

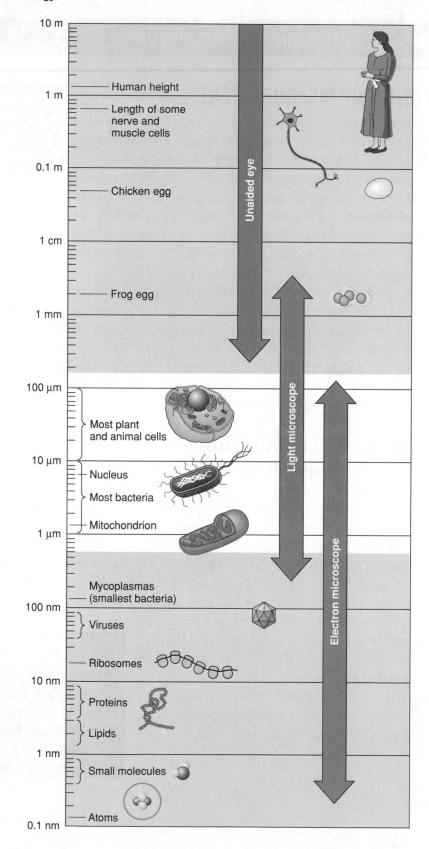

FIGURE 3.2. Relative sizes of objects.

Types of Microscopy

LIGHT MICROSCOPY

Light microscopes use light to illuminate the specimen. The most common type of light microscope is the compound light microscope (Figure 3.3). Light travels from an **illuminator**, through a **condensor lens**, which focuses the light on the specimen, and then into an **objective lens** that magnifies the image. The magnified image passes through another lens, called the **ocular lens**, in the eyepiece and is magnified further. The total magnification of the microscope is the product of the magnification of both the objective lens and ocular lens. For example, if the objective lens has a magnifying power of 40 times (40×), and the ocular lens has a magnifying power of 10 times (10×), the total magnification of the image would be 400 times (400×).

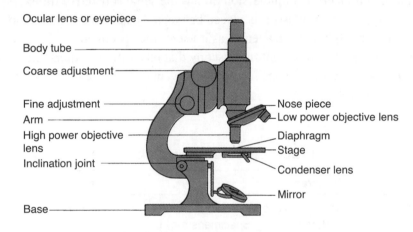

FIGURE 3.3. Parts of a compound light microscope.

The resolving power of a light microscope is determined by the quality of the lenses and by the properties of light. Visible light is a type of electromagnetic radiation and has a wavelike property as it travels. The distance between two adjacent waves is called the **wavelength** (Figure 3.4). The smaller the wavelength of light used, the greater the resolving power of the microscope.

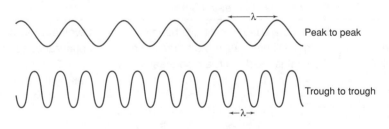

FIGURE 3.4. Wavelength.

Wavelength is measured as the distance from peak to peak
of a wave, or from trough to trough of a wave.

Another property of light that affects microscopy is its ability to bend, or **refract**. Light travels differently through different media and will refract as it passes from one medium to the next. For example, light travels differently through air than it does through water. If you stand a pencil in a glass of water, the pencil will appear broken.

> **REMEMBER**
> You must use oil to get a good image with a high-power oil immersion lens. Without oil, the light will refract and scatter, causing your image to be blurry.

This is because the light bends as it moves from the water to the air. Refraction affects light microscopy in two major ways. We can use it to our advantage when we stain specimens for the light microscope. Light will travel differently through the stain, refracting and creating greater contrast with the background. However, refraction scatters the light as it leaves the specimen. At high magnification, this becomes a problem because the objective lenses are very small. If light scatters too much from refraction, enough light will not be captured by the lens. To prevent this, immersion oil is used with high magnification lenses (100×). Immersion oil has the same **refractive index**, or light-bending ability, as that of glass. It is placed between the specimen and the objective lens to prevent the light from scattering as it leaves the specimen.

There are several types of light microscopy that use different filters and techniques to create unique effects. Each one has certain advantages. These types of microscopy are summarized in Table 3.1.

TABLE 3.1. Types of Light Microscopy

Type	Features	Advantages
Brightfield	Specimens appear dark against a light background. Specimens may be living or may be killed and stained.	Inexpensive and easy to use
Darkfield	Specimen appears light against a dark background. Light from illuminator is blocked from entering objective lens directly. Only the light that is reflected from the specimen can enter the objective.	Can be used to see specimens that cannot be seen by Brightfield microscopy or that do not stain well.
Phase contrast	Uses the differences in light from the illuminator and that diffracted from the specimen to create contrast.	Creates contrast in unstained living cells.
Differential interference contrast (DIC)	Prisms are used to split two beams of light. Differences in the refraction of the light is used to create contrast.	Gives a three-dimensional appearance to unstained cells. The refraction of the light by the prisms adds color to the specimen.

TABLE 3.1. (continued)

Type	Features	Advantages
Fluorescence	Special fluorescent stains are used along with filters that allow only certain wavelengths of light to reach the specimen. These wavelengths activate the fluorescence of the stains and cause them to glow.	Stains can be targeted to specific types of microbes that are then illuminated by the fluorescence if they are present in the sample.
Confocal	Uses computers to precisely guide lasers to illuminate the specimen. Fluorescent stains are used and activated by the lasers.	Very precise areas of the specimen can be examined at one time. Computers can then combine multiple images of the same specimen to create a very detailed image.

ELECTRON MICROSCOPY

Light microscopes have a lot of advantages. They are relatively inexpensive and easy-to-use, and you can look at living specimens. However, the smallest object that can be viewed clearly with a light microscope is about 0.2 μm in size. This means that light microscopes are good for viewing entire cells and some of the larger parts of cells, but they are not able to give detailed information about the smaller parts of cells. They are also unable to give information about viruses, which are even smaller than prokaryotes. Because resolving power of the microscope depends upon the wavelength of light used, even the best light microscope is ultimately limited by the properties of light itself.

> **REMEMBER**
> Electron microscopes have higher resolution than light microscopes, but you can't look at live samples with an electron microscope.

In the 1950s, another type of microscope was invented. Instead of using light, this type of microscope used electrons to create an image. Electrons are small, negatively charged particles that also have a wavelike property. The wavelengths of electrons, however, are even smaller than the wavelengths of light. This allows better resolving power and makes it possible to obtain clear images of much smaller structures than can be seen in a light microscope. Electron microscopes can resolve structures as small as about 2 nm. Thus, they can provide detailed information about the fine structure of eukaryotic cell parts, prokaryotic cells and can even create images of viruses and some molecules. Some disadvantages to electron microscopes are that you cannot view living specimens, and they are much more expensive and difficult to use than are light microscopes.

Electron microscopes (Figure 3.5) have an electron gun that shoots electrons at the specimen. The electrons are focused by electromagnetic lenses. Because electrons are so small and would be easily deflected by molecules in the air, the electrons must travel through a vacuum chamber from the gun to the specimen. To view a specimen on an electron microscope, it must first be killed and layered with a fine coating of metal. When the electrons strike the metal coating of the specimen, electrons from the metal are deflected away and captured either by photographic film or a computer monitor. They deflect away in a pattern that recreates the original structure. When the electrons strike photographic film, they expose the film, creating an image of the structure. These images are called **electron micrographs**.

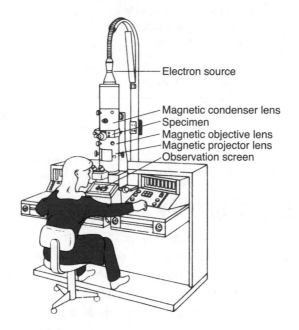

Electron source

Magnetic condenser lens
Specimen
Magnetic objective lens
Magnetic projector lens
Observation screen

FIGURE 3.5. An electron microscope.

There are two basic types of electron microscopy: **scanning electron microscoscopy (SEM)** and **transmission electron microscopy (TEM)**. In scanning electron microscopy, whole objects are layered with metal, and images are made of their surfaces. This can give three-dimensional information about the surfaces of cells and even whole organisms such as insects. In transmission electron microscopy, very thin slices called sections are made through the specimen, and each section is layered with metal. This type of electron microscopy gives information about the structure on the insides of cells and objects.

Preparing Specimens for Light Microscopy

Many cells and parts of cells are clear and would be nearly invisible if viewed directly on a light microscope. To increase contrast and make it possible to see the cells, **stains** are used. Stains are solutions that contain colored ions called **chromophores** that attach to cells or parts of cells. A summary of some common staining types and their uses is given in Table 3.2.

TABLE 3.2. Common Staining Protocols and Their Uses

Staining Protocol	Use	Commonly Used Stains
Simple stain	Stains cells evenly to reveal shape (morphology) and arrangements of cells.	Methylene blue, crystal violet
Gram stain	Determines whether bacteria are gram-positive (thick-walled) or gram-negative (thin-walled). Important for deciding upon antibiotic therapy.	Crystal violet and safranin
Acid fast stain	Determines whether bacterium has a waxy wall. Important for detecting the causative agents of leprosy (*Mycobacterium leprae*) and tuberculosis (*Mycobacterium tuberculosis*).	Methylene blue and carbol fuchsin
Flagellar stain	Detects presence and arrangement of flagella.	Basic fuchsin
Capsule stain	Detects presence of capsule.	India ink to stain the background
Endospore stain	Detects presence of endospores and location of endospores within cells.	Malachite green and safranin

SMEAR PREPARATION

To prepare microbes for staining, a few drops of liquid containing the microbes are dropped onto a glass slide and allowed to dry. This results in a cloudy film of cells on the slide called a **smear**. To help attach the cells firmly to the slide, the smear is **heat-fixed** by passing the glass slide through a flame. Once the smear is heat-fixed, it is ready to be stained.

SIMPLE STAINS

Stains that are used to color the entire cell evenly are called **simple stains**. The most commonly used simple stains have a positively charged chromophore. These types of stains are called **basic simple stains**. The positively charged chromophore of the basic stain is attracted to the negative charge on the surface of a bacterial cell. The chromophore attaches to the surface of the cell, giving the entire cell a colored appearance. Simple stains that have a negatively charged chromophore are called **acidic simple stains**.

> **REMEMBER**
> When a stain colors an entire cell evenly, it's called a simple stain.

DIFFERENTIAL STAINS

Some staining protocols will react differently with one type of cell than they will another. These stains are called **differential stains**. Because they react differently with different types of cells, they allow identification of a key characteristic of these cells.

The most important differential stain is the **Gram stain**, named for the Danish microbiologist Hans Christian Gram who invented the staining protocol in 1884. The Gram stain is used to separate most disease-causing bacteria into one of two categories, **gram-positive bacteria** and **gram-negative bacteria**. Because of differences in their cell walls, these two groups of bacteria react differently in the Gram stain. Gram-positive bacteria are stained purple by the stain, while gram-negative bacteria are stained pink. (A useful trick is to remember that there is an "N" in the word "pink," and that "negative" starts with "N.") Differences between these two groups of bacteria affect the success of antibiotic therapy, so the Gram stain is often the first piece of information that is needed about an infectious organism.

There are several steps in performing the Gram stain. After a smear is prepared, a purple stain called crystal violet is dropped onto the smear and allowed to bind to the cells. Because it is used first, this stain is called the **primary stain**. The crystal violet is then rinsed away and a solution of iodine is applied. The iodine attaches to the chromophores of the crystal violet, making a larger structure that does not easily wash out of the cell. Because the iodine helps bind the dye to the cell, it is called a **mordant**. After the iodine is rinsed away, a solution of acetone and alcohol is applied to the cell. This solution is called the **decolorizing solution** because it removes the purple color from gram-negative cells. This is also rinsed, and then a second pink stain, safranin, is applied. Because this stain is applied second, it is called the **counterstain**.

> **REMEMBER**
> Gram-negative bacteria stain pink in a Gram stain because the decolorizing solution disrupts the outer membrane of the bacteria, washing away the primary stain.

The reason the Gram stain reacts differently with gram-positive and gram-negative bacteria is due to the differences in their cell walls, which are located outside their plasma membranes. Gram-positive bacteria have a wall that is made of a thick layer of a material called **peptidoglycan** (also called murein). Gram-negative bacteria have a thin layer of peptidoglycan that is surrounded by an outer membrane made up primarily of lipids and proteins.

During a Gram stain, both types of bacteria are initially stained purple by the primary stain. In gram-positive bacteria, the iodine then reacts with the primary stain to make complexes that are too large to be rinsed back out of the thick layer of peptidoglycan. The decolorizing solution has essentially no effect on these bacteria besides a slight drying effect on the wall. The pink counterstain does bind to the cells, but it is not seen because of the darker purple stain. Thus, these cells appear purple at the end of the staining protocol (Figure 3.6). In gram-negative bacteria, however, the decolorizing solution washes away the outer membrane and allows the crystal violet–iodine complex to wash out of the thin wall, making the cells colorless. When the pink counterstain is applied, it is visible in these cells, and they appear pink (Figure 3.6).

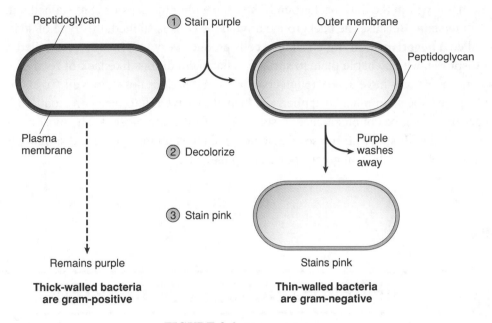

FIGURE 3.6. Gram staining.

Another important differential stain is the **acid fast stain**. This stain is important because it is used to identify certain disease-causing bacteria, including the bacterium that causes tuberculosis. These bacteria belong to a group of bacteria called mycobacteria, which have a waxy layer on their cell walls. The acid fast staining protocol is similar to that of a Gram stain, with the use of a primary stain, decolorizer, and counterstain. In this staining protocol, however, the decolorizer is a solution of **acid–alcohol**. As in the Gram stain, all cells will be stained with the primary stain, in this case the pink stain carbol fuchsin. Cells that are not mycobacteria will then be decolorized by the decolorizer and will pick up the blue color of the counterstain, methylene blue. Because of their waxy wall, however, mycobacteria will resist the decolorizer and keep the pink color of the primary stain.

SPECIAL STAINS

Special stains may be used to reveal particular structures of cells. For example, some prokaryotes are motile because they have fine whip-like structures called **flagella**. These structures are normally too small to see with a light microscope, but they can be thickened by the application of many layers of **flagellar stains** so that they can be seen. (This is analogous to thickening eyelashes with mascara.) Other cells make a slimy layer on the outside called a **capsule**, which helps them stick to surfaces and protects them from the immune system. Capsules do not stain easily, so in order to view them, cells are placed into a solution containing a dye. The dye creates a dark background all around the cells but does not stick to the capsules, which appear as clear halos within the dark background The cells within the capsules are stained with a counterstain. Because the background around the cells is stained, this type of staining is called a **negative stain**. Finally, some cells produce very tough structures called **endospores**, which enable them to survive in harsh conditions like lack of food or water. Endospores have a very tough coat on them that does not stain well. To stain endospores, the stain must be applied and heated to drive it into the spore coat. The smear must then be counterstained to stain any cells that are present. The most commonly used **endospore stain** uses malachite green to stain the spores and safranin to stain the cells, resulting in green spores and pink cells.

REVIEW EXERCISES FOR CHAPTER 3

1. Describe the relationship between meters, millimeters, and micrometers.

2. Describe the difference between magnification and resolution.

3. How does a light microscope work?

4. How do wavelength and refraction of light affect the quality of light microscopy?

5. Describe the advantage of each of the following types of microscopy: brightfield, darkfield, phase contrast, DIC, fluorescence, and confocal.

6. How does an electron microscope work?

7. What is the difference between scanning and transmission electron microscopy?

8. What are the advantages of electron microscopy over light microscopy? What are the disadvantages?

9. Distinguish between the terms smear and stain.

10. Why are stains used in light microscopy?

11. State the definition for each of the following and give an example: simple stain, differential stain, special stain.

12. Describe how the Gram stain works to distinguish between thick- and thin-walled bacteria.

SELF-TEST

1. The difference between simple and compound microscopes is that

   A. simple microscopes use light; compound microscopes use electrons.
   B. simple microscopes don't require staining; compound microscopes do.
   C. simple microscopes magnify the specimen with one lens; compound microscopes magnify the specimen with two lenses.
   D. Both A and B are correct.
   E. All of the above are correct.

2. An object measures 0.5 millimeters in length. How many micrometers long is it?

   A. 500 micrometers
   B. 50 micrometers
   C. 5000 micrometers
   D. 0.005 micrometers
   E. 0.0005 micrometers

3. All of the following are advantages of light microscopes **except**:

   A. They are easy to use.
   B. They have better resolution than electron microscopes.
   C. They are relatively inexpensive.
   D. They can be used to see living cells.
   E. All of the above are advantages of light microscopes.

4. True or False. As wavelength decreases, resolution increases.

5. True or False. The ability to distinguish two points as separate is called refraction.

6. Unstained cells could be viewed with all of the following types of microscopy **except**:

   A. Darkfield
   B. Phase contrast
   C. Differential interference contrast
   D. Fluorescence
   E. All of the above are correct.

7. Which of the following types of microscopy creates a three-dimensional effect?

   A. Brightfield
   B. Differential interference contrast
   C. Darkfield
   D. Fluorescence
   E. None of the above is correct.

8. Electron microscopes have better resolution than light microscopes because

   A. the lenses are more powerful.
   B. the images are magnified on computer screens.
   C. the specimens are coated with metal.
   D. electrons have smaller wavelengths than visible light.
   E. All of the above are correct.

9. The difference between a simple stain and a special stain is

   A. simple stains color the entire cell; special stains color a particular structure.
   B. simple stains have a positively charged chromophore; special stains have a negatively charged chromophore.
   C. simple stains color all cell types the same; special stains react differently with one cell type than another.
   D. simple stains are easy to do; special stains require special equipment.
   E. None of the above is correct.

10. After the decolorizer has been applied during a Gram stain, but before the safranin has been applied, a gram-negative cell would appear

    A. purple.
    B. pink.
    C. colorless.
    D. blue.
    E. None of the above is correct.

Answers

Review Exercises

1. A millimeter is 1/1000 of a meter. A micrometer is 1/1000 of a millimeter. In other words, if you break a meter up into a thousand equal pieces, each piece is a millimeter. If you then break a millimeter up into a thousand equal pieces, each piece is a micrometer. A micrometer is 1/1,000,000 of a meter.

2. Magnification is making things bigger. Resolution is the ability to clearly distinguish between two objects. Magnification may make it easier to see something, but ultimately the amount of detail that can be seen in any image will be limited by the resolution of the device capturing that image (the eye, a camera, or a microscope).

3. Light microscopes shine visible light through a specimen and magnify the image of the specimen by passing the light through curved glass lenses. The light is focused onto the specimen by the condenser lens and is magnified as it passes through the objective and ocular lenses. The total magnification is a product of the magnification of the objective and ocular lenses.

4. The smaller the wavelength of light used, the better the resolution of the image. At high magnifications, the refraction of light makes it more difficult to capture enough light into the objective lens. This is why immersion oil is used. However, the refraction of light can be used to our advantage when it is controlled and manipulated in phase contrast microscopy and differential interference contrast microscopy. In both of these types of microscopy, light that has been refracted differently is recombined to create an image with greater contrast. This allows better visualization of unstained specimens.

5. Brightfield microscopy is simple to do. Darkfield is useful for viewing specimens that cannot be stained or are invisible on brightfield microscopy. Phase contrast and differential interference contrast are both useful for viewing unstained specimens. An additional advantage of DIC is that it provides a three-dimensional effect and false color to the image. Fluorescence microscopy is useful when specific molecules or structures are to be targeted. This is especially powerful if used in conjunction with confocal microscopy. Confocal microscopy can be used to collect very detailed information about cells.

6. Specimens that are going to be looked at with an electron microscope are killed and layered with metal. They are then placed in the vacuum chamber of the electron microscope. Electrons are fired from an electron gun and focused with electromagnetic lenses as they travel through the vacuum chamber. When the electrons strike the specimen, they dislodge electrons from the metal coating. These electrons scatter and are collected on computer screens or photographic film. The electrons create an image of the specimen.

7. Scanning electron microscopy gives three-dimensional images of the surfaces of objects. Transmission electron microscopy is used to exam thin sections of specimens and so gives details about internal structures.

8. Electron microscopes have greater resolving power than light microscopes and thus provide more detailed information about small structures. They can be used to examine the small structures of cells and of things smaller than cells, like viruses. The disadvantages are that they cannot be used to view live specimens and that they are more expensive and more difficult to use.

9. A smear is when microbes are dried on a slide. A stain is the colored solution used to color the smear so that there will be greater contrast in the microscopic image.

10. Stains are used to provide greater contrast and allow visualization of cells and cell parts that are not colored and would be almost invisible on a brightfield microscope.

11. A simple stain colors a cell uniformly. A differential stain reacts differently with different cell types and thus allows them to be distinguished from each other. A special stain stains a certain structure in a cell.

12. In a Gram stain, both thin- and thick-walled cells are initially colored purple. When the decolorizing solution is applied, the outer membrane of thin-walled bacteria washes away, releasing the purple color. When the pink stain is applied, both cell types stain with the pink stain, but it is only visible on the thin-walled cells. The purple color that was retained by the thick-walled cells is what is seen, and they appear purple. On the thin-walled cells, the purple color washed away, so the pink can be seen.

Self-Test
1. C	5. F	9. A
2. A	6. D	10. C
3. B	7. B	
4. T	8. D	

Structure and Function of Microbial Cells

WHAT YOU WILL LEARN

This chapter explores the details of cell structure and function that are the foundation for understanding microbial growth, human disease, and defenses against disease such as antibiotics. As you study this chapter, you will:

- compare prokaryotic and eukaryotic cells;
- review the structure and function of the plasma membrane;
- learn the structure and importance of the bacterial cell wall;
- discover structures unique to prokaryotic cells;
- explore the organelles of the eukaryotic cell.

SECTIONS IN THIS CHAPTER

- Prokaryotic vs. Eukaryotic Cells
- The Plasma Membrane
- The Prokaryotic Cell
- The Eukaryotic Cell

Many microbes, like the bacterium *E. coli*, are single-celled organisms; others, like the mold you've seen growing on bread in your kitchen, are made of many cells. (Viruses, which are not cells, will be considered in a separate chapter.) Like the cells of your body, the cells of microbes do certain things to survive—they use energy, get raw materials for growth, and move. In this chapter, we will consider the parts of microbial cells and how they help microbes perform these essential functions. Many of the cellular parts discussed here also play an important role in the disease process and will be mentioned again later.

Prokaryotic vs. Eukaryotic Cells

The biggest difference among cell types is whether they are **prokaryotic** or **eukaryotic** (Figure 4.1). The cells of bacteria and archaea are prokaryotic; the cells of humans are eukaryotic (as are the cells of other animals, plants, fungi, and protists). Because the cells of disease-causing bacteria are fundamentally different from our cells, we can use **antibiotics** to fight bacterial infections: The antibiotics attack parts of bacterial cells that are not found in our cells; thus, antibiotics hurt "them" while leaving "us" alone. The major differences between prokaryotic cells and eukaryotic cells are summed up in Table 4.1.

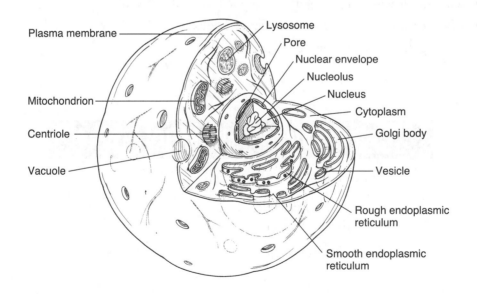

Prokaryotic cell structure

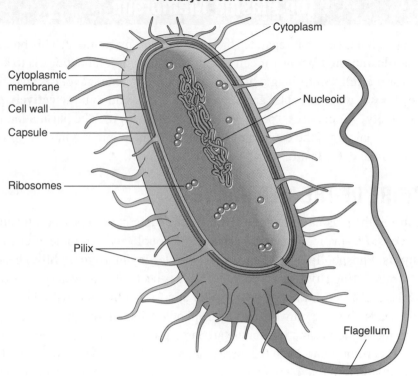

FIGURE 4.1. Prokaryotic and eukaryotic cells.

TABLE 4.1. Differences Between Prokaryotic and Eukaryotic Cells

Prokaryotic Cells	Eukaryotic Cells
DNA (genetic material) not enclosed by a membrane (the region where the DNA is located is called the **nucleoid**)	DNA enclosed by a membrane to form a **nucleus**
No membrane-enclosed organelles	Have membrane-enclosed organelles
Divide by **binary fission** (a simple process in which the DNA is copied, the cell grows, and then splits in two)	Divide by **mitosis** (a complicated process involving many movements of the chromosomes by a structure called the mitotic spindle)

The Plasma Membrane

All cells have a boundary that separates them from their environment. This boundary is called the **plasma membrane** (or cell membrane). One of its functions is to act as "customs" for the cell, controlling what can enter and what can leave. Because the membrane chooses what can enter and leave the cell, it is said to be **selectively permeable** ("selectively" for chooses; "permeable" for pass through). The plasma membrane also receives signals from the environment and passes them into the interior of the cell, which is called the **cytoplasm**.

STRUCTURE OF THE MEMBRANE

The structure of the plasma membrane is almost the same in both prokaryotic and eukaryotic cells. Plasma membranes are made of a double layer of molecules called **phospholipids**. Phospholipids have two parts: a water-loving (**hydrophilic**) head group and water-hating (**hydrophobic**) tails. Because cells live in watery environments, two layers of phospholipids get together and form a **bilayer**, with the hydrophilic heads pointing outward and hydrophobic tails pointing inward (Figure 4.2). Also found in the plasma membrane are large protein molecules: Some of these proteins are **receptors** that search the environment for signals like food molecules; others are **transporters** that help molecules cross the membrane. Eukaryotic plasma membranes also contain sterols; most prokaryotic membranes do not.

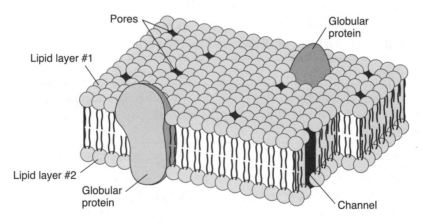

FIGURE 4.2. The plasma membrane.

The fluid mosaic model of the plasma membrane showing a lipid bilayer with embedded proteins. Phospholipids and proteins can both move within the membrane, giving it its fluid nature.

TRANSPORT ACROSS MEMBRANES

If a molecule is small enough, or if it is hydrophobic like the tails inside the phospholipid bilayer, it can scoot across the membrane all by itself. Examples of this type of molecule are carbon dioxide (CO_2), oxygen (O_2), and water (H_2O). Molecules that are hydrophilic can only cross with the help of a transport protein. For example, ions like potassium (K^+) or chloride (Cl^-) have a charge that makes them hydrophilic. Because they are attracted to the watery environment on either side of the membrane, and not to the hydrophobic tails inside the membrane itself, they would not cross the membrane without the help of a protein. Other molecules that require the help of a protein are those that are too large to cross through the phospholipids by themselves (e.g., glucose).

> **REMEMBER**
> Small, hydrophobic molecules can cross a plasma membrane without assistance, but larger and/or hydrophilic molecules need the help of a transport protein.

There are basically two types of transport proteins that help molecules cross membranes: **channel** proteins and **carrier** proteins. The names of these proteins tell you what they do. Channel proteins are folded so that they have a channel, or tunnel, through their centers (Figure 4.3A). These channels are filled with water so that small hydrophilic molecules like ions can pass through them and get across the membrane. Carrier proteins (Figure 4.3B) are folded so that they have little pockets, or binding sites, that pick up specific molecules. After a carrier protein picks up a molecule on one side of the membrane, the protein's shape changes so that it releases the molecule on the other side of the membrane. After the molecule is released, the carrier protein goes back to its original shape and can repeat the cycle all over again. Glucose, a common food molecule for cells, is transported across membranes by carrier proteins.

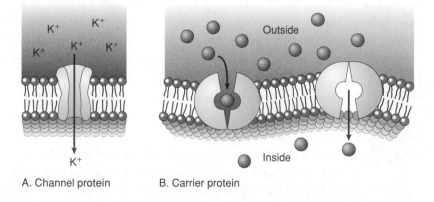

A. Channel protein   B. Carrier protein

FIGURE 4.3. Types of membrane transport proteins.

A. Channel proteins have a hollow core that allows any ions or other molecules that are small enough to fit inside the passage to cross the membrane. B. Carrier proteins have binding sites that are specific to the ion or molecule that they can transport. They bind to the molecule on one side of the membrane. This changes the shape of the carrier protein and causes it to bring the molecule to the other side of the membrane where it is released.

One thing that determines the direction of movement of molecules across membranes is their concentration, or amount, on either side of the membrane. If there is more of a molecule on one side of the membrane than the other, the concentration of the molecule will eventually become even on both sides of the membrane by the process of **diffusion**. You can demonstrate diffusion easily at home—just squirt some food coloring into a glass of water. You don't have to stir the water to disperse the food coloring, if you just leave the glass alone for a while, the food coloring will spread throughout the water. This is because molecules, including those of the water in the glass, are constantly in motion owing to their kinetic energy. Basically, even though you cannot see them, the water molecules are constantly jiggling around randomly. As they jiggle around, they bump into the dye molecules and cause them to disperse. The same thing happens around cells: If there is more of a molecule outside a cell, the molecules will randomly jiggle around and occasionally cross into the cell. Over time, the concentration of the molecule on either side of the membrane will become even. This requires *no energy* from the cell; the movement is caused by the kinetic energy of the molecules themselves. Of course, for this to happen, the molecule has to be able to cross the membrane either by itself or through a transport protein. If it crosses by itself or through a channel protein, this is called **simple diffusion**; if it crosses through a carrier protein, it is called **facilitated diffusion**.

A special case of diffusion is the diffusion of water, which is called **osmosis**. Osmosis follows the rules of diffusion: The movement of water is from an area where it is more concentrated to an area where it is less concentrated, and no energy is required from the cell. This can be tricky to think about, though, because we don't usually focus on the concentration of the water, we talk about the concentration of things that are dissolved in the water (**solutes**). If water is more concentrated (or "pure"), that means there are fewer solutes dissolved in it. If water is less concentrated, that means there are more solutes. So, water moves by osmosis from an area where there are fewer solutes toward an area where there are more solutes. An easy way to think about this is to remember that water moves toward the greater concentration of solutes.

> **REMEMBER**
> Water moves toward the area with the greatest concentration of solutes (hypertonic environment).

The movement of water is very important to cells: If cells do not have enough water inside of them, they cannot function or grow. And, because water can move across membranes by itself, solutes in the cell's environment can cause water to leave a cell.

If there are more solutes outside the cell than inside the cell, the environment is said to be **hypertonic** (*hyper* = over), and water will leave the cell. If there are more solutes inside the cell than outside the cell, the environment is said to be **hypotonic** (*hypo* = under), and water will enter the cell. This will cause some cells to burst. If the solute concentration is the same outside as it is inside the cell, the environment is said to be **isotonic**, and there will be no overall movement of water.

Another factor that can affect the movement of molecules across membranes is the cell's need for the molecule. For example, even if there is less of a food molecule outside a cell than there is inside, the cell may still need to transport the food molecules into the cell. This can be done, but the cell must spend energy to do it. Some carrier proteins can perform **active transport**, moving molecules from an area where they are less concentrated to an area where they are more concentrated. In the process, they utilize cellular energy either in the form of ATP or by tapping into a concentration gradient of hydrogen ions (H^+) across the plasma membrane, which is called the **proton motive force**. To prevent the molecule from diffusing back out again once it's been moved into the cell, the molecule might be changed slightly so that it no longer fits into the carrier protein's binding site. Many microbes achieve this through a process called **group translocation**. This is essentially a special case of active transport in which the carrier protein changes the molecule it is transporting as it moves it across the membrane. For example, as glucose is moved into the cell, the carrier protein might attach a small chemical group called a phosphate to the glucose. With the phosphate group attached, the glucose can no longer be transported by the carrier protein and so becomes locked inside the cell.

In eukaryotic cells, very large molecules or structures can be taken into the cell by **endocytosis** (Figure 4.4). During endocytosis, part of the plasma membrane reaches out and wraps around the molecule to be taken in, forming a sphere of membrane called a **vesicle**. The vesicle can be transported within the cell. Likewise, large molecules can be placed inside a vesicle and transported up to the plasma membrane for export from the cell during the process of **exocytosis**.

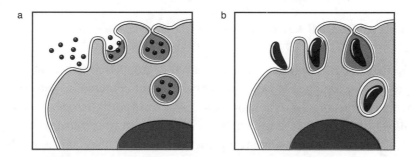

FIGURE 4.4. Endocytosis.

Two types of endocytosis are shown. A. In pinocytosis, the cell takes in substances that are dissolved in water. B. In phagocytosis, the cell takes in large particles.

The Prokaryotic Cell

MORPHOLOGY AND ARRANGEMENT

Although bacteria come in many shapes and sizes, three shapes (**morphologies**) are the most common (Figure 4.5): spherical (singular **coccus** or plural **cocci**), rod-shaped (singular **bacillus** or plural **bacilli**), and **spiral**. Within the spiral bacteria, **vibrio** bacteria are comma-shaped, **spirilla** are wavy, and **spirochetes** are twisted like corkscrews. Bacteria that are intermediate between spherical and rod-shaped are called **coccobacilli**, and those that vary in shape are called **pleomorphic**.

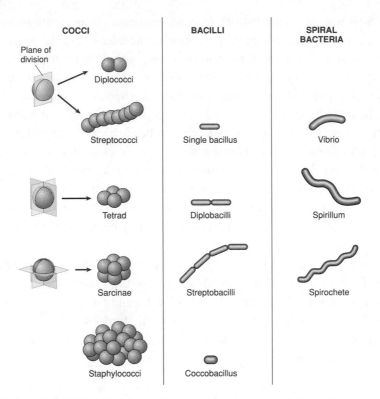

FIGURE 4.5. Morphology and arrangement of bacteria.

As bacterial cells grow and divide, cells may stay grouped together. The way the cells are grouped or arranged is called the **arrangement** (Figure 4.5). Some common arrangements are clustered (prefix *staphylo-*), in chains (prefix *strepto-*), and in twos (prefix *diplo-*). Sometimes cells are arranged in clusters of four (*tetrads*) or eight (*sarcinae*). If there is no particular arrangement of the cells, they are said to be **random** or to occur singly. The appearance of bacteria is typically described by both the morphology and arrangement together. For example, a cluster of spherical cells would be referred to as **staphylococci**, whereas a chain of rod-shaped cells would be referred to as **streptobacilli**.

> **REMEMBER**
> Most bacteria have a layer outside the plasma membrane called a cell wall. Human cells don't have a cell wall. (However, some eukaryotes—like plants, fungi, and protists—do have cell walls.)

THE CELL WALL

Most bacteria have a **cell wall**, a reinforcing layer found outside the plasma membrane. The cell wall is very important in protecting the cell from bursting if its environment becomes hypotonic. As water enters a cell that has a wall, the cytoplasm expands and presses against the wall, increasing the internal cellular pressure. This pressure provides a countermeasure against the further influx of water, and osmosis is halted, often, preventing the cell from being destroyed.

Most members of the domain Bacteria, including those that cause disease, have a cell wall that is primarily composed of a net-like molecule called **peptidoglycan** (Figure 4.6A). As the name implies, peptidoglycan has both a protein (peptido) and a sugar (glycan) component. The sugar component is a polysaccharide made of two alternating sugars, *N*-acetylglucosamine (NAG) and *N*-acetylmuramic acid (NAM). The protein component consists of short chains of amino acids that link layers of the polysaccharide together. Peptidoglycan is not found in human or other eukaryotic cells and is thus a safe target for antibiotics. For example, penicillin prevents the linking of the sugar layers by the amino acid chains and thus weakens the wall of bacteria, triggering the destruction of the cells.

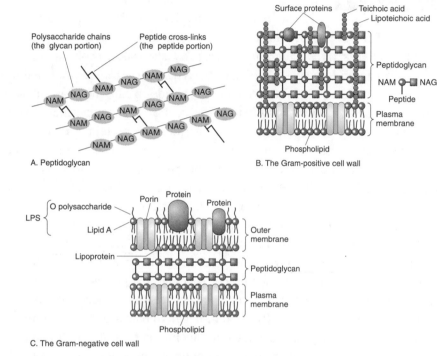

FIGURE 4.6. Bacterial cell walls.

A. Peptidoglycan. B. The gram-positive cell wall. C. The gram-negative cell wall.

The cell walls of bacteria contain other components besides peptidoglycan. Depending on these other components, and on the amount of peptidoglycan, most of the familiar bacteria can be placed into one of two categories: **gram-positive bacteria** or **gram-negative bacteria**. Because these two groups of bacteria respond differently to antibiotics, determining which type of wall a pathogen has is very important to the treatment of disease. The term "Gram" refers to a Danish microbiologist, Hans Christian Gram, who first developed a staining protocol that revealed the type of bacterial wall. That staining protocol is called the **Gram stain**, and it is usually among the first tests performed when trying to identify a bacterium.

Gram-positive bacteria include the familiar pathogens *Staphylococcus aureus* and *Streptococcus pyogenes*. These bacteria have a thick peptidoglycan layer within their walls (Figure 4.6B). In addition, their walls contain **teichoic acids**, modified alcohols that have a negative charge and may play a role in the transport of ions into the cell. The mycobacteria, which include *Mycobacterium tuberculosis*, are a group of rod-shaped bacteria that are referred to as gram-positive because of their reaction in the Gram stain, but they actually have a slightly different cell wall structure than that of other gram-positive bacteria. The most important feature of their cell wall is an outer waxy layer. This layer makes these bacteria resistant to a variety of disinfectants and is one of the reasons *Mycobacterium tuberculosis* is so hard to control.

Gram-negative bacteria, such as *E. coli* and *Neisseria gonorrhoeae*, have a thin peptidoglycan layer within their walls. Outside of the peptidoglycan, they have an additional membrane called the **outer membrane** (Figure 4.6C). The outer membrane, which is made up of phospholipids, lipopolysaccharides, lipoproteins, and channel proteins, contributes both to the disease process triggered by gram-negative bacteria and to their resistance to antimicrobial compounds. The **lipopolysaccharides (LPS)** consist of a lipid portion, called **lipid A**, and a polysaccharide portion. Lipid A is a toxin, called **endotoxin**, that can trigger fever and shock. Within the polysaccharide portion are two regions: the **core polysaccharide** and the **O polysaccharide**. The O polysaccharide is useful in identifying gram-negative bacteria. The channel proteins in the outer membrane are called **porins**. The porins are important because they help determine what passes through the outer membrane. Together, the porins and the tightly packed LPS molecules of the outer membrane are able to restrict access of some antibiotics and disinfectants to the cells of gram-negative bacteria, thus enabling the bacteria to survive. The region between the outer membrane and the plasma membrane is called the **periplasm**. This contains the peptidoglycan, enzymes, and proteins that assist in transporting substances across the membranes.

Some bacteria do not have a typical gram-positive or gram-negative cell wall. For example, the walls of mycobacteria typically have an outer waxy layer, and the mycoplasmas do not have a cell wall at all.

GLYCOCALYX

In response to certain environmental conditions, some bacteria are capable of producing yet another layer around the cell called a **glycocalyx**. For example, when you eat food that contains sucrose, or table sugar, the bacterium *Streptococcus mutans* in your mouth starts to make a glycocalyx. The glycocalyx helps *S. mutans* stick to your teeth, where it then produces acids that cause tooth decay. Another function of the glycocalyx in some bacteria is protection. Bacteria that can produce a glycocalyx can resist being eaten by **phagocytes**, white blood cells that defend our bodies by eating and breaking down bacteria and viruses. The exact chemical composition and structure of the glycocalyx depends on the type of bacterium. It may be made of polysaccharide, or protein, or a combination of the two. An organized glycocalyx that is firmly attached to the cell wall is called a **capsule**; a looser glycocalyx that is not firmly attached is called a **slime layer**.

FLAGELLA

Many bacteria can swim using **flagella**, whip-like extensions that stick out from the cell wall. The flagella rotate, propelling the bacteria forward, much like the prop on an outboard motor spins and pushes the boat through the water. Flagella have three parts: the **basal body**, the **hook**, and the **filament** (Figure 4.7). The basal body consists of disks of protein that anchor the flagellum into the cell wall and plasma membrane. The hook connects the basal body to the filament, which is the part of the flagellum that sticks out of the bacterial cell. The filament is made of a protein called **flagellin**.

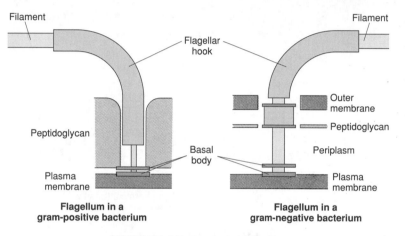

FIGURE 4.7. Bacterial flagella.

Bacteria swim toward some things, and away from others; for example, a bacterium might swim toward food or away from waste. There are variations in how bacteria accomplish this, but one well-studied way is that used by bacteria such as *E. coli*. In order to respond to something, a bacterium must be able to sense it with a receptor in its plasma membrane. In bacteria like *E. coli*, the receptor can then activate a chain of events that uses cellular energy to power the rotation of the flagellum. As the flagella rotate in one direction, the bacterium moves forward, or **runs** (Figure 4.8A). If the flagella reverse their rotation, the bacterium will spin in place, or **tumble** (Figure 4.8B). This is how bacteria like *E. coli* control their direction of movement. Random running and tumbling cause the bacterium to move randomly in an area (Figure 4.8C). If the receptors are picking up signals for something the bacterium can use, the bacterium will run more than it tumbles, continuing its movement toward the thing it can use (Figure 4.8D). However, if the receptors stop sending positive signals, or if they are picking up signals for something harmful to the bacterium, the bacterium will tumble more than it runs. The random tumbling changes the bacterium's direction until it again finds a direction that contains positive signals.

Bacterial movement is called **taxis**: When a bacterium moves toward something, that is called **positive taxis** when it moves away, that is **negative taxis**. The signal the bacterium is responding to can also be indicated by adding a prefix to the term taxis (e.g., *chemo-* for a chemical signal like food, *photo-* for light, and *magneto-* for the magnetic field of the Earth). So, movement of a bacterium toward food would be described as positive chemotaxis (Figure 4.8D).

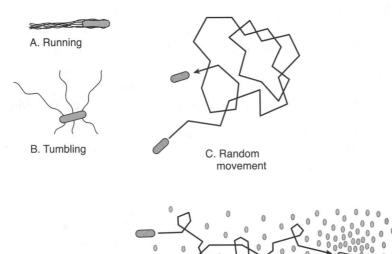

A. Running

B. Tumbling

C. Random
movement

D. Directed movement
towards a chemical (positive
chemotaxis)

FIGURE 4.8. caption

PILI AND FIMBRIAE

Another type of structure that protrudes from the cell is thin, hair-like extensions called **pili** or **fimbriae** (Figure 4.1). These structures allow bacteria to attach to things and thus may be very important for bacterial colonization and disease. For example, *Neisseria gonorrhoeae* uses fimbriae to attach to cells of the genitourinary tract so that it can grow and multiply there, causing the disease gonorrhea. Structurally, fimbriae and pili are essentially the same, and they are both made of the protein **pilin**. The distinction between the terms is usually based on function: If the structures are used by a bacterium to attach to another bacterial cell, they are called pili; if they are used to attach to surfaces or host cells, they are called fimbriae.

> **REMEMBER**
> Pili, fimbriae, and the glycocalyx all help bacteria attach to surfaces, including tissues in the human body. Thus, these structures can help pathogenic bacteria colonize the body and cause disease.

INSIDE THE CELL

Within the plasma membrane of the cell is the fluid-filled interior, or **cytoplasm**. The cytoplasm is a thick solution of molecules necessary to the cell, such as proteins, carbohydrates, lipids, and ions. Located within the cytoplasm of a prokaryote are the DNA, small structures called ribosomes, and particles of storage material called inclusion bodies.

The DNA of the cell is located in a region of the cytoplasm called the **nucleoid**. It consists of a single circular chromosome that is coiled up into many loops so that it will fit inside the cell. The DNA contains the genetic information for all necessary functions of the cell. Prokaryotes may also have smaller circular pieces of DNA called **plasmids**. Plasmids contain information for traits that may be very useful to the cell under certain circumstances (e.g., a plasmid might enable resistance to a particular antibiotic). Plasmids can be passed between prokaryotic cells, allowing the spread of these traits among bacteria.

DNA determines the traits of the cell because it contains the instructions for the formation of the proteins that perform the work of the cell. Proteins are made on structures called **ribosomes**. Ribosomes, which are made of ribosomal RNA and protein, are found in all cells, both prokaryotic and eukaryotic. However, there are differences between the ribosomes of prokaryotic and eukaryotic cells. Prokaryotic ribosomes, which are referred to as 70S ribosomes, are smaller and have slightly different components than do eukaryotic ribosomes. (70S is an indicator of their density; eukaryotic ribosomes are called 80S ribosomes.) One reason the difference between the ribosomes is important is the treatment of disease: Because the ribosomes of disease-causing bacteria are different than the ribosomes in human cells, drugs that target the bacterial ribosome can be used to slow bacterial growth without hurting human cells.

The cytoplasm of bacteria also contains **cytoskeletal proteins** that are similar to those found in eukaryotic cells. Actin-like proteins form helical bands under the cell walls of some bacteria, and tubulin-like proteins are associated with new wall formation in dividing bacterial cells. Because these cytoskeletal proteins are associated with the cell wall, scientists think they guide wall formation and thus determine the shape of bacterial cells.

Also found in the cytoplasm are various types of materials that the cell may be storing for use in its metabolism. When these materials form visible structures, they are called **inclusion bodies**. Some examples of different types of inclusion bodies are lipid, glycogen, or starch globules for energy storage; **sulfur granules** for use in certain types of metabolism; **volutin** (metachromatic granules) to provide a source of phosphate; and even crystals of magnetite, called **magnetosomes**, that allow a bacterium to detect the magnetic field of the Earth.

ENDOSPORES

If resources such as food or water become scarce, some gram-positive bacteria have the ability to form a resistant structure, called an **endospore**, that enables them to survive until conditions become better. **Sporogenesis** (Figure 4.9), or spore formation, begins when conditions become poor. First, the cell copies its DNA, so that there are two chromosomes within the cell. The plasma membrane grows inward and wraps around the new DNA and some cytoplasm, forming a double membrane. The space between the two membranes fills in with peptidoglycan; then, a tough, resistant protein forms a **spore coat** over the peptidoglycan. The original cell dies, releasing the spore to the environment, where it can survive adverse conditions for long periods of time. When conditions become better, the spores can then **germinate**, or grow into cells again.

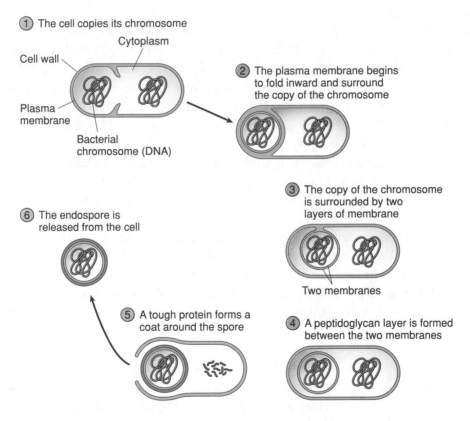

FIGURE 4.9. Sporogenesis, the process of endospore formation.

In repsonse to stress, such as starvation or lack of water, some bacteria form highly resistant cells called endospores, which can survive harsh conditions for long periods of time.

Endospores have been known to remain alive for thousands of years and to survive almost 20 hours in boiling water! In fact, endospores are so good at surviving, scientists have speculated that endospores might be able to remain alive in outer space. Because the endospores are so difficult to kill, spore-forming bacteria can be hard to control. If low-acid foods like vegetables are canned improperly, spores of *Clostridium botulinum* may survive in the food. *C. botulinum* produces a powerful neurotoxin, causing botulism in people who eat the canned goods. Another spore-forming species is *Bacillus anthracis*, which causes the disease anthrax. Spores of this species have been distributed in bioterrorist attacks. When the spores get into the body, they germinate, and the bacteria begin to multiply and produce toxins. If the spores get into a cut in the skin, a black lesion develops. This type of anthrax is easily treated with antibiotics. However, if the spores are inhaled into the lungs, a pneumonia-like illness results that is often fatal.

> **REMEMBER**
> Endospores are highly resistant structures that enable a bacterial cell to survive harsh conditions.

The Eukaryotic Cell

Some microbes, such as yeast, molds, and many pond water organisms, are eukaryotes just like us. Their cells are larger than those of prokaryotes and contain more internal structures (Figure 4.1). Many of these structures are surrounded by one or more layers of membranes. Cellular structures that have a membrane border are called **organelles** (little organs). Just like the organs of your body, each of these structures has a special job to perform for the cell. Some structures are involved in manufacturing molecules, others in producing energy for the cell, and still others in transporting materials around the cell.

THE NUCLEUS AND RIBOSOMES

The DNA of a eukaryotic cell is contained within a membrane-bounded structure called the **nucleus** (Figure 4.10). The nucleus is actually surrounded by a double membrane consisting of two phospholipid bilayers. This double membrane is called the **nuclear envelope**. Because it houses the DNA, the function of the nucleus is information storage. Information from the DNA that has been encoded in molecules of RNA can move from the nucleus to the rest of the cell via small holes in the nuclear envelope called **nuclear pores**. Also found in the nucleus is a dense region called the **nucleolus**, at which eukaryotic ribosomes are made. Once the ribosomes are made at the nucleolus, they travel through the nuclear pores to the cytoplasm of the cell to assist with protein synthesis. Some ribosomes remain in the cytoplasm as **free ribosomes**, whereas others will attach temporarily to the surfaces of membranes and become **bound ribosomes**. Proteins that are going to remain in the cytoplasm of the cell are made on free ribosomes. Proteins that will either be part of membranes or that will travel out of the cell are made on bound ribosomes.

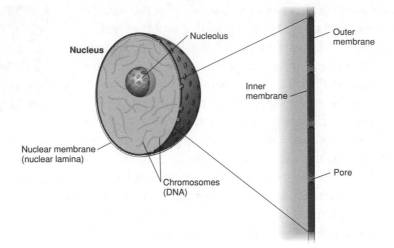

FIGURE 4.10. The nucleus.

THE ENDOMEMBRANE SYSTEM

One of the distinctive features of eukaryotic cells is the amount of membrane within the cell. A large part of this is the result of the **endomembrane system** (*endo* = inside) within the cell (Figure 4.11). The endomembrane system is a system that makes and transports proteins and lipids. It begins at the surface of the nucleus, where the **endoplasmic reticulum (ER)** develops off of the nuclear envelope. Another major component is the **Golgi apparatus**, a stack of flattened membranous structures. Lipids and proteins move between the ER, Golgi, and plasma membrane within **transport vesicles**, little spheres of membrane that travel around the cell and join with other membranes.

There are two kinds of endoplasmic reticulum: **rough endoplasmic reticulum (RER)**, which has ribosomes attached to it (making it look "rough"), and **smooth endoplasmic reticulum (SER)**, which has no ribosomes. Proteins that are going to be part of membranes or that will leave the cell are synthesized on the ribosomes of the RER. As they are made, the proteins are pushed into the **lumen** or interior of the RER. They are then packaged into vesicles, where they travel to the Golgi. The vesicles fuse with the membranes of the Golgi, depositing the proteins inside the lumen of the Golgi. Inside the Golgi, the proteins may be changed slightly and tagged with chemical groups so that they are transported to their proper destination. Similarly, lipids are synthesized at the SER and may travel to the Golgi in vesicles. Once proteins and lipids are modified at the Golgi, they can then travel to the plasma membrane in vesicles and either become part of the plasma membrane or be released to the outside of the cell.

> **REMEMBER**
> Membrane proteins and exported proteins are made on the rough endoplasmic reticulum. They travel in a vesicle to the Golgi, where they are modified, then travel in a vesicle to the plasma membrane.

Another type of vesicle that is formed from the Golgi is the **lysosome**. These vesicles contain a variety of digestive enzymes and other compounds that are used to break down cellular debris and foreign objects such as bacteria.

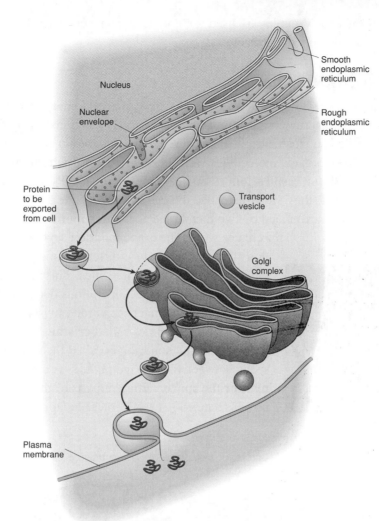

FIGURE 4.11. The endomembrane system.

Proteins that will be part of membranes or that will be exported from the cell are made at the rough endoplasmic reticulum. They travel by transport vesicles to the Golgi apparatus where they are modified and chemically marked for their destination. Then, they travel by transport vesicle to their destination, such as the plasma membrane.

Peroxisomes are similar to lysosomes in structure; however, they do not arise from the Golgi, but rather from the division of other peroxisomes. They break down certain types of molecules, like alcohol. When they break molecules down, hydrogen peroxide (H_2O_2) is produced. H_2O_2 is toxic to cells, but peroxisomes also produce catalase, an enzyme that can neutralize the H_2O_2. By neutralizing the H_2O_2, the peroxisomes protect the rest of the cell.

Some cells also contain **vacuoles**, large fluid-filled structures that are surrounded by a membrane. In particular, plant cells contain very large vacuoles that take up most of the volume of the cell. Vacuoles are used for storage by some cells and for intake of food for some single-celled eukaryotes.

ENERGY ORGANELLES

All cells require energy to grow and move. All cells can access energy for cellular processes by breaking down organic molecules (like food) and transferring that energy to a molecule called adenosine triphosphate (ATP). In a eukaryotic cell, most of the energy transfer from organic molecules to ATP occurs in the **mitochondrion** as part of a process called **cellular respiration**. Because usable energy (ATP) for the cell is made in mitochondria, they are often called the *powerplants* of the cell.

Mitochondria are surrounded by two membranes: an **outer membrane** and an **inner membrane** (Figure 4.12). This creates within the mitochondrion several different locations that can be used for different functions during the breakdown of organic molecules. This is analogous to a factory that has different operations occurring in different locations. The inner membrane twists back and forth, forming wrinkles called **cristae**. The innermost part of the mitochondrion is called the **matrix**. The space between the two membranes is called the **intermembrane space.** Different parts of the energy transfer process happen in each of these locations.

> **REMEMBER**
> Eukaryotic cells use mitochondria to transfer energy from food to ATP. Prokaryotic cells don't have mitochondria; they transfer energy from food to ATP in the cytoplasm and plasma membrane of their cells.

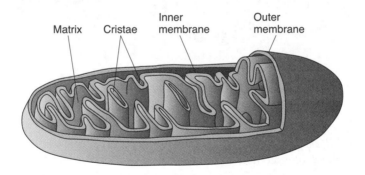

FIGURE 4.12. Mitochondrion.

Many eukaryotes, like humans, meet their energy needs by eating other organisms. These eukaryotes rely on their mitochondria to extract the energy from their food. Some eukaryotes, however, have the ability to make their own food. These organisms use energy from the sun to make sugar out of carbon dioxide and water in a process called **photosynthesis**. Once the carbon dioxide and light energy has been stored in sugar, photosynthetic eukaryotes then rely upon their mitochondria to get the energy back out of storage and transfer it to ATP to fuel the cell. Thus, virtually all eukaryotes have mitochondria. Photosynthetic eukaryotes, however, have an additional organelle that allows them to peform photosynthesis. This organelle is called the **chloroplast**.

Like mitochondria, chloroplasts are surrounded by two membranes: an **outer membrane** and an **inner membrane** (Figure 4.13). They also contain small sacs of membrane called **thylakoids**. The thylakoids are piled into stacks called **grana** within the interior of the chloroplast. The space between the two membranes is called the **intermembrane space**. The fluid within the chloroplast that surrounds the thylakoids is called the **stroma**. There is also the space within the thylakoids themselves. All of these membranes create various spaces that can be specialized for different tasks during photosynthesis.

> **REMEMBER**
> Eukaryotes like plants and algae use chloroplasts to do photosynthesis, making food by capturing energy from the sun, and carbon dioxide and water from the environment. Prokaryotic cells that do photosynthesis don't have chloroplasts, so they do photosynthesis in the cytoplasm and plasma membrane of their cells.

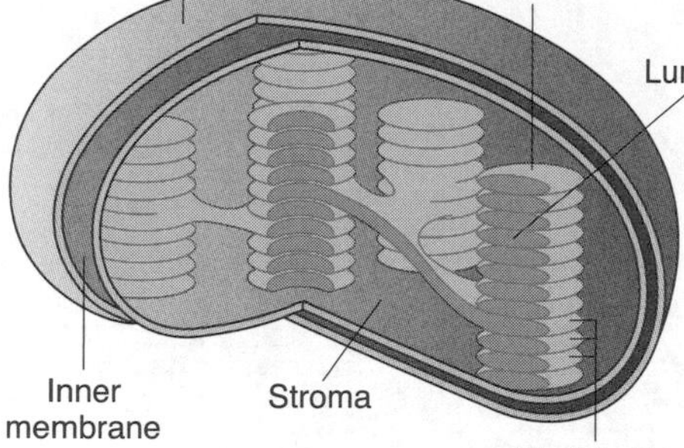

FIGURE 4.13. A chloroplast.

One of the most fascinating stories in the study of eukaryotic cells surrounds the origin of the mitochondria and chloroplasts. In the early 1900s, scientists who were looking at eukaryotic cells on a light microscope noticed that these organelles looked very similar to some types of bacteria. In particular, the French scientist P. Portier hypothesized that chloroplasts might be related to green bacteria, and the Russian K. C. Mereschovsky speculated on the relationship between bacteria and mitochondria. However, these ideas were virtually ignored until the 1960s, when it was discovered that chloroplasts and mitochondria both had their own DNA and ribosomes, and that the ribosomes were prokaryotic ribosomes! This led the scientist L. Margulis to propose that chloroplasts and mitochondria were the descendents of free-living bacteria that had long ago moved into the cytoplasm of another cell. Her idea, that cells living inside of other cells became the chloroplast and mitochondrion, is called the **serial endosymbiotic theory** (*endo* = inside; *symbiosis* = living together). At first, this idea was completely rejected by other scientists. However, when scientists began comparing the genetic code of the DNA in mitochondria and chloroplasts with that of free-living bacteria today, they indeed found that there are more similarities between the organelles and free-living bacteria than there are between the organelles and the DNA in the nucleus of the cells they were living in! This evidence finally persuaded most scientists that the serial endosymbiotic theory of the origin of chloroplasts and mito-chondria is correct.

THE CYTOSKELETON

Just as a building needs the support of beams and girders, so does a cell need support. In a cell, support is provided by protein cables that are part of the cell's **cytoskeleton** (Figure 4.14). Proteins of the cytoskeleton also act as *railroad tracks* for movement of materials around the cell and allow movement of cells by eukaryotic flagella (Figure 4.15). There are three categories of proteins within the eukaryotic cytoskeleton. The distinguishing characteristics of these proteins are summarized in Table 4.2.

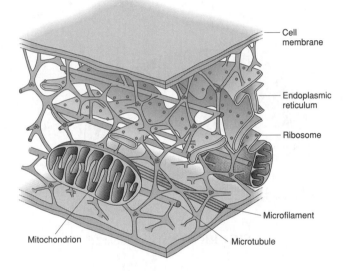

FIGURE 4.14. The eukaryotic cytoskeleton.

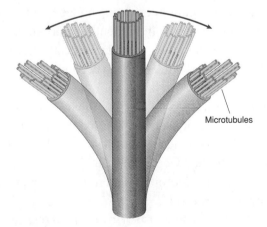

FIGURE 4.15. Cilia and eukaryotic flagella.

Flagella and cilia have the same basic structure and contain microtubules. The microtubules slide past each other, causing the cilia and flagella to bend. Some cells use these structures for swimming. Others, like cells in the human airway, use them to move mucus.

TABLE 4.2. Categories of Cytoskeletal Proteins

Category	Major Protein Component	Functions
Microtubules	Tubulin	Provide structure within cilia and eukaryotic flagella. Bending of tubulin causes movement of these structures. Provide tracks within cells. Allows movement of organelles and vesicles within cells and chromosomes during cellular division.
Microfilaments	Actin	Provide tracks within cells. Allows contractile movement of muscles and amoeboid movement of amoebae and white blood cells (phagocytes). In some cells, allows movement of organelles within cells. Provides reinforcement of plasma membrane.
Intermediate	Various proteins	Provides reinforcement of cells. For example, keratin reinforces skin cells, and lamin reinforces the nuclear envelope.

EXTERNAL STRUCTURES

Some eukaryotic cells have cell walls that are external to the plasma membrane. The walls of eukaryotes, however, are made of different molecules than those of prokaryotes. Plants and algae have cell walls that contain large amounts of the polysaccharide **cellulose**. The cell walls of fungi are reinforced with **chitin**. Cell walls support cells and help keep cells of multicellular organisms together.

Animal cells do not have cell walls, but they do produce a sticky mass of molecules that surrounds their cells. This mass of molecules is called the **extracellular matrix (ECM)**. It helps connect animal cells together and contains molecules that are important in signaling from the outside of cells to the inside.

REVIEW EXERCISES FOR CHAPTER 4

1. Compare and contrast prokaryotic and eukaryotic cells. How are they the same? How are they different?

2. Describe the structure of the plasma membrane. What functions do plasma membrane proteins perform?

3. Redraw and fill in the following table:

Type of Transport	Type of Molecules Moved? (small, large, hydrophobic, hydrophilic?)	Direction of Movement? (high to low concentration, or low to high?)	Energy Input by Cell?	Notes
Simple diffusion				
Facilitated diffusion				
Osmosis				
Active transport				
Group translocation				

4. Compare and contrast the gram-positive and gram-negative cell wall. Which is more susceptible to antibiotics? Why?

5. What molecules make up the glycocalyx? What is its function?

6. What are the three parts of the prokaryotic flagellum and the function of each? How would movement toward a sugar be generated? How would you describe it?

7. Describe the structure and function of pili and fimbriae.

8. Compare and contrast the bacterial chromosome with plasmids.

9. Describe the structure and function of the prokaryotic ribosome.

10. Describe the different types of inclusion bodies.

11. Why do bacteria make endospores? Why are spore-forming bacteria difficult to control?

12. Describe the structure of the nucleus. What is the function of the nucleolus?

13. Distinguish between free and bound ribosomes.

14. Distinguish between the rough and smooth endoplasmic reticulum.

15. For a protein that is to be exported from the cell, describe the pathway in the cell the protein would follow from the time it was made until it was released from the cell.

16. Distinguish between transport vesicles, lysosomes, peroxisomes, and vacuoles.

17. Compare and contrast mitochondria with chloroplasts.

18. Describe the three categories of cytoskeletal proteins.

19. Compare and contrast eukaryotic cell walls with prokaryotic cell walls.

20. What is the extracellular matrix? What is its function?

SELF-TEST

1. Which of the following is not true of prokaryotic cells?

   A. They have a plasma membrane.
   B. They have DNA.
   C. They divide by binary fission.
   D. They have nuclei.
   E. They have a cell wall.

2. Which of the following types of membrane transport requires energy to be input from the cell?

   A. Simple diffusion
   B. Active transport
   C. Osmosis
   D. Facilitated diffusion
   E. None of the above

3. Which of the following types of membrane transport occurs only in prokaryotic cells?

   A. Simple diffusion
   B. Osmosis
   C. Facilitated diffusion
   D. Group translocation
   E. Active transport

4. Which of the following types of membrane transport occurs only in eukaryotic cells?

   A. Simple diffusion
   B. Endocytosis
   C. Osmosis
   D. Facilitated diffusion
   E. Active transport

5. Which of the following types of membrane transport can occur without the help of a membrane protein?

   A. Simple diffusion
   B. Facilitated diffusion
   C. Active transport
   D. Group translocation
   E. None of the above

6. Which of the following correctly describes a cluster of spherical cells?

   A. Streptococci
   B. Staphylococci
   C. Streptobacilli
   D. Vibrio
   E. Diplococci

7. Which type of cell wall contains endotoxin?

   A. Gram-positive cell wall
   B. Gram-negative cell wall

8. Which type of cell wall contains a thick layer of peptidoglycan?

   A. Gram-positive cell wall
   B. Gram-negative cell wall

9. Which type of cell wall contains teichoic acids?

   A. Gram-positive cell wall
   B. Gram-negative cell wall

10. Which component of the gram-negative cell wall is responsible for resistance to some antibiotics and disinfectants?

   A. Peptidoglycan
   B. O polysaccharide
   C. Porins
   D. Periplasmic gel
   E. Teichoic acids

11. Which of the following is a function of the glycocalyx?

    A. Attachment
    B. Motility
    C. Protection
    D. Both A and B are correct.
    E. Both A and C are correct.

12. Which of the following correctly describes movement away from light?

    A. Positive chemotaxis
    B. Positive phototaxis
    C. Negative chemotaxis
    D. Negative phototaxis
    E. Negative magnetotaxis

13. True or False. Pili are used for attachment, while fimbriae are used for motility.

14. True or False. If a bacterial cell loses its plasmids, it will always die.

15. True or False. The nucleoid is not surrounded by a membrane.

16. Which of the following is an incorrect pair?

    A. Volutin—phosphate
    B. Starch—energy
    C. Magnetosome—energy
    D. Lipid—energy
    E. Sulfur—metabolism

17. Cells of the bacterium *Bacillus thuringiensis* generate a toxin that kills lepidopterans (moths). The bacterium can be sold in a dried state for agricultural use. Which of the following structures would make this possible?

    A. Glycocalyx
    B. Endospore
    C. Cell wall
    D. Flagellum
    E. Fimbriae

18. Consider the statement "Sugar rots your teeth." Is this statement true or false? Explain your answer.

19. Mitochondria, which have their own DNA and ribosomes, are believed to be the descendants of bacterial cells. What evidence to support this theory would you expect to find from an examination of mitochondrial DNA and ribosomes?

20. Penicillin, which prevents the linking of peptidoglycan layers by amino acid side chains, is more effective on gram-positive cells than gram-negative cells. Why might this be so? (*Hint:* Penicillin is a hydrophilic molecule.)

21. How does placing fish between layers of salt protect it from being spoiled by bacterial degradation?

22. The DNA of a eukaryotic cell is located in the

    A. nucleolus.
    B. nucleoid.
    C. nucleus.
    D. ribosome.
    E. rough endoplasmic reticulum.

23. Which of the following represents the pathway a plasma membrane protein would take in the cell from the time it is made to its arrival at the plasma membrane?

    A. Smooth endoplasmic reticulum to Golgi to plasma membrane
    B. Rough endoplasmic reticulum to Golgi to plasma membrane
    C. Free ribosome to Golgi to plasma membrane
    D. Nucleus to rough endoplasmic reticulum to plasma membrane
    E. Rough endoplasmic reticulum to lysosome to plasma membrane

24. True or False. Foreign objects are digested in lysosomes.

25. The cristae

    A. are made of membrane.
    B. are found in the mitochondrion.
    C. are found in chloroplasts.
    D. Both A and B are correct.
    E. Both A and C are correct.

26. True or False. Plants have chloroplasts instead of mitochondria.

27. True or False. The inner membrane of the mitochondrion is called the grana.

28. True or False. The central area of a chloroplast is called the stroma.

29. True or False. Microfilaments are made of tubulin.

30. Which of the following reinforces the plasma membrane of eukaryotic cells?

    A. Microtubulues
    B. Microfilaments
    C. Intermediate filaments
    D. Both A and B are correct.
    E. A, B, and C are correct.

31. True or False. Animal cells have cell walls.

Answers

Review Exercises

1. All cells have a plasma membrane, DNA, ribosomes, and a cytoplasm. In prokaryotic cells, the DNA is not enclosed by a membrane, nor are there any membrane-bounded organelles. In eukaryotic cells, the DNA is enclosed by a membrane to form a nucleus, and there are membrane-bounded organelles. Prokaryotes divide by binary fission; eukaryotes divide by mitosis.

2. The plasma membrane is made of a bilayer of phospholipids with proteins embedded in it. Eukaryotic membranes also contain sterols. The plasma membrane controls what enters and exits the cell. It also receives signals from the environment and passes them to the inside of the cell.

3.

Type of Transport	Type of Molecules Moved? (small, large, hydrophobic, hydrophilic?)	Direction of Movement? (high to low concentration, or low to high?)	Energy Input by Cell?	Notes
Simple diffusion	Small or hydrophobic	High to low	No	Molecules cross by themselves or through channel proteins
Facilitated diffusion	Small or larger hydrophilic	High to low	No	Move through carrier proteins
Osmosis	Water	High water to low water, or water moves toward greater solute	No	Water moves from a more hypotonic environment toward a more hypertonic environment
Active transport	Small or larger	Low to high	Yes	Carrier proteins utilize ATP during transport
Group translocation	Small or larger	Low to high	Yes	Carrier proteins utilize ATP during transport and modify molecule being transported

4. Both the gram-positive and gram-negative cell walls contain peptidoglycan. However, the gram-positive cell wall contains a thick layer of peptidoglycan, whereas the gram-negative cell wall contains a thin layer of peptidoglycan. Gram-positive cell walls contain teichoic acids; gram-negative cell walls do not. Gram-negative cell walls have an additional membrane outside the peptidoglycan; this is called the outer membrane. It contains transporter proteins called porins, and lipopolysaccharides. The lipopolysaccharides contain lipid A, a toxin, and an O polysaccharide. Gram-positive cell walls are more susceptible to certain antibiotics because the porins in the gram-negative outer membranes can exclude some of these agents, preventing them from reaching the cell.

5. The glycocalyx is made of polypeptides and polysaccharides. It functions in attachment and protection from phagocytosis.

6. The basal body attaches the flagellum to the wall and membrane. The filament propels the bacterium, and the hook attaches the basal body to the filament. Movement toward a sugar would require that the bacterium had a receptor for that sugar. Once the sugar was sensed by a receptor, a signal would be sent to affect flagellar rotation. This requires energy from the cell. As long as the bacterium was receiving positive signals from the presence of the sugar, it would run more often than it would tumble. If the signal stopped, the bacterium would start to tumble more, until it was reoriented toward the sugar. This is called positive chemotaxis.

7. Pili and fimbriae are fine projections made of pilin. They function in attachment.

8. Both the chromosome and plasmids are circular and made of DNA. The chromosome, however, is much larger than plasmids and contains the DNA necessary for essential cellular functions. Plasmids contain the code for useful functions that are not necessary to the cell under all circumstances.

9. Prokaryotic ribosomes are made of rRNA and protein. They are smaller than eukaryotic ribosomes. Their function is protein synthesis.

10. Lipid, starch, and glycogen may function as energy reserves. Sulfur granules are present for use in certain types of energy metabolism. Volutin provides a source of phosphate. Magnetosomes allow detection of the magnetic field of the Earth.

11. Bacteria make endospores in order to survive unfavorable conditions. Spore-forming bacteria are difficult to control because endospores are very resistant to control methods (e.g., some endospores can survive up to 20 hours of boiling).

12. The nucleus is surrounded by a double membrane called the nuclear lamina. Within the nucleus is the DNA of the cell and a region called the nucleolus at which ribosomes are made. Information travels out of the nucleus through the nuclear pores, small holes in the nuclear lamina.

13. Free ribosomes are in the cytoplasm of the cell. Bound ribosomes are associated with the rough endoplasmic reticulum. Both function in protein synthesis. Proteins that will remain in the cytoplasm are made on free ribosomes; proteins that will be part of membranes or that will leave the cell are made on bound ribosomes.

14. Both the rough and smooth endoplasmic reticulum are made of membranes. The rough endoplasmic reticulum has ribosomes bound to it; the smooth endoplasmic reticulum does not. The RER functions in protein synthesis; the SER functions in lipid synthesis.

15. An exported protein would be synthesized at the RER. As it was made, it would be pushed into the interior of the RER. It would then be packaged into a transport vesicle and travel to the Golgi. The vesicle would fuse with the Golgi, and the protein would be deposited inside the lumen of the Gogli. The protein would travel through the Golgi and be tagged with chemical groups that marked it for export from the cell. It would leave the Golgi in another transport vesicle and travel to the plasma membrane. The vesicle would fuse with the plasma membrane, and the protein would be deposited outside the cell.

16. All of these structures are surrounded by membrane, but their functions are different. Transport vesicles move things around the cell. Lysosomes contain digestive enzymes and function to break down foreign objects or other materials the cell needs to recycle. Peroxisomes detoxify compounds like alcohol and produce catalase to neutralize hydrogen peroxide. Vacuoles are storage organelles for some types of cells and may be involved in capture of food for other types of cells.

17. Both mitochondria and chloroplasts are involved in energy transfer. Mitochondria are involved in cellular respiration, a process by which energy from organic molecules is transferred to ATP. Chloroplasts are involved in photosynthesis, in which sunlight energy is transferred into organic molecules. Both are surrounded by two membranes, an inner membrane and an outer membrane. The inner membrane of mitochondria is very wrinkled, forming structures called cristae. Chloroplasts have additional sacs of membrane called thylakoids.

18. Microtubules are made of tubulin. They function in the movement of eukaryotic cells and in the movement of structures within eukaryotic cells. Microfilaments are made of actin. They reinforce the plasma membrane, enable contractile movement of muscle cells, enable amoeboid movement, and help things move within some types of cells. Intermediate filaments can be made of various proteins. They help reinforce cells.

19. Both types of cell walls surround the cell and reinforce it. They help protect the cell from bursting when placed in a hypotonic (lower solute) environment. Prokaryotic cell walls are made of peptidoglycan. Eukaryotic cell walls are not made of peptidoglycan. They may contain large amounts of cellulose (plants) or chitin (fungi).

20. The extracellular matrix is a mass of sticky molecules produced by animal cells and released outside the cell. It helps attach animal cells to each other and contains molecules important in signaling between cells.

Self-Test

1. D	7. B	13. F
2. B	8. A	14. F
3. D	9. A	15. T
4. B	10. C	16. C
5. A	11. E	17. B
6. B	12. D	

18. True or False as long as the explanation is correct. Sugar itself does not rot your teeth. However, it does trigger glycocalyx formation in oral bacteria. This allows the bacteria to attach to your teeth. Once attached, acids from their metabolism create holes in the tooth enamel. So, the sugar itself doesn't do it, but it is necessary to the process.

19. Mitochondrial DNA is circular, like prokaryotic DNA. Mitochondrial ribosomes are 70S ribosomes.

20. Because penicillin is hydrophilic, it cannot cross the outer membrane by itself. To gain access to the peptidoglycan, it would have to pass through the porin proteins. These proteins restrict the ability of penicillin to pass through the outer membrane.

21. The salt around the fish creates a hypertonic environment for any bacteria that might be present. Water leaves the bacterial cells, and they can no longer grow or divide.

22. C	26. F	30. B
23. B	27. F	31. F
24. T	28. T	
25. D	29. F	

Microbial Metabolism

WHAT YOU WILL LEARN

This chapter introduces the fundamentals of metabolism, especially as they apply to microbial cells. As you study this chapter, you will:

- learn about the ways humans use microbial metabolism;
- review how cells transfer energy and electrons;
- explore the process of cellular respiration;
- compare cellular respiration and fermentation;
- discover how microbes make food by photosynthesis and chemolithoautotrophy.

SECTIONS IN THIS CHAPTER

- Fundamentals of Metabolism
- An Overview of Microbial Metabolism
- Fundamental Metabolic Pathways of Heterotrophic Metabolism
- Fundamental Metabolic Pathways of Autotrophic Metabolism
- Anabolic Pathways

Cells, whether yours or a microbe's, are living, growing things. They require energy and materials that can be used as building blocks. We feed our cells by taking food into our mouths and then breaking it down in our digestive system. Once the food is broken down into protein, carbohydrate, and fat molecules, our cells can take in those molecules and use them. Similarly, the cells of microbes may take in food molecules from their environment. If the cells break the molecules down further in order to obtain energy, it is called **catabolism**. If a cell uses a molecule as a building block to make a larger molecule, it is called **anabolism**. All of these processes together, the breaking down and the building up, is called **metabolism**.

Metabolism is an amazing thing. At any one time, a cell is performing hundreds of different chemical reactions, breaking down molecules for energy and building molecules it needs to do certain jobs. And when we look at all the types of life on planet Earth, the metabolic champions are the microbes. As a group, microbes are incredibly diverse in their metabolism. They can break down more types of molecules than any other group and produce many different types of molecules as waste.

This metabolic diversity of microbes has been used by humans since ancient times. Anything you've ever eaten that's been fermented, like yogurt, wine, or beer, was made with microbial metabolism. Breads and cheeses too require the action of microbes, as do soy sauce, kim chee, sausages, sauerkraut, and kefir. Our use of microbial metabolism goes beyond food. We use microbes to clean up our sewage and other types of pollution like oil spills. In the future, we may be able to make plastics that microbes can eat, eliminating a problem substance in our landfills. Microbes also produce enzymes that are used in laundry detergents and to make products like paper and corn syrup. Microbes make vitamins and are a source of the amino acids used to make artificial sweeteners (aspartame) and MSG. Recently, we've even begun to modify microbes by introducing pieces of DNA so that they can make human proteins like insulin and interferon that are used to fight disease.

> **REMEMBER**
> Humans use microbial metabolism to produce fermented foods and products like paper and stone-washed jeans, and to clean up our waste.

As you can see from these examples, microbial metabolism is incredibly diverse and important to humans. Yet understanding metabolism, because of its complexity and diversity, is often a real challenge. To try and keep the subject manageable, this chapter will present in detail only a selection of the different types of metabolic processes that occur in microbes.

Fundamentals of Metabolism

ENZYMES

Nothing happens in metabolism without the help of an **enzyme** (Figure 5.1). Enzymes are proteins that **catalyze**, or speed up, chemical reactions. Every enzyme has a pocket called an **active site** that can bind certain chemicals, called **substrates**, and make it easier for those chemicals to react. Because the active site of an enzyme can only bind its particular substrate, enzymes are specific. So, virtually every chemical reaction that happens in a cell requires an enzyme to help it happen. The enzymes themselves are not typically changed by the reaction and can continue to do their jobs over and over again.

> **REMEMBER**
> Enzymes bind substrates in their active sites, helping them to react and speed up chemical reactions.

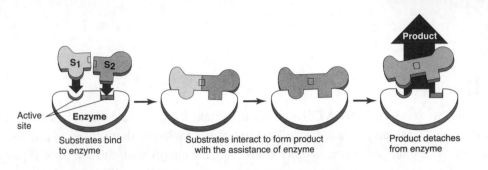

FIGURE 5.1. Enzyme structure and function.

An enzyme has a specific active site where substrate molecules bind. In this case, two substrate molecules bind to the active site, and the enzyme catalyzes bond formation between them. The product is released, and the enzyme may participate in another reaction.

METABOLIC PATHWAYS

During chemical reactions, molecules interact with one another. Bonds may be broken, and others may be formed. Energy is transferred from one source to another. When cells perform chemical changes during metabolism, they do so in very small steps. Take, for example, the reaction that we considered in Chapter 2:

$$C_6H_{12}O_6 \quad + \quad 6O_2 \quad \rightarrow \quad 6CO_2 \quad + \quad 6H_2O$$

glucose   oxygen   carbon dioxide   water

This process, the breaking down of glucose in the presence of oxygen, is done all the time in a wide variety of cells. A cell, however, would not do this conversion all at once. If it did, the energy released from the reaction could not be controlled and would damage the cell. When cells perform this process, they break it down into a series of small steps.

A **metabolic pathway** is a series of these small steps: several chemical reactions strung together, where the product of one becomes the substrate of the next. A metabolic pathway might be represented like this:

$$A \rightarrow B \rightarrow C \rightarrow D \rightarrow E$$

In this reaction, A represents the initial **substrate** and E represents the **product** of the pathway. Each arrow represents a chemical reaction that would be catalyzed by a unique enzyme. B, C, and D represent **intermediates** in this pathway, molecules that are the product of one reaction and the substrate of the next. This pathway could also intersect with many other pathways in the cell. For example, molecule D could be the substrate for a pathway that branches off of this one.

$$A \to B \to C \to D \to E$$
$$\downarrow$$
$$X$$
$$\downarrow$$
$$Y$$

ATP

Adenosine triphosphate (ATP) is a very important molecule to cells. It is sometimes called the energy currency of the cell because it is a molecule that is used to provide energy to cellular processes. In other words, it is the energy cash the cell uses to pay for certain energy requiring reactions. When ATP is used to supply energy to a reaction, it is hydrolyzed (broken down) into its component parts, **adenosine diphosphate (ADP)** and **phosphate (P)**. And, just as people do not have an unlimited supply of cash, neither does a cell have an unlimited supply of ATP. They must constantly replenish their supply of ATP by using energy from an outside source to reform the ATP from ADP and P. Cells like ours obtain this outside energy from organic molecules (food). Cells of other organisms may obtain this energy from light or other chemicals. Thus, there is a constant ATP cycle (Figure 5.2) that exists in cells and is a fundamental part of metabolism.

> **REMEMBER**
> When ATP is hydrolyzed to ADP + P, energy becomes available to the cell. The phosphorylation of ADP to ATP requires energy from the cell.

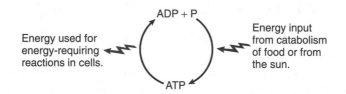

FIGURE 5.2. The ATP cycle.

Energy from catabolic processes is transferred into ATP molecules as they are formed from ADP + P. ATP molecules are then used to transfer energy into energy-requiring reactions in cells, generating ADP + P.

NAD/NADH + H⁺

Oxidation and **reduction** are very important processes during metabolism. Recall that oxidation is the loss of electrons and reduction is the gain of electrons. The process of transferring energy from one source to another during metabolism is tightly linked to the transfer of electrons from one molecule to another. In other words, to transfer energy to ATP, cells have to move electrons around.

During metabolism, electrons are often removed from one molecule and temporarily given to a carrier molecule called **nicotinamide adenine dinucleotide (NAD$^+$)**. Two hydrogen atoms, each with one proton (H$^+$) and one electron, are removed from the molecule being oxidized. The NAD$^+$ accepts the two electrons and one of the protons but releases the other proton. Thus, when NAD$^+$ accepts electrons, it is reduced to NADH + H$^+$. The NADH + H$^+$ carries the electrons to another molecule and then gives them to that molecule. In this reaction, NADH + H$^+$ is oxidized back to NAD$^+$, and the electron acceptor is reduced. You can think of NAD$^+$ as a kind of electron shuttle bus: NAD$^+$ is the empty bus that picks up its electron passengers; NADH + H$^+$ is the full bus. NAD$^+$, the empty bus, is the oxidized form of the molecule; NADH +H$^+$, the full bus, is the reduced form of the molecule. The transfer of electrons by NAD$^+$/NADH + H$^+$ is a very important part of the energy transfer processes of cells. Just as there is an ATP cycle that is part of metabolism, there is also an NAD$^+$/NADH + H$^+$ cycle (Figure 5.3).

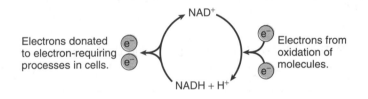

FIGURE 5.3. The NAD$^+$/NADH + H$^+$ cycle.

Many cellular processes involve oxidation and reduction reactions. As molecules are oxidized, their electrons are transferred to NAD$^+$, reducing it to NADH + H$^+$. NADH molecules can then provide electrons to electron-requiring processes in cells, such as the synthesis of macromolecules and the generation of ATP by chemiosmosis.

An Overview of Microbial Metabolism

NUTRITIONAL CLASSIFICATION

All cells need three basic things in order to function: a source of carbon, a source of energy, and a source of electrons. All cells need carbon because carbon is the fundamental building block for all organic molecules (e.g., the carbohydrates, proteins, lipids, and nucleic acids that make up the structure of the cell). All cells need energy in order to do energy-requiring processes like transport, movement, and synthesis of cellular components. All cells need a source of electrons because electron transfer is an essential part of energy metabolism and because electrons are needed during the synthesis of cellular components. How and where a cell gets these three fundamental things is the basis for the cell's metabolism.

When we know the source of energy, carbon, and electrons for a cell, we have an overview of that cell's metabolic strategy. Humans, for example, are very boring. We get everything from our food. We have metabolic pathways that break down our food, transfer electrons to electron carriers, transfer energy to ATP, and give us carbon-containing building blocks for our cells. A plant, on the other hand, gets carbon from carbon dioxide in the atmosphere, energy from sunlight, and electrons from water. As you can imagine, plants have metabolic pathways that do not exist in humans.

A certain terminology is used to describe the sources of energy, electrons, and carbon for cells. The suffix –*troph*, which means to eat, is used with prefixes that indicate the energy, electron, and carbon sources. These prefixes are shown in Table 5.1. If we return to the earlier examples, then, a human could be described as a chemoorganoheterotroph, whereas a plant would be a photolithoautotroph. Often, when a group of cells is being described by its nutritional classification, the source of electrons is omitted. In this system, humans would be chemoheterotrophs, and plants would be photoautotrophs. In the microbial world, there are many different types of metabolic strategies, from microbes that are similar to humans, to those that are similar to plants, to even more that are very different from either.

TABLE 5.1. Prefixes Used for Nutritional Classification

	Source	Prefix
Energy	Chemical	*Chemo-*
	Light	*Photo-*
Electrons	Organic molecules	*Organo-*
	Inorganic molecules	*Litho-*
Carbon	Organic molecules	*Hetero-*
	Inorganic molecules	*Auto-*

Fundamental Metabolic Pathways of Heterotrophic Metabolism

Heterotrophs are organisms that obtain their carbon from organic molecules like sugars, proteins, and fats. In other words, they obtain their carbon by eating other organisms (*hetero* = other; *troph* = to eat). Organisms that are chemoorganoheterotrophs, like humans and *E. coli*, oxidize these organic molecules to obtain their energy and electrons. In the process, many different carbon-containing intermediates are generated; they can be used by the cells to build cellular components during anabolic processes.

CELLULAR RESPIRATION

Cellular respiration refers to the oxidation of glucose to carbon dioxide. When oxygen is used for this process, it is called **aerobic respiration**. The overall equation for aerobic respiration is:

$$C_6H_{12}O_6 \quad + \quad 6O_2 \quad \rightarrow \quad 6CO_2 \quad + \quad 6H_2O$$

glucose oxygen carbon dioxide water

However, as stated before, cells do not do this reaction all at once. Instead, the oxidation of glucose by aerobic respiration happens in many small steps. We can study these steps in three main stages: glycolysis, the Kreb's cycle, and the electron transport chain (Figure 5.4).

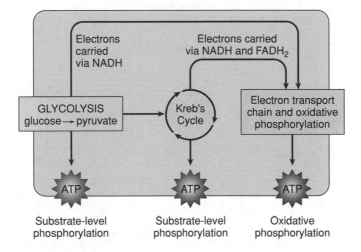

FIGURE 5.4. An overview of cellular respiration.

Cellular respiration consists of three phases: the initial oxidation of organic molecules by glycolysis, the additional oxidation of organic molecules in the Kreb's cycle, and the geneneration of ATP using an electron transport chain. ATP is generated by substrate-level phosphorylation during both glycolysis and the Kreb's cycle, and by oxidative phosphorylation at the electron transport chain.

GLYCOLYSIS

All cells on planet Earth need to manipulate carbon-containing molecules to access energy and to form the building blocks of living cells. The most common metabolic pathway used for this purpose by cells on Earth is **glycolysis**, the oxidation of glucose to two molecules of pyruvate (Figure 5.5). During glycolysis, which is also known as the Embden–Meyerhof Pathway of glycolysis, the six-carbon backbone of glucose is split into two three-carbon molecules. These three-carbon molecules are then oxidized, ultimately yielding two molecules of pyruvate, which also has three carbons. The electrons from the glucose are transferred to two NAD^+ molecules, which are reduced to $2NADH + 2H^+$. During the beginning of glycolysis, two ATP molecules are used to transfer energy to the pathway, but in the second half of glycolysis four ATP molecules are made as energy becomes available from the oxidation of the carbon backbone. Thus, the net amount of ATP produced by glycolysis is two ATP per glucose molecule; overall, glycolysis is a process that yields usable energy to the cell.

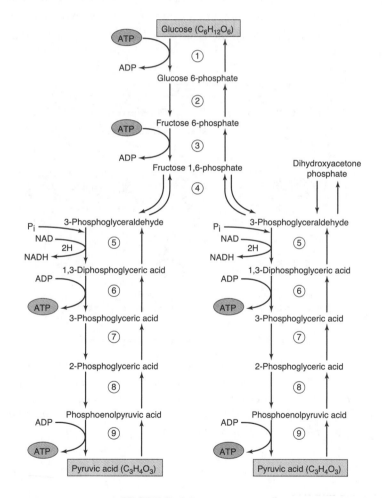

FIGURE 5.5. Glycolysis.

Glycolysis is a multistep biochemical process in which a molecule of glucose is converted to two molecules of pyruvic acid. Note that, in the process, two molecules of ATP are used, and four molecules of ATP are produced, for a net gain of two ATP molecules. Two molecules of NADH + H⁺ are also formed.

The ATP made during glycolysis is made by a process called **substrate-level phosphorylation**. (To phosphorylate is to add a phosphate to something; in this case, the phosphate is being added to ADP to make ATP.) During substrate-level phosphorylation, an enzyme binds a phosphate-containing substrate and an ADP molecule in its active site and then catalyzes the transfer of the phosphate from the substrate to ADP (Figure 5.6).

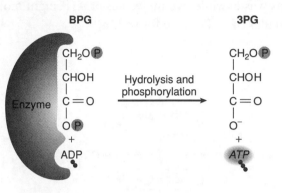

FIGURE 5.6. Substrate-level phosphorylation.

An enzyme transfers a phosphate from a substrate,
bisphosphoglyceric acid, to ADP, forming ATP.

In addition to glycolysis, some microbes oxidize carbon-containing molecules by either the Enter–Doudoroff pathway or the pentose phosphate pathway. Both of these pathways generate some ATP, although the individual steps of the pathways and the enzymes involved are somewhat different. Most microbial cells use either Embden–Meyerof or Entner–Doudoroff glycolysis, as these pathways are very similar. Many microbes use the pentose phosphate pathway in addition to one of these other pathways in order to generate the five-carbon intermediates necessary for the formation of amino acids and nucleic acids. All of these metabolic pathways allow cells to generate the ATP, carbon-containing intermediates, and reduced electron carriers that are necessary for survival.

KREB'S CYCLE

After glucose is oxidized in glycolysis, there is still a lot of energy in pyruvate. To obtain this energy, pyruvate is oxidized. First, a series of reactions that converts pyruvate into a molecule called acetyl-CoA occurs (Figure 5.7), by removing carbon and hydrogen and attaching co-enzyme A. This series of reactions has many names, the simplest of which is **pyruvate oxidation** (Figure 5.7). After pyruvate oxidation occurs, acetyl-CoA is oxidized in a series of

> **REMEMBER**
> Pyruvate oxidation and the Kreb's cycle provide cells with carbon in the form of 2-, 4-, 5-, and 6-carbon intermediates, energy in the form of ATP, and electrons carried by NADH and $FADH_2$.

reactions called the **Kreb's cycle** (Figure 5.7), which is also referred to as the citric acid cycle or the tricarboxylic acid (TCA) cycle. As the intermediates of this pathway are oxidized, electrons are transferred to electron carriers (NAD^+ and a related molecule called FAD), and carbon dioxide is released as waste. Energy is transferred to ATP by substrate-level phosphorylation. For every two molecules of pyruvate that enter pyruvate oxidation and the Kreb's cycle from glycolysis, six molecules of CO_2 are released as waste, while two molecules of ATP, eight molecules of NADH + 8H$^+$, and two molecules of $FADH_2$ are formed.

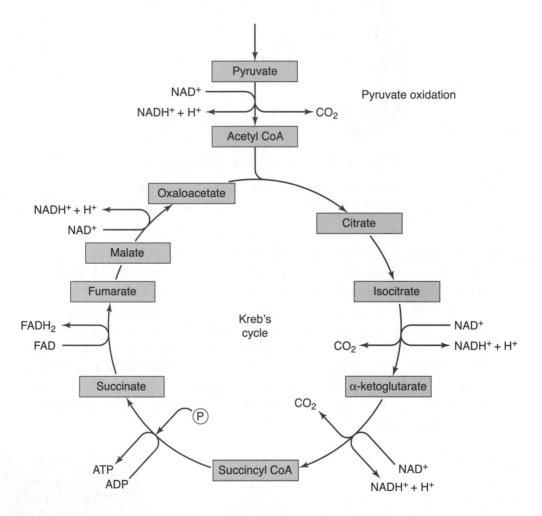

FIGURE 5.7. Pyruvate oxidation and the Kreb's cycle.

Pyruvate, or pyruvic acid, is oxidized and coverted to acetyl-CoA. Acetyl-CoA is oxidized further during the Kreb's cycle, generating the reduced electron carriers NADH + H$^+$ and FADH$_2$. Carbon dioxide is released as waste, and ATP is produced by substrate-level phosphorylation.

ELECTRON TRANSPORT CHAIN
AND OXIDATIVE PHOSPHORYLATION

By the end of the Kreb's cycle, the carbon backbone that began as glucose has been fully oxidized to CO_2. Some of the energy in the original glucose molecule has been transferred to ATP, but much of the energy remains with the electrons that are being carried by NADH + H$^+$ and FADH$_2$. This energy can be transferred from the electron carriers to ATP by a process called the **chemiosmotic theory of oxidative phosphorylation**.

This process involves a set of large protein complexes, called an **electron transport chain** (Figure 5.8), which is embedded in a membrane. The proteins of the electron transport chain accept electrons from the reduced carriers NADH + H$^+$ and FADH$_2$. In a series of oxidation and reduction reactions (**redox reactions**), they pass the electrons down the chain to a final electron acceptor. In the case of aerobic respiration, the final electron acceptor is oxygen. When oxygen accepts electrons, it is reduced to water, which is released as waste. Some cells are able to perform **anaerobic respiration** and use a different electron acceptor, such as nitrate (NO_3^-) or sulfate (SO_4^{2-}), which are reduced to nitrite (NO_2^-) and hydrogen sulfide (H_2S), respectively. Thus, the electrons that were originally part of glucose are now part of these waste molecules (H_2O, NO_2^-, or H_2S).

> **REMEMBER**
> Anaerobic respiration includes glycolysis, the Kreb's cycle, and oxidative phosphorylation, but the final electron acceptor is an inorganic molecule other than oxygen.

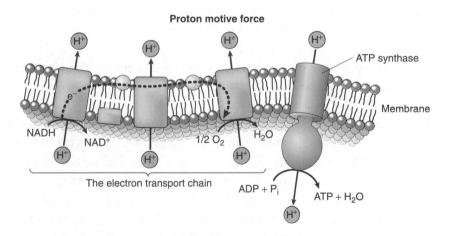

FIGURE 5.8. The electron transport chain and oxidative phosphorylation.

The electron transport chain consists of several large protein complexes and associated molecules that are situated in a membrane. Electrons are transferred to the electron transport chain by reduced electron carriers such as NADH + H$^+$. The electrons are pulled through the chain as they move to substances that have greater pull for electrons, or electronegativity. As the electrons are pulled through the chain, released energy is used to pump protons (H$^+$) across the membrane. This creates a source of potential energy called the proton motive force. Proton motive force is used by ATP synthase to generate ATP from ADP and P.

The electrons are transferred from one carrier in the electron transport chain to another based on the pull each carrier has for electrons. The degree of pull a molecule has for electrons is called **electronegativity**. To create a standard reference, called the **redox potential** *(E)*, the electronegativity of substances such as atoms, ions, and molecules are all compared to that of hydrogen. The redox potential of hydrogen is set at zero ($E = 0$). Substances that are more electronegative than hydrogen have positive redox potentials ($E > 0$). Substances that are less electronegative than hydrogen have negative redox potentials ($E < 0$). As electrons move through electron transport chains, they move from carriers that are less electronegative, or have smaller redox potentials, to those that are more electronegative, or have greater redox potentials. The degree of difference in redox potential between two substances is proportional to the amount of energy that is released as the electrons are transferred between them. Because oxygen is very electronegative, there is a large difference between its redox potential and that of NADH + H$^+$. Thus, when oxygen is used as the final electron acceptor in cellular respiration, a great deal of energy is made available to the cell.

> **REMEMBER**
> Cells use oxidative phosphorylation to transfer the energy stored in NADH and FADH$_2$ to ATP. During this process, the electrons from NADH and FADH$_2$ are transferred to final electron acceptors, regenerating the oxidized form of the carriers, NAD$^+$ and FAD.

Cells utilize the energy released as electrons flow from NADH + H$^+$ to a final electron acceptor such as oxygen. As the electrons are pulled through the chain toward oxygen, the protein complexes transfer some of their energy to pump hydrogen ions (H$^+$) across the membrane. This creates a concentration gradient of H$^+$ across the membrane called the **proton motive force**. The proton motive force is a source of potential energy, energy that originally came from glucose. Generation of proton motive force is an important energy source for microbial cells and is used directly for several processes, such as transport of solutes, flagellar rotation, and generation of NADH for anabolism.

During cellular respiration, potential energy from the proton motive force can be transferred to ATP by another protein in the membrane called **ATP synthase**. ATP synthase allows the H$^+$ to diffuse back across the membrane, using the energy of their movement to catalyze the formation of ATP from ADP and P. The process of making ATP using the flow of H$^+$ is called **chemiosmosis**. It is by this process that much of the energy from glucose ultimately gets transferred to ATP. When chemiosmosis occurs during cellular respiration, it is called **oxidative phosphorylation**. For every glucose molecule that is broken down by cellular respiration, up to thirty-eight molecules of ATP can be made, four by substrate-level phosphorylation in glycolysis and Krebs and up to thirty-four by oxidative phosphorylation at the electron transport chain. A summary of the key events in the process of cellular respiration is presented in Table 5.2.

TABLE 5.2. A Summary of the Key Events in Cellular Respiration

Phase	Location in Cell	Initial Substrate	Products per Glucose	Major events
Glycolysis	Cytoplasm	Glucose	2Pyruvate 2ATP 2NADH + 2H$^+$	Oxidation of glucose. Reduction of NAD$^+$. Substrate-level phosphorylation.
Pyruvate oxidation and Kreb's cycle	Prokaryotes: cytoplasm Eukaryotes: matrix of mitochon-drion	Pyruvate (pyruvate oxidation) then acetyl-CoA (Kreb's cycle)	6CO$_2$ 8NADH + 8H$^+$ 2FADH$_2$ 2ATP	Oxidation of pyruvate and acetyl-CoA. Reduction of NAD$^+$ and FAD. Substrate-level phosphorylation.
Electron Transport Chain	Prokaryotes: plasma membrane Eukaryotes: inner membrane of the mitochondrion	10NADH + 10H$^+$ 2FADH$_2$ O$_2$ (aerobic respiration)	Up to 32 ATP H$_2$O (aerobic respiration) 10NAD$^+$ 2FAD	Oxidation of NADH + H$^+$ and FADH$_2$. Creation of proton motive force. Oxidative phosphorylation. Reduction of final electron acceptor (O$_2$, NO$_3^-$, SO$_4^{2-}$).

FERMENTATION

Not all cells can perform cellular respiration. Some cells simply lack the necessary enzymes. Other cells can perform cellular respiration under certain circumstances but sometimes need to switch to a different method of ATP production. Human muscle cells, for example, can perform aerobic respiration when oxygen is available but, if there isn't enough oxygen, will switch to **fermentation**. Many microbial cells perform fermentation as well. Some, like human muscle cells, can switch between respiration and fermentation depending on their growth conditions. Others never perform respiration but rely solely upon fermentation. There are many different kinds of microbial fermentation, and several of them have economic importance to humans.

The simplest fermentations are essentially glycolysis, plus an additional step to oxidize the NADH + H$^+$ (Figure 5.9). NADH + H$^+$ transfers the electrons it accepted during glycolysis to an organic molecule. This regenerates the NAD$^+$ needed to repeat glycolysis. Recall that there is an oxidation step during glycolysis, and NAD$^+$ is required to pick up the electrons. Without NAD$^+$, this step cannot happen, and glycolysis stops. The organic molecule that accepts electrons from NAD$^+$ is reduced and forms a waste product called a **fermentation product**. The key differences between cellular respiration and fermentation are summarized in Table 5.3.

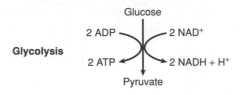
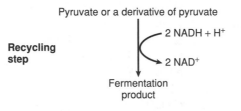

FIGURE 5.9. An overview of a simple fermentation.

TABLE 5.3. Key Differences Between Fermentation and Cellular Respiration

	Fermentation	Cellular Respiration
Final electron acceptor	Organic molecule (e.g., pyruvate)	Inorganic molecule (e.g., O^2, NO^{3-}, SO^{4-})
Amount of ATP per glucose	Net 2ATP	38ATP
Method of ATP production	Substrate-level phosphorylation	Substrate-level phosphorylation and oxidative phosphorylation

The simplest form of fermentation is **lactic acid fermentation** (Figure 5.10). First, glycolysis occurs, producing two pyruvate, two ATP, and two NADH + H⁺. Then, the NADH + H⁺ transfers its electrons back to pyruvate. Pyruvate is acting as the **electron acceptor** and is reduced to the fermentation product lactic acid. NAD⁺ is regenerated and glycolysis can occur again.

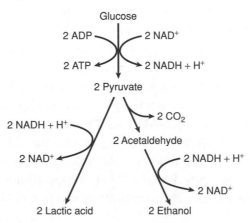
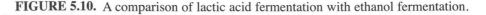

FIGURE 5.10. A comparison of lactic acid fermentation with ethanol fermentation.

Another type of fermentation is **ethanol fermentation** (Figure 5.10). Again, glycolysis occurs producing two pyruvate, two ATP, and two NADH + H$^+$. A carbon is then removed from pyruvate as CO_2. Thus, pyruvate is **decarboxylated** (carbon is removed) by a non-oxidative reaction and becomes **acetaldehyde**. The CO_2 is released as waste. This is what causes bread to rise or bubbles to form in beer. Acetaldehyde now serves as the electron acceptor, taking electrons from NADH + H$^+$. The acetaldehyde is reduced, becoming **ethanol**.

There are many types of fermentation in the microbial world. Microbes can ferment a wide variety of organic molecules by a diverse array of catabolic pathways, generating many different fermentation products. As stated in the beginning of this chapter, many of these fermentation products are useful to humans. Examples of different fermentations and their products are summarized in Table 5.4.

> **REMEMBER**
> Fermentation and anaerobic respiration are both anaerobic processes, but they are very different. Fermentation produces much less ATP because it consists of just glycolysis plus a recycling step, whereas anaerobic respiration includes the Kreb's cycle and oxidative phosphorylation.

TABLE 5.4. Examples of Microbial Fermentation Products

Fermentation Products	Examples
Lactic acid	*Lactobacillus* to make yogurt
Ethanol	Yeast to make beer and bread
Acetone and butanol	*Clostridium* to make these products for industrial uses
Propionic acid and carbon dioxide	*Propionibacteria* to make swiss cheese

CATABOLISM OF LIPIDS AND PROTEINS

The processes of cellular respiration and fermentation are always written as if the starting molecule is glucose. However, it is important to remember that this is just a convenient starting point and that these pathways are actually just part of a larger interconnected metabolic web. For example, you know that you get energy from all types of food, not just sugar. If you were to eat a peanut butter and jelly sandwich, you wouldn't just get energy from the sugar in the jelly, you'd get energy from the protein and fat in the peanut butter and also from the starch in the bread. This is because you have the enzymes necessary to break down the starch, protein, and fat. The starch would be broken down into its building block, glucose, and the glucose would get catabolized in cellular respiration. Likewise, the protein and fat would also get broken down into their building blocks, and these building blocks would be processed through cellular respiration. Thus, cells can catabolize all sorts of macromolecules as long as they have the necessary enzymes.

LIPID CATABOLISM

To break down lipids like fats, cells must have **lipases**. Lipases are enzymes that break fats down into their components, glycerol and fatty acids. Glycerol is a three-carbon molecule that can easily be converted into an intermediate in glycolysis. Fatty acids are further broken down into the two-carbon molecule acetyl-CoA, the starting molecule for the Kreb's cycle. Thus, the energy stored in fats is transferred to ATP by cellular respiration after the fats have been partially catabolized into intermediates in this pathway (Figure 5.11).

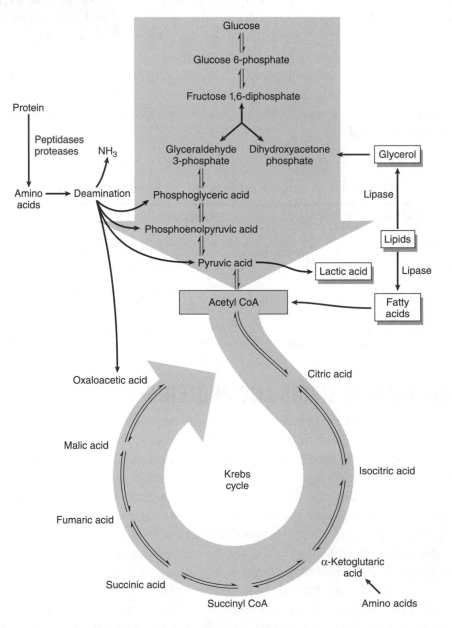

FIGURE 5.11. Catabolism of molecules other than glucose.

PROTEIN CATABOLISM

Proteins can also be catabolized by cellular respiration after they have been converted into intermediates in this pathway. Proteins are broken down into their building blocks, amino acids, by enzymes called **proteases** and **peptidases**. Amino acids, however, cannot be immediately converted into intermediates in cellular respiration. Recall that amino acids have a nitrogen-containing amino group. This group must first be removed before the remaining part of the molecule can be catabolized in cellular respiration. The removal of the amino group is called **deamination**, and it often results in release of ammonia (NH_3) by microbial cells. After the amino acids have been deaminated, the remaining parts of the molecules are converted into intermediates in the Kreb's cycle (Figure 5.11).

Fundamental Metabolic Pathways
of Autotrophic Metabolism

Autotrophs are organisms that obtain their carbon from CO_2 and convert it into the organic building blocks needed by the cell. Because they do not have to eat other organisms to obtain their building blocks, they are considered self-feeders (*auto =* self). The importance of autotrophs really can't be overstated. Without autotrophs capturing CO_2 and converting it into organic molecules, there would be no other kinds of life. Autotrophs are at the base of every food chain. They are the grass in the field, the algae in the ocean, the bacteria living around the thermal vents in the ocean floor.

> **REMEMBER**
> Photosynthesis is important because it enables cells to store energy, electrons, and carbon in food. Oxygen is just a waste product of the process.

PHOTOSYNTHESIS

The most familiar autotrophs are the photosynthetic organisms. Because they are big and green, you probably automatically think of plants. However, over half of the photosynthesis on the planet is not done by plants at all, but by microbes such as the eukaryotic green, brown, and red algae or the prokaryotic blue-green, purple, and green bacteria.

Photosynthesis essentially occurs in two major phases, **photophosphorylation** and **carbon fixation**. During the photophosphorylation phase of photosynthesis, light energy is transferred to ATP, and electrons from an inorganic source are energized and transferred to an electron carrier ($NADP^+$). Then during carbon fixation, the energy from the ATP and the electrons from the electron carrier are used to reduce CO_2 to sugar. Because photophosphorylation requires light, these reactions are sometimes called the light-dependent reactions of photosynthesis. Likewise, because carbon fixation does not directly require light, these reactions are sometimes called the light-independent reactions of photosynthesis.

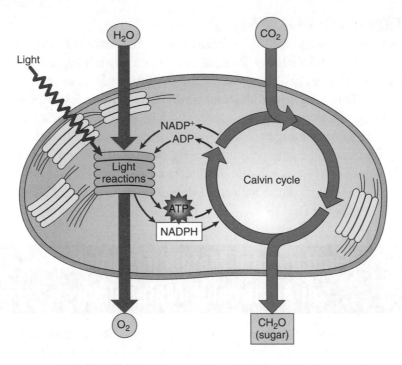

FIGURE 5.12. An overview of photosynthesis.

PHOTOPHOSPHORYLATION (LIGHT-DEPENDENT REACTIONS)

Photophosphorylation takes place in a membrane and involves an electron transport chain (Figure 5.13). The electron transport chain is connected to **photosystems**, protein complexes that contain a **pigment** molecule capable of absorbing light energy. In plants and blue-green bacteria (cyanobacteria), the primary photosynthetic pigment is **chlorophyll a**. In green and purple bacteria, it is **bacteriochlorophyll**. The light energy excites electrons in the pigment, and the electrons are then transferred via redox reactions through the electron transport chain. As the electrons move through the chain, some of their energy is used to pump H^+ and create a **proton motive force**. **Chemiosmosis** occurs using **ATP synthase**, and the energy from sunlight is ultimately transferred to ATP. In eukaryotes, photophosphorylation occurs in the thylakoid membranes within the chloroplast. In prokaryotes, it occurs in the plasma membrane or membranes within the cell that are formed from the plasma membrane.

> **REMEMBER**
> During the light reactions, light energy is transformed into chemical energy. Electrons from an inorganic molecule like water are transferred to $NADP^+$, reducing it to NADPH.

Two types of photophosphorylation occur in microbes, **noncyclic photophosphorylation** and **cyclic photophosphorylation**. In noncyclic photophosphorylation (Figure 5.14), the electrons from the electron transport chain are transferred to an electron carrier called $NADP^+$, which is very similar to the carrier NAD^+. $NADP^+$ is reduced to

NADPH + H$^+$. In cyclic photophosphorylation (Figure 5.14), electrons cycle from the pigment molecule through the electron transport chain and then back to the pigment molecule, making a cycle. This type of electron flow generates proton motive force, which can be used to make ATP, but no NADPH + H$^+$. Cells that perform cyclic photophosphorylation must use additional steps to obtain the NADPH to supply the electrons necessary to convert CO_2 to sugar. In noncyclic photophosphorylation, electrons are constantly being transferred to NADP$^+$, generating the necessary NADPH + H$^+$.

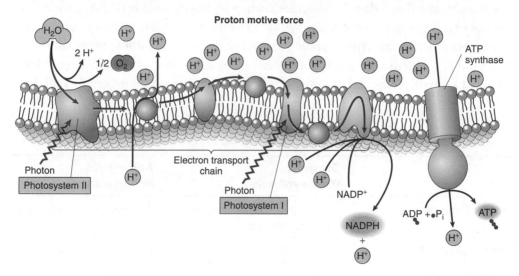

FIGURE 5.13. Photophosphorylation.

Light energy is used to energize electrons that move through an electron transport chain and are transferred to the electron carrier NADP$^+$, reducing it to NADPH + H$^+$. As the electrons move through the chain, energy is transferred and used to pump protons (H$^+$) across the membrane. ATP synthase uses the proton motive force to synthesize ATP from ADP and P.

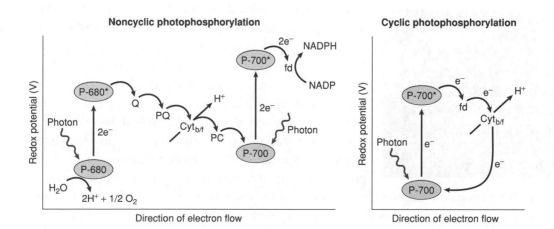

FIGURE 5.14. A comparison of cyclic and noncyclic photophosphorylation.

Because electrons are constantly being transferred from the pigment molecules to $NADP^+$, noncyclic photophosphorylation requires electron donors to resupply the chain with electrons. In green algae and blue–green bacteria (cyanobacteria), the electron donor is water. Light energy splits the water, and its electrons are transferred to the electron transport chain. In the process, oxygen (O_2) and hydrogen ions (H^+) are released as waste. Because oxygen is released, this type of photosynthesis is called **oxygenic photosynthesis** (*gen* = genesis, so this is oxygen-generating photosynthesis). Green and purple bacteria use molecules such as hydrogen sulfide (H_2S) as their electron donor, releasing molecules like sulfur (S) as waste. Because no oxygen is released during this process, this type of photophosphorylation is called **anoxygenic photosynthesis** (*an* is a negative, so this is not oxygen-generating photosynthesis). The differences between oxygenic and anoxygenic photosynthesis are summarized in Table 5.5.

TABLE 5.5. A Comparison Between Oxygenic and Anoxygenic Photosynthesis

	Oxygenic Photosynthesis	Anoxygenic Photosynthesis
Types of organisms	Plants, green algae, blue-green bacteria (cyanobacteria)	Green and purple bacteria
Primary pigment	Chlorophyll a	Bacteriochlorophyll
Source of electrons	H_2O	H_2S, H_2

CARBON FIXATION (LIGHT-INDEPENDENT REACTIONS)

Photophosphorylation provides cells with energy in the form of ATP derived from proton motive force and electrons in the form of the reduced carrier, NADPH. However, it does not provide the cell with the carbon building blocks it needs to grow. To get their carbon, autotrophs must take in CO_2 and convert it to an organic form. This process is called **carbon fixation**, and it occurs in pathways such as the Calvin–Benson cycle (Figure 5.15). In prokaryotes, the Calvin–Benson cycle occurs in the cytoplasm. In eukaryotes, it occurs in the stroma of the chloroplast. Some prokaryotes can fix carbon by pathways other than the **Calvin–Benson** cycle.

> **REMEMBER**
> During the Calvin–Benson cycle, oxidized, energy-poor carbon (CO_2) is converted into reduced, energy-rich carbon (sugar = food). In order to do this conversion, cells must supply energy (from ATP) and electrons (from NADPH).

Although the Calvin–Benson cycle contains many steps, it can be simplified into four basic processes. The first process is carbon fixation. An enzyme catalyzes the attachment of carbon dioxide to an existing five-carbon sugar, called ribulose bisphosphate. This creates an unstable six-carbon compound that immediately breaks down into two three-carbon compounds, called phosphoglyceric acid. The second process is the reduction of phosphoglyceric acid using the products of photophosphorylation.

Energy from the ATP and electrons from the NADPH are transferred to the phospho-glyceric acid, converting them into three-carbon sugars, called glyceraldehydes 3-phosphate. Third, some of the glyceraldehydes 3-phosphate molecules may now be used for the synthesis of glucose. The fourth process is the rearrangement of bonds in some of the molecules of glyceraldehydes 3-phosphate in order to recreate the five-carbon sugar, ribulose bisphosphate, that started the cycle. The regeneration of ribulose bisphosphate, which also requires energy from ATP, allows the Calvin–Benson cycle to continue.

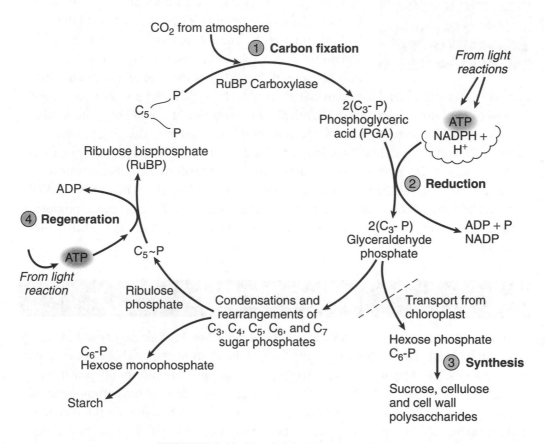

FIGURE 5.15. The Calvin–Benson cycle.

Many cells use the Calvin–Benson cycle to do carbon fixation, the reduction of carbon dioxide into sugar. Carbon dioxide is first joined with a five-carbon sugar; then, the carbon intermediates of the pathway are reduced using energy (ATP) and electrons (from NADPH) that were stored during photophosphorylation. After the carbon intermediates have been reduced, they can be used to synthesize sugar. To continue the pathway, some intermediates are rearranged to recreate the five-carbon sugar needed at the beginning of the pathway.

CHEMOLITHOAUTOTROPHY

The diversity and uniqueness of microbial metabolism is perhaps best demonstrated by the chemolithoautotrophs. *Litho* means rock, so these are literally rock-eaters. Chemolithoautotrophs can also be found at the Earth's surface in the iron-rich streams near mines, or oxidizing methane in the soil. Because of their ability to oxidize metals, they have practical value in cleaning up crude ore to obtain desired minerals like copper and gold. Down in the depths of the ocean, where no light can penetrate, chemolithoautotrophs form the basis for food chains in communities clustered around the hydrothermal vents. These communities do not rely on carbon fixation by photosynthetic organisms and exist only because of the ability of the chemolithoautotrophic

> **REMEMBER**
> Chemolithoautotrophs allow life to flourish on Earth, even in places where there is no light—from the deepest, darkest caves to the bottom of the ocean.

bacteria to fix carbon and to obtain energy and electrons from the hot chemicals spewing out of the Earth's core. These chemicals are things like H_2S and Fe^{2+}, chemicals that would be completely useless to the cells of most organisms. The chemolithoautotrophs, however, oxidize these compounds and use their electrons to power an electron transport chain, generate proton motive force, and form ATP by chemiosmosis. They also use some of the electrons to reduce NAD^+ and $NADP^+$. Then with the ATP and NADPH, they fix carbon using pathways like the Calvin–Benson cycle, producing organic molecules for their own growth and the growth of the other organisms in the community.

Anabolic Pathways

A living cell must constantly produce new molecules. We have already seen how glucose can be synthesized from CO_2 using the Calvin–Benson cycle. This is an example of an anabolic pathway used by autotrophs. Heterotrophs must also build new molecules, but they cannot build them from CO_2. Instead, they rely on the carbon-containing molecules in their food. Heterotrophs are constantly catabolizing food molecules and generating the intermediates within their catabolic pathways. To build molecules, they simply use some of these intermediates and rearrange them into the molecules they need. Recall that some autotrophs also break glucose down by catabolic pathways. Thus, they also can use intermediates in this pathway as precursors for their macromolecules. In addition to carbon, anabolism requires energy in the form of ATP and electrons from electron carriers. The intermediates that are commonly used to synthesize carbohydrates, lipids, proteins, and nucleic acids are shown in Figure 5.16.

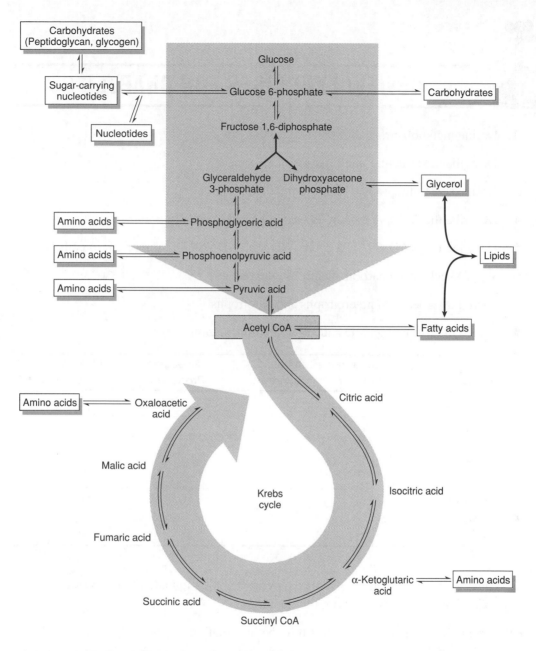

FIGURE 5.16. Interconnections between catabolic and anabolic pathways.

REVIEW EXERCISES FOR CHAPTER 5

1. Define metabolism, anabolism, and catabolism.

2. Describe the structure and function of enzymes.

3. Define substrate, product, and intermediate.

4. Describe the ATP cycle. Why is ATP imporant to cells?

5. Describe the NAD^+/NADH cycle. Why is NAD^+/NADH important to cells?

6. Why do cells need carbon, electrons, and energy?

7. Distinguish between heterotrophs and autotrophs.

8. Create the following table and fill in the information.

Phase	Location in Cell?	Initial Substrate	Molecules Produced?	Links to Other Phases?
Glycolysis				
Pyruvate oxidation and Kreb's cycle				
Electron transport chain				

9. State the major events in glycolysis, pyruvate oxidation and the Kreb's cycle, oxidative phosphorylation, and fermentation.

10. Create the following table and fill in the information:

	Aerobic Respiration	Anaerobic Respiration	Fermentation
Amount of ATP produced per glucose			
Types of phosphorylation			
Electron acceptor			
Oxygen required (Y/N)			

11. Compare and contrast lactic acid fermentation and ethanol fermentation.

12. Compare and contrast photophosphorylation and oxidative phosphorylation.

13. Distinguish between oxygenic photosynthesis and anoxygenic photosynthesis.

14. Describe the major events of the Calvin–Benson cycle.

15. Describe how lipid catabolism and protein catabolism fit into cellular respiration.

16. Describe chemolithoautotrophy.

17. Describe the relationship between cellular respiration and anabolism.

SELF-TEST

1. In which metabolic pathways do the following processes occur? Choose all the pathways in which the processes occur. In the blank next to the process, write G for glycolysis, P for pyruvate oxidation, K for Kreb's cycle, E for electron transport chain, and F for fermentation.

    _____ a. ATP hydrolysis (breakdown of ATP)
    _____ b. NAD+ reduction
    _____ c. Substrate-level phosphorylation
    _____ d. Production of CO_2
    _____ e. $FADH_2$ oxidation
    _____ f. O_2 reduction
    _____ g. Generation of proton motive force
    _____ h. Oxidative phosphorylation
    _____ i. Oxidation–reduction reactions
    _____ j. Production of lactic acid

2. Which metabolic pathway is common to both fermentation and cellular respiration?

    A. Kreb's cycle
    B. Electron transport chain
    C. Glycolysis
    D. Synthesis of acetyl CoA from pyruvate
    E. Reduction of pyruvate to lactate

3. In a prokaryotic cell, most the enzymes of the Kreb's cycle are located in the

    A. plasma membrane.
    B. cytoplasm.
    C. inner mitochondrial membrane.
    D. mitochondrial matrix.
    E. intermembrane space.

4. The light reactions of photosynthesis supply the light-independent reactions (Calvin–Benson cycle) with

    A. light energy.
    B. CO_2 and ATP.
    C. H_2O and NADPH.
    D. ATP and NADPH.
    E. sugar and O_2.

5. Which sequence correctly portrays the flow of electrons during photosynthesis?

    A. NADPH → O_2 → CO_2
    B. H_2O → NADPH → Light-independent reactions (Calvin–Benson cycle)
    C. NADPH → Chlorophyll → Light-independent reactions (Calvin–Benson cycle)
    D. H_2O → ATP → Glucose
    E. NADPH → Electron transport chain → O_2

6. A fat molecule with radioactively labeled carbon is fed to a cell. Of the following molecules, which is the first place you would expect to find the radioactively labeled carbons?

    A. Glucose
    B. NADH + H^+
    C. ATP
    D. Acetyl-CoA
    E. CO_2

7. True or False. Fermentation and anaerobic respiration produce approximately the same amount of ATP for a cell.

8. True or False. Neither fermentation nor anaerobic respiration uses an electron transport chain.

9. True or False. Both photophosphorylation and oxidative phosphorylation generate a proton motive force.

10. During lactic acid fermentation, which is the final electron acceptor?

    A. Lactic acid
    B. NAD^+
    C. ATP
    D. Pyruvate
    E. NADH + H^+

11. A farmer notices that the nitrates (NO_3^-) from his fertilizer are disappearing rapidly from his soil. This could be due to

   A. aerobic respiration.
   B. anaerobic respiration.
   C. fermentation.
   D. photosynthesis.
   E. chemolithoautotrophy.

12. Which of the following do chemolithoautotrophs and photosynthetic organisms have in common?

   A. Electron transport chains
   B. Calvin–Benson cycle
   C. Light as a source of energy
   D. Both A and B are correct.
   E. Both A and C are correct.

Answers

Review Exercises

1. Metabolism is all the chemical reactions in a cell. Catabolism is the reactions that break things down. Anabolism is the reactions that build things up.

2. Enzymes are proteins that are folded up to create pockets called active sites. The active site binds the enzyme's substrate. Enzymes speed up chemical reactions involving their substrate. The enzymes themselves are not changed in the process.

3. A substrate is a reactant in an enzyme-catalyzed reaction. An intermediate is a molecule that is a product of one reaction and a substrate of another reaction. A product is a molecule that is produced by a reaction.

4. Cells obtain energy by the catabolism of organic molecules, by the capture of energy from light, or by the oxidiation of inorganic molecules. They use this energy to form ATP from ADP and P. The ATP can then be used to fuel energy requiring reactions for the cell such as transport, movement, and synthesis of molecules. When ATP is used, it is hydrolyzed back to ADP and P. ATP is important for cells because it is a useful form of chemical energy that can be used for energy-requiring reactions.

5. Cells obtain electrons from the oxidation of molecules. These electrons are transferred to carriers like NAD^+. When NAD^+ accepts electrons, it is reduced to $NADH + H^+$. $NADH + H^+$ can then supply electrons to electron transport chains for ATP synthesis or to anabolic reactions that build molecules. When $NADH + H^+$ gives up its electrons, it is oxidized back to NAD^+. $NAD^+/NADH + H^+$ is

important to cells because it is the carrier for electrons obtained in certain reactions that are necessary to run other reactions. In other words, it is a kind of middleman between electron-yielding and electron-requiring reactions.

6. Cells need carbon because it is the building block for all organic molecules. They need energy so they can move, transport materials, and build molecules. They need electrons to produce ATP and to build molecules.

7. Heterotrophs obtain their carbon from organic molecules, in other words, by eating other organisms. Autotrophs can obtain their own carbon from CO_2.

8.

Phase	Location in Cell?	Initial Substrate	Molecules Produced?	Links to Other Phases?
Glycolysis	Prokaryotes: cytoplasm Eukaryotes: cytoplasm	Glucose	2pyruvate 2ATP 2NADH + H$^+$	Pyruvate to pyruvate oxidation NADH to electron transport chain
Pyruvate oxidation and Kreb's cycle	Prokaryotes: cytoplasm Eukaryotes: matrix of mitochondrion	Pyruvate (pyruvate oxidation) Acetyl-CoA (Kreb's)	6CO$_2$ 8NADH + H$^+$ 2FADH$_2$ 2ATP	NADH, FADH$_2$ to electron transport chain
Electron transport chain	Prokaryotes: plasma membrane Eukaryotes: inner membrane of mitochondrion	NADH FADH$_2$ O$_2$ (aerobic respiration)	34ATP NAD$^+$ FAD H$_2$O	NAD$^+$ and FAD back to glycolysis and Kreb's

9. Glycolysis and fermentation both result in oxidation of glucose to produce pyruvate and NADH + H$^+$. ATP is produced by substrate-level phosphorylation. Fermentation also results in oxidation of NADH and reduction of an organic electron acceptor. Pyruvate oxidation results in oxidation of pyruvate to produce NADH + H$^+$ and acetyl-CoA. CO_2 is released as waste. The Kreb's cycle results in oxidation of acetyl-CoA to produce NADH + H$^+$ and FADH$_2$. ATP is produced by substrate-level phosphorylation. CO_2 is released as waste. The electron transport chain begins with oxidation of NADH + H$^+$ and FADH$_2$. Electrons move through the chain via redox reactions and are accepted by an inorganic electron acceptor like O_2 or NO_3^-. Energy from the electrons is used to pump H$^+$ across the membrane creating a proton motive force. The H$^+$ flow back through ATP synthase, and the energy is used to synthesize ATP from ADP and P.

10.

	Aerobic Respiration	Anaerobic Respiration	Fermentation
Amount of ATP produced per glucose	Up to 38	Up to 38	2 (varies somewhat but is much less than that produced by respiration)
Types of phosphorylation	Substrate level and oxidative	Substrate level and oxidative	Substrate level
Electron acceptor	Inorganic, always O_2	Inorganic such as NO_3^-, SO_4^{2-}	Organic such as pyruvate
Oxygen required (Y/N)	Yes	No	No

11. Both lactic acid fermentation and ethanol fermentation begin with glycolysis, resulting in oxidation of glucose to produce pyruvate and $NADH + H^+$. ATP is produced by substrate-level phosphorylation. In lactic acid fermentation, pyruvate acts as the electron acceptor, taking electrons from $NADH + H^+$ and being reduced to lactic acid. In ethanol fermentation, pyruvate is first decarboxylated (carbon is removed), resulting in the production of CO_2 (the removed carbon) and acetaldehyde (a two-carbon molecule). The acetaldehyde then acts as the electron acceptor and is reduced to ethanol when it accepts electrons from $NADH + H^+$.

12. Both photophosphorylation and oxidative phosphorylation use chemiosmosis to produce ATP. In oxidative phosphorylation, the source of electrons is $NADH + H^+$ and $FADH_2$. In photophosphorylation, the source of electrons is H_2S or H_2O. In oxidative phosphorylation, the electron acceptor is O_2 or another inorganic molecule. In photophosphorylation, the electron acceptor is $NADP^+$. In oxidative phosphorylation, the source of energy is chemical (ultimately from glucose). In photophosphorylation, the source of energy is light.

13. During oxygenic photosynthesis, electrons are supplied to the electron transport chain by H_2O. When water is split to provide electrons, oxygen is produced. During anoxygenic photosynthesis, other electron donors, such as H_2S, are used. This produces waste products like sulfur instead of oxygen.

14. During the Calvin–Benson cycle, carbon is fixed when an enzyme attaches carbon dioxide to an existing five-carbon sugar. Then energy from ATP and electrons from NADPH (both from the photophosphorylation) are used to reduce the carbon-containing molecules and create sugars. Finally, some of the carbon-containing molecules are rearranged to recreate the initial five-carbon sugar needed at the beginning of the pathway. These reactions require ATP.

15. Lipids and proteins are both partially broken down by enzymes to create intermediates that fit into the process of cellular respiration. Fats are broken down by lipases into glycerol and fatty acids. The glycerol is converted into a three-carbon intermediate in glycolysis; the fatty acids are broken down into two-carbon molecules that are converted into acetyl-CoA. Proteins are broken down by proteases into amino acids. The amino acids are then deaminated (nitrogen-containing amino groups removed) and the remaining carbon-containing molecules are converted into intermediates in the Kreb's cycle.

16. Chemolithoautotrophs oxidize inorganic molecules such as H_2S and H_2 in order to supply an electron transport chain. They use chemiosmosis to generate ATP. They then use ATP and electrons from electron carriers to fix CO_2 via the Calvin–Benson cycle.

17. As cells break down organic molecules via cellular respiration, they generate many different types of carbon-containing intermediates. These intermediates can then be used as substrates in anabolic pathways leading to the synthesis of lipids, carbohydrates, nucleic acids, and proteins.

Self-Test

1. a: G, F; b: G, F, P, K; c: G, F, K; d: P, K;
   e: E; f: E; g: E; h: E; i: G, F, P, K, E; j: F

2. C   6. D   10. D
3. B   7. F   11. B
4. D   8. F   12. D
5. B   9. T

6

Microbial Growth and Reproduction

WHAT YOU WILL LEARN

This chapter explains what microbes need in order to grow, and how their environment can affect their growth. As you study this chapter, you will:

- learn to calculate the number of bacteria in a sample after a given time period;
- examine the growth pattern of bacteria in the laboratory;
- explore how environmental conditions such as temperature, pH, and salinity can affect microbial growth;
- discover how oxygen can be harmful to cells;
- get an overview of how microbiologists grow microbes in the lab.

SECTIONS IN THIS CHAPTER

- Bacterial Growth
- Conditions That Affect Growth
- Growth in the Laboratory

Imagine the discovery of a planet that has temperatures near boiling, acidic waters, and no oxygen. Could life as we know it exist on such a planet? You might be tempted to say no, but maybe you haven't considered all the kinds of life that exist on our own planet. There are cells here that are very different from your own, cells that *can* survive high temperatures, acidic conditions, or even a lack of oxygen. They are the **extremophiles** (*phile* = love), a diverse group of bacteria and archaea that live in hot springs, at ocean vents, in salt lakes, in soda lakes, and in acid rivers. These prokaryotes have strategies that enable them to survive conditions that would kill our own cells and the cells of most of the familiar bacteria. In this chapter, we will consider the growth requirements of prokaryotes and the diversity of strategies they employ to inhabit such a wide range of environments.

Bacterial Growth

Growth of bacteria refers to increase in the number of cells in the population, not an increase in cell size. Bacteria multiply by a process called **binary fission** (Figure 6.1). During binary fission, the bacterial cell copies its chromosome and other materials and increases in size. Then, cell membrane and wall form between the two chromosomes, splitting the cell in two (**cytokinesis**).

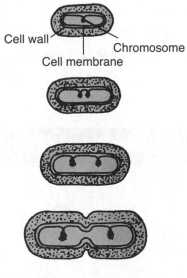

Cell wall

Chromosome

Cell membrane

FIGURE 6.1. Binary fission.

The cell copies its chromosome and doubles in size. Membrane and wall material grow inward and separate the cell into two.

GENERATION TIME

The time it takes for a bacterial cell to divide, or for a population to double, is called the **generation time**. Generation time for bacteria can range from as little as 10 minutes, to 24 hours or even longer. Many of the familiar disease-causing bacteria divide in 20–30 minutes.

> **REMEMBER**
> Generation time is the time it takes a bacterial population to double.

Because generation times are so short, it does not take long for disease-causing organisms to multiply to dangerous levels. For example, the generation time for *Streptococcus pneumoniae* is 30 minutes. If a population of two *S. pneumoniae* cells multiplied for 4 hours, this would result in 512 cells. To figure this out, you could just figure out how many generations in 4 hours, which is two per hour for a total of eight, then double the number of cells eight times. In other words, two doubled is four, four doubled is eight, eight doubled is sixteen, sixteen doubled is thirty-two, and so on up to 512. If you are dealing with large numbers or many generations, this method becomes impractical. To solve more difficult problems, the following formula is used:

Initial number of cells $\times\ 2^{\text{number of generations}}$ = Number of cells

In the preceding example, this calculation would be

2 cells $\times\ 2^8$ = 512 cells

The number of generations is determined by dividing the total time by the time per generation (in this case, 240 minutes ÷ 30 minutes/generation = 8 generations).

BACTERIAL GROWTH CURVE

When bacteria are introduced into a new environment that has fixed resources, like a culture dish in the laboratory, they will exhibit a pattern of growth that can be plotted as a **bacterial growth curve** (Figure 6.2). The bacterial growth curve shows the changes in numbers of bacteria over time. Initially, when bacteria are placed into a new environment, there will be no increase in numbers in the population. This phase is referred to as the **lag phase**. During the lag phase, the bacteria are adjusting to the new environment and synthesizing necessary enzymes to break down new food sources. After the bacteria have made the necessary enzymes, they begin growing at a rapid rate, and the population doubles at regular intervals. This is called the **exponential phase**, or **log phase**, of growth. It is the phase during which the bacteria are most metabolically active. As nutrients begin to run out and wastes begin to accumulate, growth slows, and the numbers of new cells begins to equal the number of dying cells. This is the **stationary phase** of growth. As growth slows even further

> **REMEMBER**
> Bacterial enzymes are most active during log phase, so bacteria are most sensitive to antibiotics when they are in log. Log phase is also the best time to measure the activity of a particular enzyme.

due to these conditions, the number of dying cells becomes greater than the number of new cells, and the population enters the **death phase**.

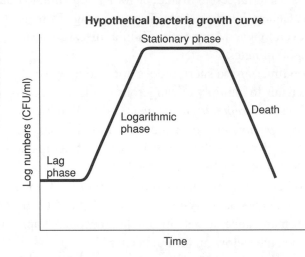

Hypothetical bacteria growth curve

FIGURE 6.2. Bacterial growth curve.

The log number of bacteria are shown over time. In the lag phase, bacteria are synthesizing cellular components but not dividing. In the logarithmic, or exponential phase, bacteria divide at regular intervals. During the stationary phase, the generation of new bacteria is balanced by the number of dying bacteria. During the death phase, the number of dying bacteria is greater than the number of new bacteria.

CONDITIONS THAT AFFECT GROWTH

TEMPERATURE

All cells have a range of temperatures at which they will grow. This range is defined by the lowest survivable temperature, or **minimum temperature**, and the highest survivable temperature, or **maximum temperature**. The temperature at which a cell grows best is the **optimum temperature**. Many bacteria and most eukaryotes survive best at moderate temperatures. Some prokaryotes, however, have the ability to tolerate very low or very high temperatures.

Based on their temperature preferences, cells are divided up into three main categories (Figure 6.3): **psychrophiles** (cold-loving organisms), **mesophiles** (moderate-loving organisms), and **thermophiles** (heat-loving organisms). Additionally, the psychrophiles and thermophiles can be further divided into organisms that are more extreme and ones that are less extreme. The approximate optimum temperatures for each group of organisms is given in Table 6.1. Strategies used by psychrophiles and thermophiles to survive in extreme environments are given in Table 6.2.

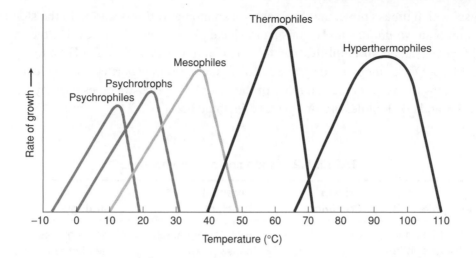

FIGURE 6.3. The effect of temperature on bacterial growth.

Bacteria can be categorized by their ability to grow at different temperatures. The beginning of each curve in the figure represents the minimum temperature at which a particular category of bacteria can grow. The peak of the curve represents the optimum temperature for that type of bacteria. The end of each curve represents the maximum temperature at which that type of bacteria grows.

TABLE 6.1. Temperature Ranges of Different Types of Cells

Category	Optimum Growth Temperature ($^\circ$ C)
Psychrophile	−5 to 15
Psychrotroph	20 to 30
Mesophile	25 to 45
Thermophile	45 to 70
Hyperthermophile	70 to 110

pH

Most bacteria grow best at a neutral pH of about 7. Fungi are more tolerant of slightly acidic conditions and grow well at a pH of 5–6. Some prokaryotes, called **acidophiles**, can tolerate conditions of very low pH. For example, *Helicobacter pylori* survives in the stomach, where pH is around 2. Some people think *H. pylori* is able to survive these conditions because it secretes an enzyme called

REMEMBER
Extremes of temperature and pH can denature enzymes, thus killing cells. Bacteria that live in extreme conditions have strategies to protect their enzymes.

urease, which breaks down urea and releases ammonia. If this occurred in the stomach, the ammonia would pick up H^+, raising the local pH to a survivable level. Other prokaryotes, called **alkalophiles**, can tolerate conditions of very high pH up to a pH of about 12.5. One strategy used by these prokaryotes is to transport H^+ into the cell, which helps to maintain the neutrality of the cytoplasm. The strategies used by acidophiles and alkalophiles are summarized in Table 6.2.

TABLE 6.2. Strategies of Extremophiles

Extremophile	Growth Condition	Potential Problem	Strategy
Thermophiles and hyperthermophiles	High temperatures	High temperatures cause plasma membranes to become too fluid and may denature (unfold) enzymes.	More rigid plasma membrane (saturated fatty acids), enzymes that maintain structure in high temperatures.
Psychrophiles and psychrotrophs	Low temperatures	Low temperatures cause plasma membranes to become too rigid. Lower kinetic energy results in less molecular motion.	More flexible plasma membrane (unsaturated fatty acids), enzymes that maintain flexibility in cold.
Acidophiles	Acidic conditions	Low pH may denature enzymes.	Ability to produce compounds that buffer local pH. Protect intracellular pH by pumping H^+ out of the cell.
Alkalophile	Basic conditions	High pH may denature enzymes, cause RNA to break down	Import H^+ into the cell to maintain neutrality of the cytoplasm.
Halophiles	High salt environment	A high salt environment is hypertonic to the cell. Water will leave the cell, causing the cell to shrink (plasmolyse).	Balance intracellular osmotic conditions with those outside the cell by concentrating organic molecules such as amino acids in their cytoplasm. Proteins may be modified to withstand high salt conditions.

OSMOTIC PRESSURE

Most cells require conditions of about 98 percent water to function. If cells are placed in a hypertonic environment, water will leave the cell by osmosis. In bacterial cells, this causes the cell to collapse, drawing the plasma membrane away from the wall. This condition is called **plasmolysis** (Figure 6.4). When cells are plasmolysed, they may not be dead, but they cannot grow. Prokaryotes that grow in high salt environments are called **halophiles**. **Facultative halophiles** grow in typical environments but can also adjust and survive in higher salt environments. **Extreme halophiles** have protein and membrane modifications that allow them to live in high salt environments such as salt lakes and are not capable of surviving in low salt environments. The strategies used by halophiles to survive in extreme environments are summarized in Table 6.2.

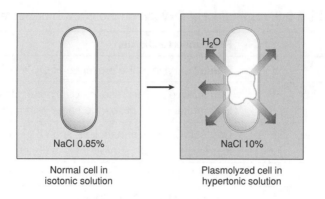

FIGURE 6.4. Plasmolysis.

When cells are placed in a hypertonic solution, water leaves the cell by osmosis. The cell wall maintains the cell shape, but the plasma membrane pulls away from the wall as the cytoplasm collapses.

OXYGEN

Oxygen has a strong pull for electrons (it is **electronegative**). This characteristic makes it both very useful and very dangerous to cells. If a cell can use oxygen to perform aerobic respiration, it is very useful to the cell because it allows the cell to make lots of ATP from its food molecules. However, because molecular oxygen (O_2) accepts electrons, it can easily be converted to more dangerous **free radicals**, such as **superoxide (O_2^-)** and **hydrogen peroxide (H_2O_2)**. Superoxide and hydrogen peroxide are both very reactive and will damage organic molecules by stealing electrons from them. This cellular damage can kill cells.

> **REMEMBER**
> Oxygen benefits cells because it enables them to fully oxidize their food, maximizing the transfer of energy to ATP. Oxygen is dangerous to cells because it can form free radicals that damage cellular molecules such as DNA.

In order for a cell to survive in the presence of oxygen, it must have protective enzymes that allow it to detoxify oxygen free radicals. One such enzyme is **superoxide dismutase**, which converts two superoxide radicals into hydrogen peroxide and water:

$$O_2^- + O_2^- + 2H^+ \rightarrow H_2O_2 + O_2$$

Hydrogen peroxide may be neutralized if the organism has **catalase** or **peroxidase**, which catalyze the following reactions:

$$\text{Catalase: } 2H_2O_2 \rightarrow 2H_2O + O_2$$

$$\text{Peroxidase: } H_2O_2 + 2H^+ \rightarrow 2H_2O$$

The effect of oxygen on a cell depends on what metabolic pathways the cell uses (aerobic respiration, anaerobic respiration, or fermentation) and whether the cell can make the enzymes to detoxify free radicals. Based on their response to oxygen, organisms can be placed in the categories shown in Table 6.3.

TABLE 6.3. The Effect of Oxygen on the Growth of Bacteria.

Category	Response to Oxygen
Obligate anaerobe	Cannot survive in the presence of O_2
Aerotolerant anaerobe	Can survive in the presence of O_2, but does not use O_2 in its metabolism; grows the same with or without O_2
Facultative anaerobe	Can use O_2 in its metabolism and so grows better if O_2 is present, but can survive if O_2 is absent
Microaerophile	Grows best at low concentrations of O_2
Obligate aerobe	Can only grow if O_2 is present

NUTRITIONAL REQUIREMENTS

All cells need essential elements in order to grow and make necessary molecules. The elements that make up the bulk of the necessary lipids, carbohydrates, proteins, and nucleic acids are required in large amounts by cells. These elements are called **macronutrients**. The macronutrients can be remembered by the following phrase: "See Hopkins Café. Mighty Good." This phrase represents the element symbols CHOPKNS Ca Fe Mg, which stand for carbon, hydrogen, oxygen, phosphorous, potassium, nitrogen, sulfur, calcium, iron, and magnesium. Microbes obtain these elements from food or from inorganic salts that are present in the soil and water. As was discussed in Chapter 5, one major difference between organisms is how they get their carbon. Heterotrophs obtain carbon from organic molecules (food), while autotrophs obtain carbon from CO_2 in the environment.

Elements that are required in very small amounts are called **micronutrients** or **trace elements**. These include elements such as zinc, copper, cobalt, manganese, and molybdenum. Most of the elements are required for enzyme function. Sufficient quantities of trace elements are usually found in water.

Growth in the Laboratory

To grow a particular microorganism in the laboratory, all of its growth requirements for temperature, pH, oxygen, and nutrient conditions must be met. This is easier to do for some organisms than for others. For example, obligate anaerobes must be grown in environments that contain no oxygen. To culture a hyperthermophile, temperatures near boiling must be maintained. Some organisms have very particular nutritional requirements, including complex **growth factors**, large molecules such as amino acids or vitamins that they cannot make for themselves. It can be very difficult to figure out the exact combination of nutrients and growth factors that a particular cell requires for its growth. In fact, it has been estimated that less than 10 percent of all the prokaryotes on the planet have ever been grown in a laboratory culture.

CULTURE MEDIA

To grow organisms in the laboratory, nutrients are combined with water to form **culture media**. Routine culture media commonly include digested yeast or proteins (peptone) that contain a variety of nutrients. These media are called **complex media** because their precise chemical composition is not known. In contrast, **chemically defined media** are made by mixing precise combinations of known chemical elements. Culture media may be liquid (**broths**), or they may be solidified by the addition of agar. Solid media may be poured into dishes called **petri plates** or into glass tubes to form **agar slants** and **deeps** (Figure 6.5).

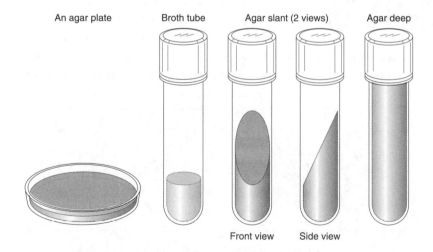

FIGURE 6.5. Common laboratory culture vessels showing the relative volumes of media contained in each (shadowed).

Special types of culture media may favor the growth of one type of organism over another, or help identify certain characteristics of organisms. These types of media and their definitions are summarized in Table 6.4.

TABLE 6.4. Special Culture Media

Media	Purpose
Selective agar	Favors the growth of one type of organism because it contains a chemical that restricts the growth of other types of organisms.
Enrichment agar	Favors the growth of one type of organism because it contains a chemical that is necessary for their growth. Used to encourage growth of organisms that might be rare in a population.
Differential agar	Contains dyes or other compounds that react if certain metabolic processes are performed by the organism. Helps to identify an organism because it reveals certain traits.

PURE CULTURE TECHNIQUES

To determine the growth characteristics of a bacterium, it is necessary to have a **pure culture** of that bacterium. In a pure culture, all of the cells are descended from a single cell; in other words, they are all the same kind of bacteria. In nature, bacteria exist in mixed populations of all different types of organisms. One way to obtain a pure culture of just one kind is to dilute the mixed population until you can collect single cells.

> **REMEMBER**
> A pure culture is a culture that's descended from a single cell. An isolated colony is a colony that isn't touching any other colonies.

A common and simple way to obtain a pure culture is to perform a **pure culture streak**. In this method, a mixed population of bacteria are dragged over the surface of a solid media until they are diluted enough that single cells are being deposited (Figure 6.6). The cells are allowed to grow until they form visible spots of growth called **colonies**. Each colony is the result of the division of a single cell. An **isolated colony** can be picked up and transferred to new growth media to create a pure culture.

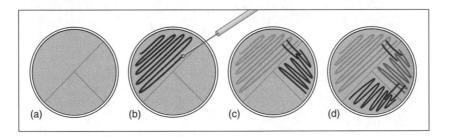

FIGURE 6.6. Pure culture streaking.

The pattern of lines indicates the path the inoculating loop would travel on the surface of the agar. After incubation, growth would be visible along these streaks, with decreasing amounts of growth in each section of the plate. In the final section, the amount of growth would be the least, resulting in the formation of separate colonies. Note: The inoculating loop is sterilized between each section.

MEASURING BACTERIAL GROWTH

There are many reasons why someone might want to know the exact number of microbes in a population. For example, microbiologists monitor the numbers of bacteria in our food and water to make sure they are safe for human consumption. A physician might need to monitor the number of bacteria in a patient's blood to determine whether antibiotic therapy is effective. There are two basic methods for determining the number of bacteria in a population: **direct counts** and **indirect counts**. Direct counts involve actually counting the cells or colonies that arise from cells. Indirect counts estimate the population size by measuring some property of the population such as mass or turbidity (cloudiness) of a solution.

> **REMEMBER**
> Direct counts determine the number of cells in a culture by actual counting of cells or colonies. Indirect counts estimate the number of cells by measuring a property of growth, such as weight.

Direct counts of bacterial cells can be done by placing a very small sample of the population into a special microscope slide that contains a counting chamber. The number of bacterial cells visible in the chamber is counted. This is called a **direct microscope count**. Bacterial cells can also be counted electronically in machines called **coulter counters**. In this method, a sample from the population is placed in the machine. The sample is passed between two electrodes. Every time a cell passes between the electrodes, it causes a disturbance in the electrical field, and the cell is counted.

Direct counts can also be done by the **dilution plate method**. In this method, samples from the population are diluted and then inoculated onto an agar plate. The plate is incubated to allow bacterial growth. Every cell on the plate will divide and produce a visible colony that can then be counted. To achieve a reasonable amount of colonies on the plate, the **serial dilution** technique is employed: A series of dilutions of the original population is made, and samples from each dilution are spread over agar plates. The plate that has the appropriate number of colonies is counted, and the count is multiplied by the **dilution factor** of the plate in order to determine the number of bacteria in the original population.

For example, a food microbiologist might want to know the number of bacteria in some juice. If 1 milliliter of the juice were directly poured onto an agar plate, the growth that resulted would be too dense to count. So, the microbiologist prepares a series of dilutions (Figure 6.7). If 1 milliliter of juice is added to 9 milliliters of sterile water, the juice has been diluted by a factor of ten (the juice is only 1/10 of the sample). If 1 milliliter of the 1/10 (10^{-1}) dilution is added to 9 milliliters of sterile water, the juice has now been diluted by a factor of 100 (1/10 times 1/10 is 1/100, or 10^{-2}). This method is repeated until several dilutions have been made. A 1-milliliter sample from each dilution is spread onto an agar plate. The more diluted the sample, the fewer the colonies that will result from growth of the cells in the juice. The microbiologist chooses the *most countable* plate, one that has between approximately 30 and 300 colonies and counts the colonies on that plate. The number of colonies on the plate is multiplied by the dilution factor of that plate to determine the number of cells in the original juice sample. For example, if there were thirty-one colonies on the 1/10,000 (10^{-4}) plate, 31 would be multiplied by 10,000. This would indicate that there were 310,000, or 3.1×10^5, cells per milliliter (ml) in the original juice sample.

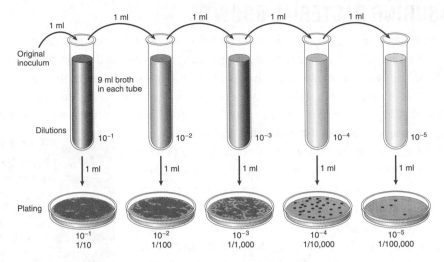

FIGURE 6.7. The dilution plate method.

In order to count the number of bacteria in samples that have great numbers of cells, the samples are first diluted, then placed in petri plates with agar. The dilute samples are allowed to grow; then the number of colonies are counted. The number of colonies from the diluted sample is multiplied by the dilution factor to determine the number of bacteria in the original sample. For example, if there were thirty-one colonies on the 1/10,000 (10^4) plate, 31 would be multiplied by 10,000. This would indicate that there were 310,000, or 3.1×10^5, cells per milliliter (ml) in the original sample.

Indirect counts do not count cells directly but rather assess a property of the population, such as mass or turbidity (cloudiness), that is proportional to the number of cells in the population. For example, filamentous organisms such as molds may be dried and their **dry weight** determined. For bacterial populations, increase in cell number is sometimes assessed indirectly by measuring the **turbidity**, or cloudiness, of the sample. The greater the number of bacterial cells in a solution, the more the light rays entering the solution are refracted (bent) by the cells, giving the solution a cloudy appearance. Turbidity can be measured with an instrument called a **spectrophotometer**.

REVIEW EXERCISES FOR CHAPTER 6

1. Describe the process of binary fission.

2. Define generation time.

3. Name and describe the phases of bacterial growth curve.

4. Name and describe the categories that organisms are placed into based on their temperature requirements.

5. What is the optimum pH for bacteria? For fungi? What is the name for acid-tolerant organisms?

6. Define plasmolysis. What causes plasmolysis? What are salt-tolerant organisms called?

7. What are free radicals? How are they formed? How do cells deal with them?

8. Name and describe the categories that organisms are placed into based on their tolerance for oxygen.

9. What are the macronutrients? Why do cells need them?

10. What are micronutrients? Why do cells need them?

11. Distinguish between complex media and chemically definied media.

12. Name and describe the different types of special culture media.

13. Describe how pure culture streaking can be used to obtain a pure culture.

14. Distinguish between direct and indirect counts of bacteria.

15. Describe the dilution plate method.

SELF-TEST

1. *Salmonella* has a generation time of 30 minutes. If four cells are present on a turkey that is left on the counter, how many cells will be present after 3 hours?

   A. 12
   B. 24
   C. 64
   D. 256
   E. 1256

2. Penicillin blocks cell wall synthesis. During which phase of bacterial growth would a population of *Streptococcus* be most sensitive to penicillin?

   A. Lag phase
   B. Log phase
   C. Stationary phase
   D. Death phase
   E. It would be equally sensitive at all times.

3. An organism that has an optimum temperature of 50° Celsius is a

   A. mesophile.
   B. psychrophile.
   C. psychrotroph.
   D. thermophile.
   E. extreme thermophile.

4. An organism that grows at a pH of 4 is a(n)

   A. thermophile.
   B. halophile.
   C. acidophile.
   D. psychrophile.
   E. None of the above are correct.

5. Which enzyme peforms the following reaction: $2H_2O_2 \rightarrow 2H_2O + O_2$?

   A. Superoxide dismutase
   B. Catalase
   C. Peroxidase
   D. Both B and C are correct.
   E. All of the above are correct.

6. An organism that uses oxygen if it is available, but can survive without it if it is not available is a(n)

   A. obligate aerobe.
   B. microaerophile.
   C. aerotolerant anaerobe.
   D. facultative anaerobe.
   E. obligate anaerobe.

7. True or False. An aerotolerant anaerobe cannot grow in the presence of O_2.

8. Which of the following is **not** a macronutrient?

   A. Carbon
   B. Sulfur
   C. Nitrogen
   D. Phosphorous
   E. Zinc

9. True or False. In chemically defined media, the exact amount of every chemical present is known.

10. Which would be the best choice to encourage the growth of a rare organism in the blood?

    A. Selective agar
    B. Differential agar
    C. Enrichment agar
    D. Both A and B are correct.
    E. None of the above are correct.

11. If a 10^{-5} dilution yielded forty-nine colonies on a plate, what was the concentration of cells in the original sample?

   A. 4.9×10^{6}
   B. 4.9×10^{5}
   C. 4.9×10^{-6}
   D. 4.9×10^{-5}
   E. None of the above.

Answers

Review Exercises

1. During binary fission, cells copy their DNA and other cellular components and increase in size. Cell membrane and wall form in the middle of the cell, splitting the original cell into two identical copies.

2. Generation time is the time it takes a cell to divide or a population to double.

3. Lag phase occurs immediately after the culture is transferred to new media. During this time, the number of cells is not increasing. Cells are synthesizing the enzymes they need to utilize the new media for growth. Log phase or exponential phase occurs when cells are dividing at regular intervals. This is the most rapid phase of growth. Stationary phase is when the number of new cells equals the number of dying cells. In other words, the population is no longer increasing. This occurs as food decreases and wastes increase. Finally, death phase is when the number of dying cells exceeds any new cells. The population is in decline owing to the lack of food and amount of waste materials.

4. Psychrophiles are very cold-loving organisms. Their optimum temperatures are around 10°C. Psychrotrophs are cold-tolerant. Their optimum temperatures are around 20°C. Mesophiles like moderate temperatures and grow best around 37°C. Thermophiles like warm temperatures and grow best around 60°C. Extreme thermophiles can tolerate temperatures above boiling and have optimums around 90°C.

5. Bacteria have an optimum pH near neutral (7). Fungi have a pH optimum of 5–6. Organisms that tolerate acid conditions are called acidophiles. Organisms that tolerate alkaline conditions are called alkalophiles.

6. Plasmolysis occurs when the cytoplasm of the cell collapses owing to water loss. The plasma membrane pulls away from the cell wall. This happens when cells are placed in a hypertonic environment. Salt-tolerant organisms are called halophiles.

7. Free radicals are dangerous forms of oxygen. Free radicals are very reactive and can destroy other molecules. They are formed during metabolic processes in cells. Cells must have enzymes that convert these free radicals into less harmful substances in order to survive in the presence of oxygen.

8. Obligate anaerobes are killed by oxygen. Aerotolerant anaerobes can survive in oxygen, but they do not use it in their metabolism. Facultative anaerobes will use oxygen if it is available, but they can also survive when it is not available. Microaerophiles like low concentrations of oxygen. Obligate aerobes require the presence of oxygen to survive.

9. The macronutrients are carbon, hydrogen, oxygen, phosphorous, potassium, nitrogen, sulfur, magnesium, calcium, and iron. They are needed for the construction of macromolecules.

10. Micronutrients are elements such as copper and zinc that are needed as partners for enzymes.

11. Complex media are made from digested cells such as yeast. Their exact chemical composition is not known. Chemically defined media are mixed from known pure chemicals. The exact amount of each chemical is known.

12. Selective media allow some organisms an advantage because they restrict the growth of other organisms. Enrichment media provide special nutrients to enhance the growth of rare organisms. Differential media show a visible reaction to certain metabolic processes. This allows identification of certain traits of the organisms growing on the plate.

13. Pure culture streaking is when a sample of bacteria is dragged repeatedly over the surface of a plate. This results in fewer and fewer cells being deposited at one time. If the sample is diluted enough, single cells will be deposited. These single cells will grow into isolated colonies that can be seen. Each colony represents growth from a single cell, which is the definition of a pure culture. The colony can be picked off the plate and used to inoculate sterile media.

14. Direct counts actually count the bacteria, for example, either individually in a microscope or after they form colonies on a plate. Indirect counts measure some property of the bacteria like weight or turbidity that is used to figure out the number of cells.

15. In the dilution plate method, a dense sample of cells is diluted. This is done by taking a small sample of the original and placing it in sterile media. A sample of the first dilution is further diluted, and so on, until a series of dilutions has been made. A sample from each dilution is placed onto a plate and allowed to grow. The plate that has a countable number of colonies is counted, and the number is multiplied by the dilution factor to figure out the number of cells in the original sample.

Self-Test

1. D	5. B	9. T
2. B	6. D	10. C
3. D	7. F	11. A
4. C	8. E	

Microbial Genetics

7

WHAT YOU WILL LEARN

This chapter dives into the complexities of microbial genetics, from how cells copy and read their DNA to the mechanisms that control when and how genes are used. As you study this chapter, you will:

- get the details on how cells use DNA replication to copy DNA for cell division;
- learn the steps of RNA synthesis (transcription) with protein synthesis (translation);
- explore how cells respond to changes in their environment by controlling their genes;
- investigate the effects of mutations on cells and how mutations can be useful to microbiologists;
- discover the ways in which bacteria can exchange genes.

SECTIONS IN THIS CHAPTER

- DNA Replication
- RNA Synthesis
- Protein Synthesis
- Gene Regulation
- Mutation
- Gene Transfer

If someone says *E. coli*, what do you think of? Chances are you think of a scary, disease-causing organism, one that can make people very sick and even kill them. And for some kinds of *E. coli,* you would be absolutely right. But did you know you have *E. coli* living in your intestines right now? The *E. coli* in your intestines is not only harmless, it actually helps you by making vitamins and keeping other "bad bugs" (disease-causing bacteria) from taking over. The headline-grabbing *E. coli* that causes severe diarrhea is different from your normal *E. coli*. It has different abilities and can make certain toxins that lead to severe disease. The abilities, or **traits**, of the two types of *E. coli* are different because their genetic material (DNA) is different. To understand these differences and where new types, or **strains**, of bacteria come from, we need to understand how the information stored in DNA is used by cells, and how cells can transfer that information among them. This involves learning several complicated processes that occur in cells (Figure 7.1), but these processes are important because they are absolutely fundamental to every kind of cell on Earth.

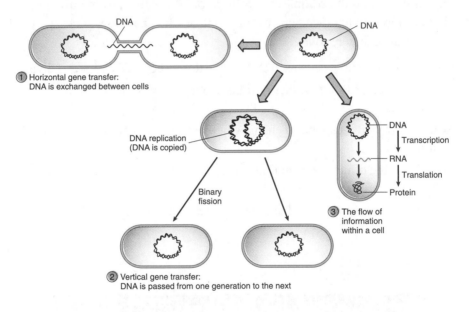

FIGURE 7.1. Overview of bacterial genetics.

DNA stores the information for the traits of a cell. 1. During horizontal gene transfer, DNA is exchanged between cells, passing traits from one cell to another. 2. During vertical gene transfer, a cell passes its traits on to the next generation by copying its DNA. 3. Cells use the information in their DNA to build proteins that determine the cells' characteristics.

DNA Replication

To replicate something is to copy it. **DNA replication**, then, refers to the copying of DNA. Cells copy their DNA when they are getting ready to divide because each cell needs its own copy of the genetic information. Recall that most bacterial cells have a single, circular chromosome (Figure 7.2). Like all chromosomes, it is made of a double-stranded DNA molecule. Each half of the strand is **complementary** to the other: Adenine (A) pairs with thymine (T), guanine (G) pairs with cytosine (C) (Figure 7.3). When DNA is copied, the two partner strands are separated from each other, and each is used as a pattern for the synthesis of a new strand (Figure 7.3). When DNA replication is finished, two double-stranded DNA molecules result, each containing one of the original strands and one new strand. Because half of the original DNA molecule is conserved, or saved, in the new molecule, DNA replication is said to be **semiconservative**.

> **REMEMBER**
> Cells use DNA replication to copy their DNA during cell division.

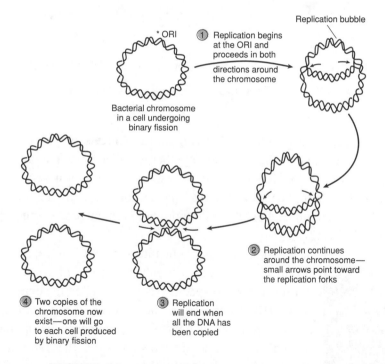

FIGURE 7.2. Replication of a bacterial chromosome.

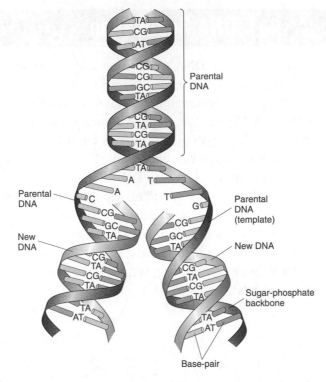

FIGURE 7.3. Semiconservative replication of a DNA molecule.

Each strand of the double helix is used as a template for the synthesis of a new strand.

The details of the process of how DNA is replicated in cells can be confusing. It involves a group of enzymes that all work together, with each one having a specific role in the process. Each one functions a certain way based on its structure, and none of them can do the whole job by themselves. Because of this, there are many steps in the process. In order to understand it, it is important to remember the specific job of each enzyme (Table 7.1).

> **REMEMBER**
> In bacteria, DNA replication proceeds simultaneously in both directions, away from the origin of replication (ORI).

The first step in the replication of DNA is to separate the two original stands from each other. This occurs within the bacterial chromosome at a specific sequence called the **origin of replication (ORI)** (Figure 7.2). The enzymes of replication assemble at the origin, and enzymes called **helicases** disrupt the hydrogen bonds that hold the two strands together. This creates a separation, or **replication bubble**, in the chromosome (Figure 7.2).

Two teams of enzymes begin to copy the DNA, moving away from each other around the bacterial chromosome. The helicases are part of the team and continue to separate the strands as progress around the chromosome continues. As the two teams approach each other again, replication slows and is complete when all of the DNA has been copied (Figure 7.2).

TABLE 7.1. Enzymes Involved in DNA Replication

Enzyme	Role in Replication	How It Works
Helicase	Separates the DNA strands	Disrupts hydrogen bonds that hold the two strands together
Primase (RNA polymerase)	Provides a starting place for DNA polymerase III	Synthesizes short strands of RNA (primers) that are complementary to the DNA
DNA polymerase III	Synthesizes the new strands of DNA complementary to the original	Must have a template (the original strand) Must add new nucleotides to the 3' end of the growing strand Cannot begin a strand by itself
DNA polymerase I	Cleans RNA out of the new DNA	Removes RNA primers and replaces them with DNA
DNA ligase	Seals the breaks in the backbone of the new DNA	Catalyzes the formation of covalent bonds along the sugar phosphate backbone between the fragments

To see the details of how the enzyme teams work together, we need to take a closer look at what is happening within the replication bubble. On either side of the bubble, there is an area where the original DNA has been newly separated and replication is just beginning. This area is called a **replication fork** (Figure 7.4). At the replication fork, the original DNA is single-stranded and available for the synthesis of a new partner strand.

The enzyme that synthesizes the new DNA strands is **DNA polymerase III** (Figure 7.4). Its name indicates its function as an enzyme that makes polymers of DNA. To synthesize new DNA, DNA polymerase III requires certain things. It must have a template for the ordering of the new nucleotides, which is provided by the original DNA strand. Also, it cannot start a strand on its own; it must have an existing nucleotide as a starter to which it can attach new nucleotides. Finally, it only adds new nucleotides to the 3' end of existing nucleotides. If we consider these last two requirements of DNA polymerase III, we can see that the enzyme needs a little help before it can get started copying DNA at the replication fork.

The enzyme that helps DNA polymerase III get started is called **primase** (Figure 7.4). Unlike DNA polymerase III, primase can start new chains of nucleotides on its own. However, primase is an RNA polymerase, and the chains it synthesizes are RNA. So, after helicases open the original DNA strands, primase puts down short chains of RNA that are complementary to the original strand. These short chains are called **primers** because they provide starting points for DNA polymerase III to begin DNA replication.

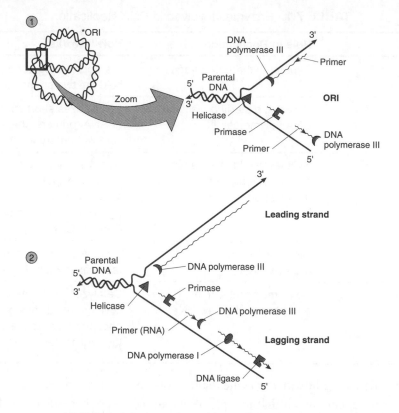

FIGURE 7.4. The progression of DNA replication.

1. DNA is unwound at the origin. Primase puts down RNA primers on both parental strands. DNA polymerase III begins to replicate DNA. 2. On the leading strand, DNA polymerase III has continuously synthesized DNA. DNA polymerase I has removed the primer and DNA ligase has sealed the strand. On the lagging strand, DNA polymerase III is starting synthesis of a new fragment. On previously synthesized fragments, DNA polymerase I is replacing RNA primer with DNA. DNA ligase is forming covalent bonds between fragments. Primase is starting a new primer close to the fork.

After the primers are in place, DNA polymerase III has everything it needs to synthesize new DNA strands complementary to the original, parental strands. Using the parental strands as a template, it brings in complementary nucleotides and attaches them first to the 3' end of the primers and then to the 3' end of the growing chains (Figure 7.4). However, because DNA polymerase III can only add to 3' ends, the chains of new DNA on opposite sides of the fork will grow in opposite directions. (Recall that when nucleotide strands pair, they always pair in an antiparallel fashion; see Chapter 2.) The 3' end of one of the new strands points toward the replication fork. As the enzyme team moves around the DNA circle, this strand continues to grow in the same direction that the team is moving. Thus, once DNA polymerase III has started to synthesize this

> **REMEMBER**
> The lagging strand is the new strand of DNA that has its 3′ end pointing away from the replication fork.

strand, it can just keep going. This strand is built in one continuous piece and is called the **leading strand**. On the other side of the fork, however, the 3' end of the new strand points away from the fork. As DNA polymerase III adds new nucleotides to this 3' end, the strand grows away from the fork, in the opposite direction of the movement of the enzyme team around the DNA circle. Because of this, DNA polymerase III synthesizes this strand in a bunch of small pieces called **Okazaki fragments**. It synthesizes a short stretch of DNA; then, it stops and begins again on a new primer closer to the fork (Figure 7.4). By synthesizing this strand, called the **lagging strand**, in a bunch of small pieces, DNA polymerase can stay with the enzyme team as it moves around the circle.

After the work of primase and DNA polymerase III are completed, other enzymes play a part in the finishing of the new DNA. Because primase makes polymers of RNA, the primers that exist in the new DNA strands need to be removed and replaced with DNA. This is the job of **DNA polymerase I** (Figure 7.4). Also, because the new DNA is being synthesized in pieces, there are places in the DNA where fragments are next to each other but not attached to each other by covalent bonds. This is the job of **DNA ligase** (Figure 7.4), an enzyme that forms covalent bonds between the fragments of the newly synthesized DNA.

RNA Synthesis

So far, we have looked at how cells copy their DNA. When DNA is copied, the information within the DNA is transferred to a new molecule and the cell can now divide. However, we haven't yet looked at how the information in DNA is *used* by an active, functioning cell to specify the activities of that cell. Most cellular activities are controlled by the actions of proteins, and some are controlled by molecules of RNA. The information for building proteins and RNA is located in the DNA. So, when we say that DNA is the genetic material of a cell and determines a cell's traits, that is because the DNA of a cell determines the RNA and proteins that cell can make and therefore what that cell can do. The sequence of DNA that contains the information for a single protein or RNA is called a **gene**.

During the process of **transcription**, the information in the DNA is copied into a molecule of RNA. The enzyme that reads the DNA and synthesizes the complementary molecule of RNA is called **RNA polymerase**. RNA polymerase locates the beginning of the gene it is going to copy by finding the **promoter**, a specific DNA sequence that marks the beginning of the gene. A useful analogy for understanding promoter function is that if you think of all of the DNA of a cell as a set of files containing information, the promoter is like the tab on an individual file folder that helps you locate specific information. The information in a single file is a gene. In prokaryotes, the main file cabinet would be the circular chromosome. Within the chromosome would be thousands of files, or genes, that contain the information for the construction of proteins.

When RNA polymerase binds to the promoter sequence for a particular gene, it is positioned on the DNA so that it reads the correct strand in the correct direction.

Beginning at the promoter, RNA polymerase unwinds a short stretch of the double helix of DNA (Figure 7.5). It then reads one of the DNA strands and builds a complementary RNA molecule following the base-pairing rules shown in Table 7.2.

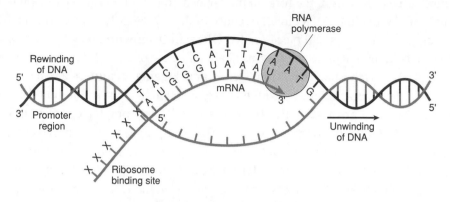

FIGURE 7.5. Transcription.

RNA polymerase began copying DNA to RNA at the promoter. It will proceed along the gene until it reaches the transcription terminator.

TABLE 7.2. Complementary Base Pairing Between DNA and RNA

DNA Nucleotide	Complementary RNA Nucleotide
C	G
G	C
T	A
A	U

As RNA polymerase moves along the gene, the growing RNA molecule detaches from the DNA, and the double helix rewinds itself behind the enzyme (Figure 7.5). RNA polymerase continues until it reaches the end of the gene, which is marked by another sequence called the **transcription terminator**. After RNA polymerase copies the terminator sequence, it detaches from the DNA and releases the RNA into the cell (Figure 7.5). Some RNA molecules, like rRNA and tRNA, directly perform a function for the cell. If the RNA molecule is mRNA, it will now go to the ribosome to be translated into protein.

In eukaryotes, each gene has its own promoter and transcription terminator, so each mRNA molecule is **monocistronic**, containing the information for just one polypeptide. In prokaryotes, however, several genes may be organized together under one promoter, forming an **operon** in which several genes are controlled together. When transcription of an operon begins, all of the genes in the operon may be transcribed into the same piece of mRNA. Thus, most mRNA molecules in prokaryotes are **polycistronic** and contain the information for several polypeptides.

Protein Synthesis

Making proteins from the DNA code involves two processes: transcription and **translation**. During transcription, the information in the DNA is copied into a molecule of RNA called **messenger RNA (mRNA)**. The mRNA message travels to the **ribosome** where it is decoded and used to build the protein. Translation of mRNA into protein involves the mRNA, the **ribosome**, which contrains rRNA, and another type of RNA called **transfer RNA (tRNA)**. Each of these plays a specific role in the process. The mRNA carries the blueprint for the protein from the DNA. The ribosome is like the factory in which translation occurs. It organizes the process and has binding sites for the mRNA and tRNA. The rRNA within the ribosome catalyzes bond formation between the amino acids building blocks of the protein. The tRNA acts as a decoder, reading the mRNA code and bringing the correct amino acids into position in the growing polypeptide chain (recall that a polypeptide chain is a chain of amino acids that folds up to form a protein).

The mRNA code is read in **codons**, groups of three nucleotides. Each codon represents one amino acid. There are four nucleotides in RNA (C, G, A, and U). If we think of all of the possible combinations of three nucleotides, there are sixty-four possible mRNA codons (Figure 7.6), yet there are only twenty amino acids found in cells. Thus, some amino acids are represented by more than one codon. Certain codons have a special function. The codon AUG is called the **start codon** because, in addition to representing the amino acid methionine, it also marks the starting point for translation of the mRNA. Three codons, UAA, UGA, and UAG, are called **stop codons** because they typically mark the end point for translation of the mRNA. These codons do not represent amino acids.

Just as there are rules for reading English, there are rules for reading mRNA. When reading English, you read from left to right. When reading mRNA, you read from 5' to 3'. When reading English, you start at the beginning of a sentence that is marked by a capitalized word, when reading mRNA you begin at the start codon closest to the 5' end. When reading English, spaces are used to indicate separate words. When reading mRNA, you divide the nucleotides into codons beginning with the start codon. Finally, when reading English, you stop when you reach a period. When reading mRNA, you stop when you reach a stop codon. Thus, to decode a molecule of RNA, you begin at the 5' end, find the start codon, break the nucleotides into codons, and stop when you reach the stop codon (Figure 7.7). The amino acids represented by the codons are listed in Figure 7.6, which shows the **genetic code**.

> **REMEMBER**
> The genetic code shows the codons in mRNA (not the anticodons in tRNA).

The Genetic Code

		2nd base in codon				
		U	C	A	G	
1st base in codon	U	Phe Phe Leu Leu	Ser Ser Ser Ser	Tyr Tyr Stop Stop	Cys Cys Stop Trp	U C A G
	C	Leu Leu Leu Leu	Pro Pro Pro Pro	His His Gln Gln	Arg Arg Arg Arg	U C A G
	A	Ile Ile Ile Met	Thr Thr Thr Thr	Asn Asn Lys Lys	Ser Ser Arg Arg	U C A G
	G	Val Val Val Val	Ala Ala Ala Ala	Asp Asp Glu Glu	Gly Gly Gly Gly	U C A G

(3rd base in codon)

Amino acids

Ala	Alanine
Arg	Arginine
Asn	Asparagine
Asp	Aspartic acid
Cys	Cysteine
Gln	Glutamine
Glu	Glutamic acid
Gly	Glycine
His	Histidine
Ile	Isoleucine
Lys	Lysine
Leu	Leucine
Met	Methionine/start codon
Pro	Proline
Phe	Phenylalanine
Ser	Serine
Thr	Thrreonine
Trp	Tryptophane
Tyr	Tyrosine
Val	Valine
Stop	termination codon

FIGURE 7.6. The genetic code.

To find out which amino acid is specified by a codon, first find the row that corresponds to the first base in the codon. Next, find the column that corresponds to the second base in the codon. Then, find the box at the intersection of that row and column. Finally, find the line in the box that corresponds to the third base in the codon. Note that the code is listed in codons from mRNA, not anticodons from tRNA. Also, the codon AUG is both the start codon and the code for the amino acid methionine. Stop codons do not code for amino acids; instead, they signal the end of translation and the arrival of the enzyme release factor.

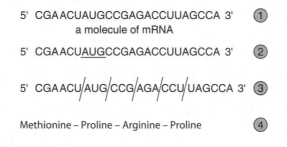

5' CGAACUAUGCCGAGACCUUAGCCA 3'   (1)
a molecule of mRNA

5' CGAACU<u>AUG</u>CCGAGACCUUAGCCA 3'   (2)

5' CGAACU/AUG/CCG/AGA/CCU/UAGCCA 3'   (3)

Methionine – Proline – Arginine – Proline   (4)

FIGURE 7.7. Decoding an mRNA molecule.

1. A molecule of mRNA. 2. Find the start codon closest to the 5' end. 3. Break the code into codons. Look up the codons, beginning with AUG and stopping at the stop codon (UAG). 4. Write the sequence of amino acids in the polypeptide.

In cells, the mRNA is "read" by the tRNA. tRNA is a nucleic acid just like mRNA, and so can bind to mRNA in a complementary fashion, just like we've seen for the DNA strands of the double helix, or the pairing of RNA to DNA during transcription.

tRNA is a single-stranded nucleic acid, but it has several folds that are held together by base pairing within the molecule (Figure 7.8). Each tRNA has a special sequence of three nucleotides called the **anticodon**. The anticodon is the part of the tRNA that binds to the codons in mRNA. The tRNA also has a special binding site for a specific amino acid. When tRNA anticodons bind to mRNA codons, the binding is specific and complementary. For example, if the mRNA codon is 5'AUG3', the only tRNA that can bind to it is the one whose anticodon is 3'UAC5'. (Recall that all nucleic acids bind in an antiparallel fashion.) This particular tRNA would also

> **REMEMBER**
> The anticodons in tRNA hydrogen bond to the codons in mRNA during translation. Like all nucleic acids, this pairing occurs when the two strands are antiparallel to each other. The specific pairing of anticodon to codon determines the sequence of amino acids in the polypeptide chain.

have a binding site for the amino acid methionine. Because this tRNA always carries methionine, and because it is the only one that can bind to the codon AUG, the amino acid methionine is always positioned in the protein when the codon reads AUG.

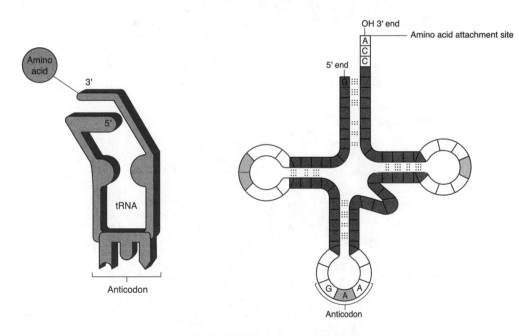

FIGURE 7.8. Transfer RNA.

Transfer RNA is made of single-stranded RNA, but it folds into a three-dimensional shape that is important to its function. The shape is held together by base pairing between nucleotides in the strand. Two important components of tRNA are the anticodon, which base-pairs with mRNA codons, and the amino acid attachment site, which carries the amino acid.

For translation to begin, the ribosome must bind to the mRNA. Recall that the ribosome consists of two subunits, a large subunit and a small subunit. These subunits are made separately and only come together to form a completed ribosome when translation begins. When a complete ribosome is assembled around the mRNA, pockets are formed within the ribosome. These pockets, called the **A site** (for acceptor), the **P site** (for peptide), and the **E site** (for exit) (Figure 7.9), serve as binding sites for tRNAs. As the polypeptide chain is synthesized, tRNAs carrying amino acids enter at the A site, are transferred to the P site where they hold the growing polypeptide, and then leave by the E site.

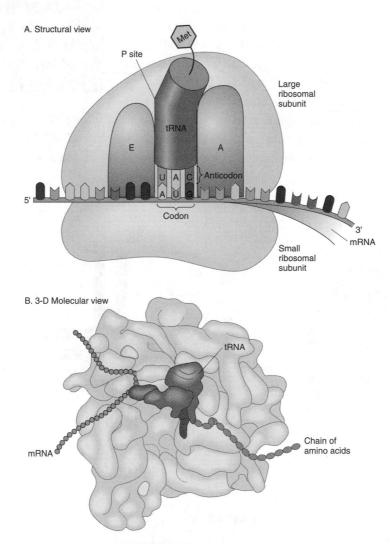

FIGURE 7.9. The ribosome.

The ribosome consists of two subunits, a large subunit and a small subunit. When it is assembled around an mRNA molecule, internal binding sites for tRNAs are formed.

The process of stringing the amino acids together to form the polypeptide involves several steps. Initiation of translation begins when the small subunit of the ribosome binds to the mRNA (Figure 7.10). After this, the tRNA that has the anticodon to the start codon attaches to the mRNA in what will become the P site of the ribosome. Finally, the large subunit of the ribosome binds to the small subunit, and translation is initiated.

① Initiation: First, the small subunit of the ribosome binds the mRNA. Second, the first tRNA with the anticodon to the start codon binds to the mRNA. Then, the large subunit binds. The second codon is now available in the A site.

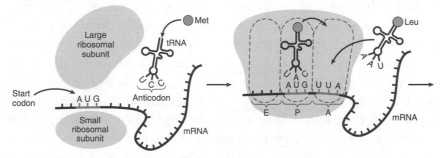

② Elongation: tRNAs carrying amino acids enter the A site. An enzyme in the ribosome catalyzes bond formation between the adjacent amino acids. The ribosome-mRNA complex shifts. The first tRNA exits from the E site. A new tRNA enters the A site, and the process repeats.

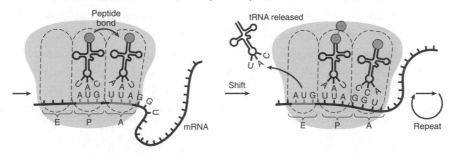

③ Termination: When a stop codon is present in the A site, the enzyme release facter enters the ribosome and releases the polypeptide chain. After that, the ribosomal subunits separate from the mRNA.

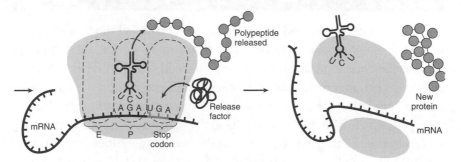

FIGURE 7.10. Translation.

After translation is initiated, the Met-tRNA is positioned in the P site over the start codon and the next codon in the mRNA is exposed in the A site of the ribosome. To begin elongation of the polypeptide chain, the tRNA with the complementary anticodon enters the A site and binds to the mRNA. This positions the amino acid of the incoming tRNA next to the amino acid of the first tRNA (Figure 7.10). At this point, a **ribozyme**, an RNA molecule that acts as an enzyme, catalyzes bond formation between the two amino acids, shifting the amino acid from the tRNA in the P site onto the amino acid of the tRNA in the A site. The ribozyme is part of the ribosome. There are now two amino acids joined to form the start of the polypeptide chain. To continue the process and bring in another tRNA with its amino acid, the mRNA with its attached tRNAs shifts relative the ribosome, effectively moving the tRNAs into new sites. The tRNA that is no longer carrying an amino acid moves into the E site, from which it will exit the ribosome and go have another amino acid attached to it. The tRNA holding the polypeptide chain is now positioned in the P site, and a new codon is exposed in the A site. Now the tRNA with the correct anticodon can enter the A site, and the elongation process can be repeated so that amino acids are added to the chain one at a time.

Elongation continues until a stop codon is exposed in the A site. Recall that the stop codons do not represent amino acids. There are no tRNAs that have anticodons to stop codons. Instead, an enzyme called release factor recognizes stop codons. When a stop codon is exposed in the A site, release factor enters the ribosome and releases the polypeptide chain (Figure 7.10) from the tRNA, terminating the process of translation. The polypeptide can now fold up into a functional protein and perform a job for the cell. Other proteins release the ribosome and mRNA so that they can restart the process of translation.

Gene Regulation

The processes of transcription and translation are occurring all the time in cells to manufacture the many proteins a cell needs to function. However, although a cell's DNA contains the code to make all the proteins the cell will ever need, at any one time a cell only needs certain proteins and not others. Cells can control which proteins are being used by regulating the activity of the proteins or by controlling production of the proteins (Figure 7.11). To help you understand the reasons cells control protein production, consider the example of your own body. You are a multicellular organism, with over 200 functionally distinct types of cells. You have skin cells that protect your body from microbial invasion, heart cells that reliably contract all day long in order to keep your blood flowing, and pancreatic cells that tell your liver and fat cells when sugar is in the blood. Each of these cells serves a very different function and thus requires very different tools. A skin cell needs lots of the protein keratin, which is very strong and helps make your skin resistant to penetration. Heart cells need lots of the proteins actin and myosin to keep the muscles contracting. A pancreatic cell needs to make lots of the protein insulin, so that it can release it into the blood to signal that sugar is available. All three types of cells have the genes for all the proteins mentioned, but they don't all need the proteins to the same degree. As you can imagine, it would be very inefficient if your cells made proteins they didn't need. And in fact, they don't. Cells have the ability to select the genes they will transcribe at any one time. This is called **gene regulation**. If a gene is being transcribed and the mRNA is then translated so that the protein is present in the cell, the gene is said to be **expressed**, or "on."

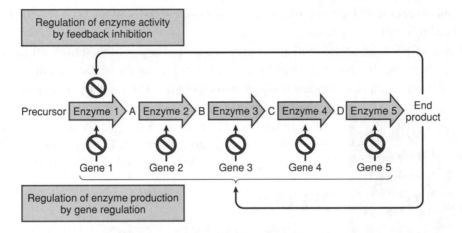

FIGURE 7.11. Two ways to regulate enzymes.

A metabolic pathway that converts a precursor to an end product is shown. A, B, C, and D are intermediates in the pathway. One way a cell might regulate this pathway is by feedback inhibition. When enough of the end product is available in the cell, it binds to a regulatory site on an enzyme in the pathway, making it inactive. Another way to regulate this pathway is at the level of transcription. In this case, end product could bind to a regulatory protein that would then bind to regulatory DNA near the gene for one of the enzymes. This may block transcription and ultimately the synthesis of that enzyme.

Microbes also need to be able to regulate their genes. Most bacteria are not multicellular and so don't have specialized cell types like humans do, but they need to be able to respond to changes in their environment. If growing conditions are bad and food becomes unavailable, a spore-forming bacterium might form an endospore. The poor growing conditions would signal the cell to turn on the genes necessary for sporulation. Also, as food sources change in the environment, bacteria need to adjust their enzymes to take advantage of what is available. To do so, they turn the genes that contain the code for the necessary enzymes on and turn the genes for unnecessary enzymes off.

Cells need some proteins all the time. Essential proteins include RNA polymerase, which is required for all protein synthesis, and certain enzymes, which are fundamental to metabolism. The genes for proteins like this are sometimes called housekeeping genes and are on all the time. Genes that are on all the time are also called **constitutive genes**.

Although it is often useful to frame discussions of gene regulation in terms of what a cell needs in a given circumstance, it is important to remember that cells can't think or make conscious choices about which genes to turn on and which genes to turn off. Cellular processes are accomplished by the chemical interactions of molecules. Signaling molecules bind to receptor molecules causing a change that relays through a series of molecules, ultimately leading to change in the function of the cell. Most gene regulation occurs at the level of transcription: Genes are turned on and off when proteins bind to the DNA near the genes and either help or block RNA polymerase from being able to read the gene. If RNA polymerase cannot read the gene and no mRNA is made, the protein will not be made. Although there are differences in the details of how this is accomplished in prokaryotes such as *E. coli* or humans such as you, the essential principle is the same.

Much of what we understand about gene regulation began with studies on *E. coli*. In *E. coli*, there are two basic systems by which genes are regulated. Some genes are typically off, unless an environmental signal turns them on. These genes are called **inducible genes**, and the process of turning the genes on is called **induction**. Other genes are typically on, unless an environmental signal turns them off. These genes are called **repressible genes** and the process is called **repression**.

> **REMEMBER**
> Inducible genes are turned on in response to environmental signals called inducers. Repressible genes are turned off in response to environmental signals called co-repressors.

OPERONS

To understand induction and repression, we must first consider the organization of genes within *E. coli* and other bacteria. In eukaryotes, each gene has its own promoter, its own sequence that marks the beginning of the gene. In prokaryotes, however, several genes can share a common promoter. The region of DNA that contains genes, whether there is one or several, plus the regulatory DNA associated with the genes is called an **operon** (Figure 7.12). One example of a regulatory region that might be part of an operon is a section of DNA that helps regulate transcription called an **operator.** When RNA polymerase binds to the promoter of an operon and begins transcription, all the genes in the operon are transcribed at the same time.

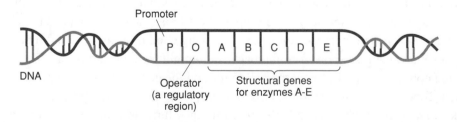

FIGURE 7.12. An operon.

In prokaryotes, several genes may share a single control region. The control region contains the promoter, which is recognized by RNA polymerase, and regulatory regions that affect whether transcription can occur. The genes for polypeptides are called structural genes.

INDUCTION

The lactose operon, or *lac* **operon**, is an example of an inducible operon. This operon consists of a promoter, an operator, and three genes that code for enzymes involved in the catabolism of lactose, or milk sugar. One of these genes is the

lacZ gene that codes for the enzyme β-galactosidase. This enzyme breaks the disaccharide lactose into two monosaccharides, glucose, and galactose.

Lactose is not always available in the environment of *E. coli*. If lactose is unavailable, *E. coli* does not make β-galactosidase and its companion enzymes. If lactose becomes available, then *E. coli* can produce β-galactosidase. Whether or not the genes of the *lac* operon are on depends on the activity of a protein called the lactose repressor, or **Lac repressor**. It is coded for by a gene outside of the *lac* operon and is constitutively expressed. If lactose is not present in the environment, the Lac repressor is active and binds to the operator of the *lac* operon (Figure 7.13). When the Lac repressor is bound, it acts like a roadblock, blocking access of RNA polymerase to the lac promoter. RNA polymerase cannot bind, and thus the genes for β-galactosidase and its companion enzymes are not transcribed. However, if lactose becomes available, it binds to the Lac repressor and inactivates it, changing its shape so it can no longer bind the lac operator (Figure 7.13). The roadblock is no longer there, and RNA polymerase can bind to the lac promoter.

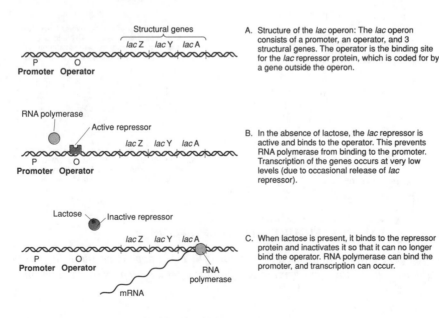

A. Structure of the *lac* operon: The *lac* operon consists of a promoter, an operator, and 3 structural genes. The operator is the binding site for the *lac* repressor protein, which is coded for by a gene outside the operon.

B. In the absence of lactose, the *lac* repressor is active and binds to the operator. This prevents RNA polymerase from binding to the promoter. Transcription of the genes occurs at very low levels (due to occasional release of *lac* repressor).

C. When lactose is present, it binds to the repressor protein and inactivates it so that it can no longer bind the operator. RNA polymerase can bind the promoter, and transcription can occur.

FIGURE 7.13. The *lac* operon.

Even when lactose is available and the Lac repressor is inactive, the ability of RNA polymerase to bind to the promoter is also affected by a separate regulatory system that detects the presence of glucose. This effect is called **catabolite repression** or the glucose effect. If glucose is available, *E. coli* will use the glucose first before turning to other food sources. This regulation is achieved through a signaling molecule called cyclic AMP, or **cAMP**. When cAMP is present in the cell, it will bind to its receptor protein, the catabolite activator protein (CAP). When bound together, cAMP-CAP bind near the *lac* promoter and help RNA polymerase to bind (Figure 7.14). If glucose is available in the environment, levels of cAMP are low in the cell. If cAMP is unavailable to bind to CAP, they do not help RNA polymerase bind to the *lac* promoter. Thus, in the presence of glucose, RNA polymerase will not bind efficiently to the *lac* promoter and transcription of β-galactosidase will occur at very low levels. This results in the glucose being catabolized before the lactose. As the glucose is used up and glucose levels decrease in the cell, the amount of cAMP increases. You can think of the cAMP as an alarm signal that the glucose is running out. As more cAMP becomes available, it binds to its receptor, CAP. Together, the cAMP-CAP bind near the promoter and help RNA polymerase bind. Transcription of β-galactosidase and its companion enzymes becomes efficient and occurs at high levels.

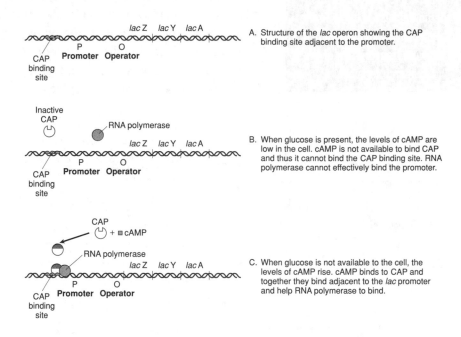

FIGURE 7.14. The glucose effect.

If glucose is present, it will be used first by the cell. This happens because the presence of glucose affects the transcription of catabolic operons such as the lac *operon.*

The two systems for regulation of the *lac* operon are separate from each other, but both should be considered when predicting the regulation of the *lac* operon. Lactose is one environmental signal affecting the operon. Its effect on the operon is through the activity of the Lac repressor. Glucose is the other environmental signal that affects the operon. Its effect on the operon is through the activity of cAMP. A useful analogy is to consider a stereo. A stereo has a power switch that turns it on and off and a volume control that increases or decreases the level of sound. The control of the *lac* operon through the Lac repressor is like the power switch. When lactose is available, the power is on. When lactose is unavailable, the power is off. The glucose effect is like the volume control. If glucose is available, the volume is turned way down. If glucose is unavailable, the volume is turned way up. So, it is possible for lactose to be present (power on) but not to have any transcription (sound) because glucose is also present (volume turned way down). If lactose is present (power on) and glucose is absent (volume turned up), then there is a high level of transcription (lots of sound). On the other hand, if glucose is absent (volume turned up), but lactose is also absent (power off), there won't be any transcription (no sound).

REPRESSION

In the case of the *lac* operon, the environmental signal lactose turns the operon on. Lactose is called the **inducer**, and the operon is an inducible operon. In the case of repressible operons, environmental signals act to turn the operon off. An example of this is the tryptophan operon, or ***trp* operon**. The *trp* operon contains the genes for enzymes necessary to synthesize tryptophan, an amino acid (Figure 7.15). A gene outside of the *trp* operon codes for a repressor protein called the Trp repressor. Like the Lac repressor, the Trp repressor is constitutively expressed.

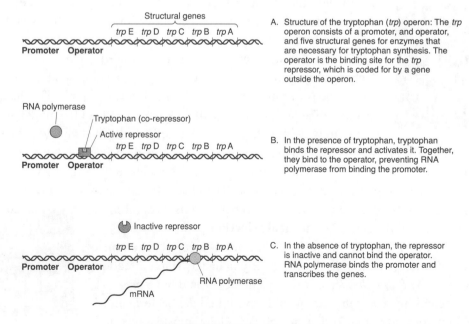

FIGURE 7.15. The *trp* operon.

If given a basic carbon source, *E. coli* has the ability to synthesize tryptophan using the enzymes that are coded for by the *trp* operon. If tryptophan is not available in the environment, the Trp repressor is inactive, and the *trp* operon is on (Figure 7.15). The genes are transcribed and translated, and the enzymes for tryptophan synthesis are made. If tryptophan becomes available in its environment, *E. coli* will use this tryptophan and stop making its own. Tryptophan binds to the Trp repressor and activates it. The activated Trp repressor binds to the Trp operator and blocks RNA polymerase from transcribing the genes (Figure 7.15). Because tryptophan acts as a partner with the Trp repressor, tryptophan is called the **co-repressor**.

Both the *lac* operon and the *trp* operon are regulated operons. The operons are switched on and off by the action of a repressor protein that binds to the operator. The difference is that in the case of the inducible *lac* operon, the inducer, lactose, inactivates the repressor protein. In the case of the repressible *trp* operon, the repressor protein is *activated* by the co-repressor, tryptophan.

Mutation

By regulating their genes, cells can change their behavior in response to environmental signals. Cells can also respond to environmental signals by regulating enzymes with feedback inhibition (Figure 7.11). However, there are other, more permanent, changes that can occur in cells when their DNA code becomes altered. These changes are called

mutations, and although they may result from exposure to certain things in the environment, a cell cannot control their effects and may even be harmed by them.

In fact, most mutations have negative consequences for cells: A change in a gene results in a change in a protein such that the protein doesn't function very well or perhaps not at all. This leads to a loss of an ability of the cell and may even lead to cell death.

Only rarely do mutations change a cell's traits in such a way that the cell functions better in a certain circumstance. When this does happen, however, that cell is more likely to survive and reproduce than other cells and so the new, beneficial trait is passed on to the next generation. The idea that organisms with advantageous traits will have more offspring than other organisms is referred to as "survival of the fittest" or **natural selection**.

The evolution of antibiotic-resistant bacteria is a perfect example of natural selection in action. In the 1940s, if you had a "staph infection," an infection by *Staphylococcus aureus*, a doctor would have prescribed penicillin. The penicillin would have killed the bacteria and cleared up your infection. Today, just 60 years later, most strains of *Staphylococcus aureus* are resistant to penicillin. Where did the antibiotic-resistant bacteria come from? How did they become the most common *Staphylococcus*? The answers to these questions are **mutation** and **selection**.

There are two types of mutations that occur in cells, **spontaneous mutations** and **induced mutations**. Spontaneous mutations happen whenever cells divide as a result of uncorrected mistakes by DNA polymerase III. When DNA polymerase III is copying DNA prior to cell division, it occasionally inserts the wrong nucleotide into the growing chain. DNA polymerase III has the ability to detect these errors and, along with DNA polymerase I, will fix most of them, but a few uncorrected errors do occur. Because of its proofreading ability, the accuracy of DNA polymerase is very high, with only roughly one mistake for every billion base pairs of DNA that is copied. However, cells may have billions of base pairs of DNA, so when a population of cells is dividing, spontaneous mutation causes changes in at least some of them.

> **REMEMBER**
> Some spontaneous mutations randomly occur every time a cell divides.

MUTAGENS

Anything that increases the error rate of DNA polymerase III is a **mutagen** (something that causes mutations). Mutations that happen in response to exposure to mutagens are **induced mutations**. Mutations may be induced by exposure to certain chemicals or to radiation. Chemical mutagens may alter the chemical properties of DNA so that complementary base pairing does not work properly (Table 7.3). Radiation can cause mutations by two different mechanisms: Ultraviolet radiation causes abnormal bonds within the DNA molecule (Figure 7.16), while ionizing radiation such as X-rays generates free radicals that damage the DNA. After DNA is damaged by radiation, repair mechanisms will attempt to fix the DNA, but these often leave errors or delete portions of the code.

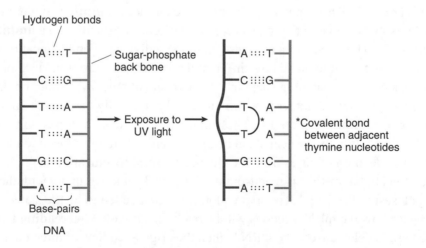

FIGURE 7.16. A thymine dimer.

Formation of thymine dimers in response to UV light can lead to increased mutations during DNA replication.

TABLE 7.3. The Effects of Some Chemical Mutagens

Type of Mutagen	Effect	Examples
Nucleoside analog	Has a shape similar to that of a nucleotide. Is incorporated into DNA instead of a normal nucleotide and doesn't follow the same base-pairing rules.	5-Bromouracil acts like thymine but may pair with guanine. 2-Aminopurine acts like adenine, but may pair with cytosine.
Intercalating agent	Planar (flat) molecules slip between the base pairs of DNA (like a piece of cardboard slipping between the rungs of a ladder). When DNA polymerase III tries to copy the DNA around the intercalating agent, it slips and may skip some DNA or copy other parts more than once.	Ethidium bromide (used to visualize DNA via fluorescence in molecular biology) Aflatoxin (made by a mold that grows on peanuts and corn) Benzpyrene (found in smoke)

TYPES OF MUTATIONS

The majority of mutations, whether spontaneous or induced, are random. The cell cannot choose or control where the changes will happen or what the changes will be. The consequences of the changes depend on the extent and location of the change. Mutations that result in a small change in a single gene are called **point mutations**. Because some amino acids are represented by more than one codon, it is possible for a point mutation in the DNA to occur without actually changing the sequence of the protein (Figure 7.17). This type of mutation is called a **silent mutation**. If a point mutation results in a change in the sequence of amino acids, it is a **missense mutation** (Figure 7.17). The effects of a missense mutation will depend on the location of the change within the protein and how different the amino acids are from each other. For example, a change in an amino acid in the active site of an enzyme might be expected to have a large effect. Another type of mutation that typically has a great effect is a **nonsense mutation** (Figure 7.17). These mutations introduce a stop codon into the mRNA and result in the protein chain being shortened. If faulty replication leads to the insertion or deletion of bases from the chain, a **frameshift mutation** (Figure 7.17) results. Recall that mRNA is read from the 5' to the 3' end, beginning with the start codon closest to the 5' end. The start codon is the first codon to be translated, and it establishes how the mRNA codons will be read. This is called the **reading frame**. If a single base is deleted from the mRNA, then the original codons of three nucleotides are all going to be shifted by one nucleotide after the site of the deletion. The reading frame will be completely different, and the protein that is produced will be completely different from what it was supposed to be. Thus, frameshift mutations generally have severe effects.

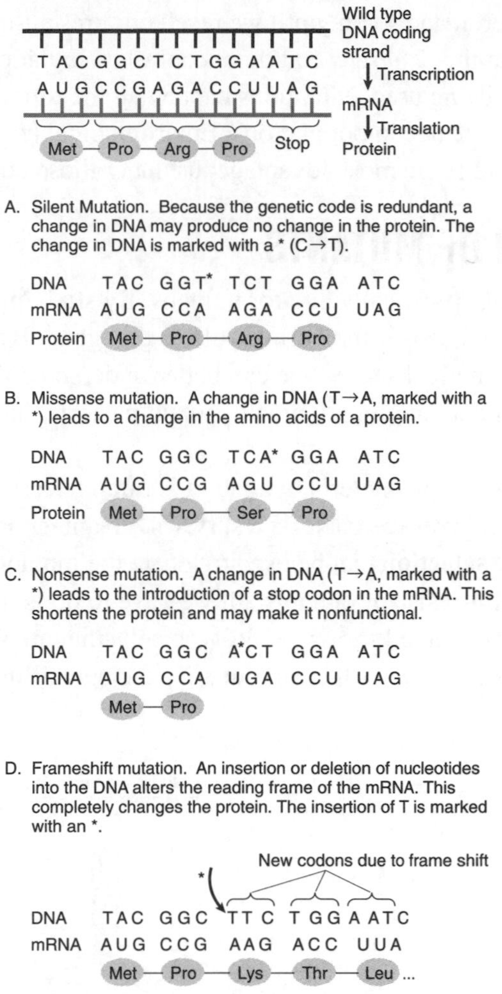

```
            ↓↓↓↓↓↓↓↓↓↓↓↓↓↓↓↓↓          Wild type
            T A C G G C T C T G G A A T C   DNA coding
                                            strand
            A U G C C G A G A C C U U A G            ↓ Transcription
            ↓↓↓↓↓↓↓↓↓↓↓↓↓↓↓↓↓          mRNA
              ⌣   ⌣   ⌣   ⌣                    ↓ Translation
            Met  Pro  Arg  Pro  Stop       Protein
```

A. Silent Mutation. Because the genetic code is redundant, a change in DNA may produce no change in the protein. The change in DNA is marked with a * (C→T).

```
DNA      TAC  GGT*  TCT  GGA  ATC
mRNA     AUG  CCA   AGA  CCU  UAG
Protein  Met — Pro — Arg — Pro
```

B. Missense mutation. A change in DNA (T→A, marked with a *) leads to a change in the amino acids of a protein.

```
DNA      TAC  GGC  TCA*  GGA  ATC
mRNA     AUG  CCG  AGU   CCU  UAG
Protein  Met — Pro — Ser — Pro
```

C. Nonsense mutation. A change in DNA (T→A, marked with a *) leads to the introduction of a stop codon in the mRNA. This shortens the protein and may make it nonfunctional.

```
DNA      TAC  GGC  A*CT  GGA  ATC
mRNA     AUG  CCA  UGA   CCU  UAG
         Met — Pro
```

D. Frameshift mutation. An insertion or deletion of nucleotides into the DNA alters the reading frame of the mRNA. This completely changes the protein. The insertion of T is marked with an *.

```
                          New codons due to frame shift
              *⌐
DNA      TAC  GGC TTC  TGG  AATC
mRNA     AUG  CCG AAG  ACC  UUA
         Met — Pro — Lys — Thr — Leu ...
```

FIGURE 7.17. Classes of point mutations.

Point mutations are small changes in the DNA code. Changes in the DNA lead to changes in RNA molecules, which can lead to changes in proteins.

All of these types of mutations generate changes in cells. Most of the time, these mutations have negative effects, and cells may die as a result. Other times, the mutants survive and are just a little bit different from the normal **wild-type** cells. If we return to our example of penicillin-resistant *Staphylococcus aureus*, we can imagine that back in the 1940s, populations of *S. aureus* were multiplying and that mutations were generating small changes in the individuals of those populations. How, then, did we get to the modern reality of most populations consisting of penicillin-resistant cells? Essentially, by using penicillin to treat *S. aureus*, humans **selected** for the antibiotic-resistant strains. When penicillin was used, the most susceptible *S. aureus* cells died first, leaving alive those that were more able to withstand the penicillin. These cells reproduced, and the resulting populations of *S. aureus* were more resistant than the original populations. More mutations resulted in more genetic change; the use of penicillin again selected for the most resistant bacteria, and so on, until we reach our present-day situation. It is important to realize that the *S. aureus* could not control its mutations nor did it mutate *because* penicillin was being used. Mutations happen all the time, and they happen randomly, creating genetic change in populations. Environmental conditions then select for the individuals whose traits are most advantageous under those conditions.

IDENTIFICATION OF MUTANTS

The study of mutants has practical value for humans. By studying cells with mutations in known genes, we can discover the normal role of that gene. By identifying mutations associated with genetic diseases, we can better understand the causes of the disease. Some scientists are attempting to create mutant bacteria that are better able to degrade oil or even plastic.

When working with mutant bacteria in a lab, it is often necessary to identify the mutants. If the mutation gives the bacteria a survival advantage in certain environmental conditions, **positive selection** can be used to locate the mutants. For example, if we are looking for penicillin-resistant mutants, all we have to do is try to grow our population on media containing penicillin. Any wild-type cells that are sensitive to penicillin will not be able to grow, and we will easily identify the penicillin-resistant mutants (Figure 7.18).

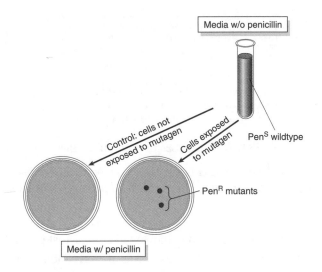

FIGURE 7.18. Positive selection.

A culture of cells that is normally sensitive to penicillin (pen^S) can be exposed to mutagens to create penicillin-resistant (pen^R) mutants. To positively select for these mutants, the mutagenized cells are inoculated into media containing penicillin. Only those cells that have gained the ability to grow in penicillin will survive. As a control, cells from the original pen^S culture are also inoculated into media containing penicillin. These cells are not expected to grow.

If the mutation does not give the mutant an advantage under certain circumstances, then we cannot use direct selection to find them. Instead, we must examine, or **screen**, our population for the mutants. This may involve spreading bacteria on plates and looking for observable changes in the colonies like changes in color. In the case of mutants that have acquired mutations in metabolic enzymes, we may be able to use **negative selection** to find them (Figure 7.19). Negative selection involves a technique called **replica plating** (Figure 7.20).

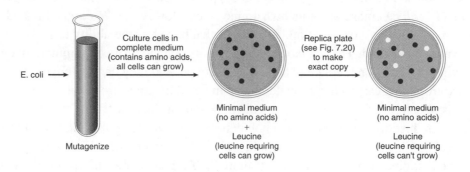

FIGURE 7.19. Negative selection.

To look for mutants that have lost an ability, negative selection is used. E. coli can normally make all of its own amino acids. If it is exposed to mutagens, some cells may have mutations in the DNA that code for enzymes necessary to synthesize an amino acid. These mutants are called auxotrophs. To grow, they must be fed the amino acid they can no longer make. In this example, leucine auxotrophs (cells that must be fed leucine) are located by negative selection. The white circles indicate locations where leucine auxotrophs failed to grow on a minimal medium. These auxotrophs can be recovered by finding the colony in the same position on the plate supplemented with leucine.

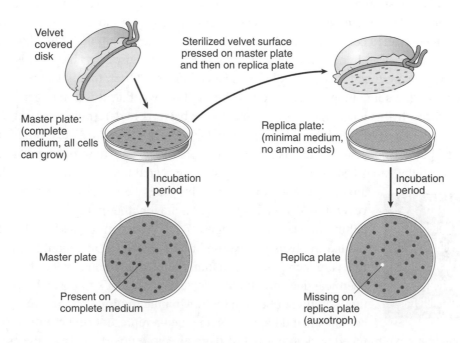

FIGURE 7.20. Replica plating.

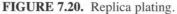
A block covered with sterile velvet is used to make exact copies of an original plate.

Replica plating can be used to look for mutants called **auxotrophs**. Auxotrophs have lost the ability to make an essential molecule and thus must be fed this molecule in order to grow. For example, wild-type *E. coli* can grow on minimal media that has a basic carbon source like glycerol but is not supplemented with any amino acids. This is because *E. coli* can make all of its own amino acids. However, if *E. coli* is mutated by exposure to chemicals or radiation, auxotrophs that have lost the ability to make an amino acid may result. These auxotrophs will not be able to grow on minimal media; they will only grow on media that contains the amino acid they can no longer make. An auxotroph that can no longer make the amino acid histidine would be called a histidine auxotroph. It will only grow on media that contains histidine.

To locate histidine auxotrophs in a population of *E. coli*, the population would first be spread onto a plate that does contain histidine. This would allow both wild-type cells and histidine auxotrophs to grow (Figure 7.20). Exact copies of this original plate would be made by pressing a piece of sterile velvet onto the surface of the plate and then pressing the velvet onto fresh plates of minimal media with and without histidine. Wherever the fibers of the velvet touched a colony on the original plate, they would have picked up a few cells. These cells would then be transferred to the same locations on the new plates. By comparing the new plates and looking for those colonies that do not grow on plates lacking histidine, the location of the histidine auxotrophs can be determined.

USING MUTANTS TO IDENTIFY CARCINOGENS

One way that histidine auxotrophs can be used is in the search for cancer-causing chemicals, called **carcinogens**. Cancer results from changes in the DNA of cells that lead to uncontrolled cell division. Thus, it should not be surprising that when known carcinogens are tested, they are found to be mutagens. Likewise, anything that is known to be a mutagen is suspected of being a carcinogen. Often the first step in determining whether something is a carcinogen is to find out whether it is a mutagen.

The **Ames test** is a test that was designed to use bacteria to determine whether a chemical is a mutagen, and therefore a potential carcinogen. The Ames test relies upon the use of auxotrophs, and often histidine auxotrophs are used. Auxotrophs are already mutated, so the Ames test checks to see whether a suspected mutagen can mutate the auxotrophs back to normal. This type of a mutation is called a **reversion**. To test a suspected mutagen, the mutagen is added to a culture of auxotrophs (Figure 7.21). As a control, a second culture of auxotrophs is not exposed to the suspected mutagen. Both cultures are then plated onto minimal media and incubated. If the chemical does not mutate the auxotrophs back to normal, then they should not be able to grow on the minimal media. Any colonies that do appear on the media represent **revertants**, auxotrophs that have mutated back to normal. If there are more revertants from the culture exposed to the mutagen than from the culture that wasn't exposed, it can be concluded that the chemical is a mutagen. Further testing, perhaps in rats, would then be done to determine whether the chemical was also a carcinogen. The Ames test is relatively fast

> **REMEMBER**
> The Ames test determines whether a chemical is a mutagen by testing its ability to cause mutations in auxotrophic bacteria.

and inexpensive and can be used to screen a wide variety of substances to identify good candidates for further testing.

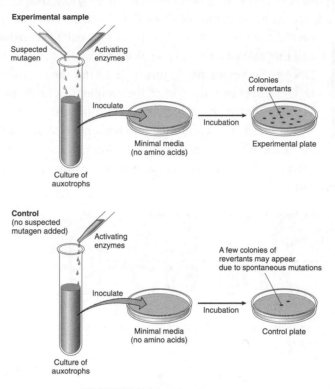

FIGURE 7.21. The Ames test.

Auxotrophs are exposed to a suspected mutagen. If the chemical is a mutagen, the auxotrophs might mutate back to normal. This type of mutation is called a reversion. After exposure to the test chemical, the auxotrophs are plated onto minimal media. If they are still auxotrophs, they will not be able to grow. Only the revertant cells that have regained the ability to make all amino acids can grow. If the plate of auxotrophs exposed to the test chemical has more colonies than a control plate, the chemical is a mutagen.

Gene Transfer

Although mutation acts to generate genetic diversity in populations constantly, other processes work to bring unique combinations of traits together in individuals. In humans, our genetic decks get shuffled and recombined when two individuals engage in sexual reproduction to produce a new, unique individual. When parents pass genes to offspring, it is called **vertical gene transfer**. In bacteria, however, reproduction is not linked to sex. Reproduction is a matter of binary fission, of one cell making an almost identical copy of itself. Yet bacteria do have the ability to shuffle their genetic decks and can even acquire new genes from organisms that are very distantly related to them. When genes are passed to cells of the same generation, it is called **horizontal gene transfer**.

TRANSFORMATION

Bacteria can acquire new genes from other bacteria in a variety of ways, the simplest of which is to pick up naked fragments of DNA that are released as neighboring bacte-

> **REMEMBER**
> Transformation occurs when bacteria pick up naked fragments of DNA from their environment.

ria die and break down. This process is called **transformation**. The cell that released DNA is the **donor** and the cell that picks up the DNA is the **recipient** (Figure 7.22). After the recipient picks up the DNA, it may become part of the recipient's chromosome. When this happens, the recipient becomes **recombinant** because it has DNA from two sources combined into one cell. Not all bacteria can take up DNA from their environment, although laboratory procedures can help enable bacteria to do so. If a cell is capable of picking up a fragment of DNA, it is called **competent**.

A. Transformation. Fragments of naked DNA are taken up by a cell.

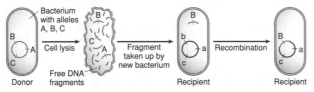

B. Conjugation. DNA is passed from one cell to another by direct contact between the cells.

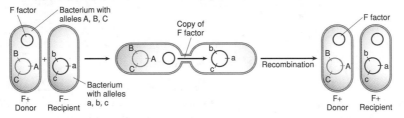

C. Transduction. Bacterial DNA is transferred from one cell to another by a virus.

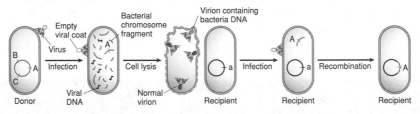

FIGURE 7.22. Gene exchange in prokaryotes.

The process of transformation got its name from the original observations of a researcher named Frederick Griffiths in 1928 (Figure 7.23). When Griffiths injected mice with a strain of *Streptococcus pneumoniae* that produced a capsule, the bacteria killed the mice. When Griffiths took cultures from the blood of the mice, he was able to grow colonies of the encapsulated bacteria. When Griffiths injected mice with a strain of *Streptococcus pneumoniae* that could not make a capsule, the mice lived, and Griffiths grew very few colonies of unencapsulated bacteria from the blood of the mice. When Griffiths used heat to kill encapsulated bacteria before he injected them into the mice, the mice also lived, and no bacteria were cultured from the blood. But when Griffiths mixed these heat-killed encapsulated bacteria with living unencapsulated bacteria and then injected the combination into the mice, the mice died and cultures from the blood grew colonies of encapsulated bacteria. Griffiths concluded that the living unencapsulated bacteria had acquired *something* from the dead encapsulated bacteria that had *transformed* them into encapsulated bacteria. Later experiments proved that the *something* was DNA.

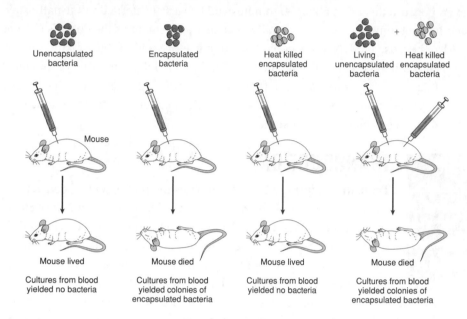

FIGURE 7.23. The Griffiths experiment.

Griffiths used two strains of Streptococcus pneumoniae. *One strain could produce a capsule and cause disease. The other strain did not produce a capsule and did not cause disease.*

CONJUGATION

Bacteria can acquire DNA from other living bacteria by a process called **conjugation**. It requires direct contact between two living bacterial cells. Many gram-negative bacteria connect to each other via **sex pili,** hollow tubes of protein that pull two attached cells together so that DNA can be passed. Gram-positive bacteria may connect to each other by sticky surface molecules. Conjugation has been most extensively studied in gram-negative bacteria like *E. coli*, and this system is usually used as an example.

> **REMEMBER**
> Conjugation transfers genes by direct contact between two living bacterial cells.

E. coli that can make a sex pilus have a special plasmid called a **fertility factor** that contains the genes for the sex pilus. Cells that can make a sex pilus are called **F+**, and they will conjugate only with cells that are **F–** and don't have the fertility factor. When an F+ cell encounters an F– cell, it makes the sex pilus and attaches (Figure 7.22). The F+ cell is the donor cell, and it copies its F factor and sends a copy of the F factor to the F– cell. The F– cell becomes an F+ cell.

In some *E. coli*, the F factor inserts itself into the main chromosome of the cell (Figure 7.24). Cells that have the F factor inserted into the main chromosome are called **Hfr** for

high-frequency recombination. Hfr *E. coli* can still initiate conjugation with F– cells, but now they have the potential to pass portions of their main chromosome through the sex pilus. After the sex pilus is attached to the recipient cell, the Hfr cell starts copying its DNA, beginning at one end of the integrated F factor. As the DNA is replicated, it is passed to the recipient cell (Figure 7.24). If the two cells remain attached long enough, the donor cell could potentially pass a copy of its entire chromosome to the recipient. However, because this takes about 90 minutes and because bacteria don't usually stay attached for that long, the recipient usually gets only a fragment of the donor's chromosome. Also, because the DNA was copied beginning at one end of the F factor, the entire F factor is not usually copied. The recipient cell gains new DNA and is recombinant, but usually remains F–. By using various Hfr strains of *E. coli* and determining the order of genes as they were transferred (Figure 7.24), scientists were able to figure out the order of many of the genes on the chromosome.

TRANSDUCTION

REMEMBER
Transduction occurs when bacteriophage transfer fragments of bacterial DNA between bacterial cells.

Transduction (Figure 7.22) involves the movement of genes from one bacterial cell to another by a virus. Viruses that attack bacteria are called **bacteriophage**. When a phage attacks a bacterial cell, it can overwhelm the cell, destroying the DNA and using the cell for the manufacture of more phage. After the components of the phage are manufactured, the new phage assemble themselves and leave the bacterial cell, destroying it in the process. They then move on to attack new bacterial cells. Sometimes, when the phage assemble themselves, they may pick up fragments of bacterial DNA rather than the genetic material of the phage (Figure 7.22). When these phage move on to new hosts, they can insert the fragments of bacterial DNA into these cells. These fragments may then incorporate into the main chromosome of the cell. The original destroyed bacteria are the donors; the newly invaded cells are the recipients and become recombinant.

GENETIC ELEMENTS

When we think about all of the ways that bacteria have for sharing DNA, we might begin to wonder what it is they are sharing. What information is contained on the plasmids and fragments of DNA that are getting passed around? Some bacteria acquire DNA that enable them to cause disease or resist antibiotics. Plasmids that contain genes for antibiotic resistance as well as the gene for the sex pilus are called **R factors**, for **resistance factors**. Some resistance factors have as many as seven genes for resistance to different antibiotics. Imagine what would happen if a significant human pathogen that is currently treatable by antibiotics were to acquire such an R factor from another bacterium!

The ability of bacteria to cause disease can be enhanced by genes for toxin production or enzymes that help a bacterium penetrate a host's tissues. These genes are sometimes located on sections of DNA called **pathogenicity islands**. A pathogenicity island may contain several genes that enable the bacteria to cause disease and that may be transferred as an entire group from one cell to another.

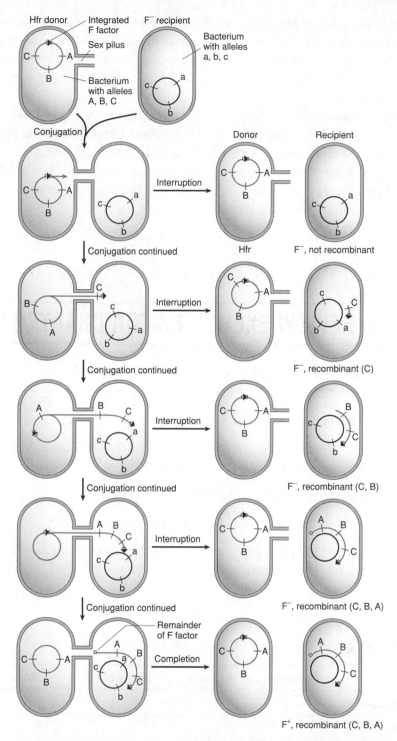

FIGURE 7.24. Conjugation between Hfr and recipient cell.

In the donor cell, the F factor has integrated into the main chromosome, forming an Hfr cell. Replication of the main chromosome begins within the integrated F factor. The longer the two cells remain connected, the more genetic information may be passed to the recipient. The entire chromosome must be transferred for the recipient to receive the complete F factor. This would take about 90 minutes and does not usually occur.

One beneficial application of gene transfer may be to identify or create bacteria that can help us clean up pollution. Some bacteria have **dissimilation plasmids** that contain genes for catabolic enzymes. These enzymes might enable the bacteria to break down unusual carbon sources such as oil, which makes these bacteria of interest for environmental cleanup.

Because all of these traits are easily shared among bacteria, we need to have a global awareness of what is happening in different populations of bacteria. For example, if the overuse of antibiotics is selecting for one type of antibiotic-resistant bacteria, it is possible that those bacteria could pass their resistance on to others. The microbial world is one vast interconnected system, and as we travel the globe, our microbes travel with us, sharing genes as they travel.

REVIEW EXERCISES FOR CHAPTER 7

1. Define semiconservative replication.

2. Draw a picture of a replication fork. At the fork, identify the enzymes involved in replication and the leading and lagging strands.

3. Describe the process of DNA replication, including the role of origins of replication, DNA polymerase III, DNA polymerase I, primase, and ligase.

4. Define antiparallel, including a definition of 5' and 3' ends.

5. Describe the characteristics of DNA polymerase III.

6. Describe the process of transcription, including the role of promoters, RNA polymerase, and transcription terminators.

7. Describe the process of translation, including the role of ribosomes, mRNA, tRNA, and the release factor.

8. Create the following table and fill in the information.

Process	Template	Product	Starts at?	Ends at?
Replication				
Transcription				
Translation				

9. State the rules for translating a molecule of mRNA using the genetic code.

10. Define feedback inhibition, gene regulation, constitutive genes, regulated genes, operon, induction, and repression.

11. Describe the process of induction, including the roles of inducers, repressors, operators, promoters, structural genes, and RNA polymerase.

12. Describe the process of repression, including the roles of co-repressors, repressors, operators, promoters, structural genes, and RNA polymerase.

13. Describe the glucose effect (catabolic repression) on inducible operons, including the role of cAMP, CAP, glucose, and the cAMP-CAP binding site.

14. Describe the gene regulation of the *lac* operon in the following conditions: lactose as the only carbon source, glucose as the only carbon source, lactose and glucose as carbon sources.

15. Describe the gene regulation of the *trp* operon in both the presence and absence of tryptophan.

16. Describe the two types of mutation.

17. Define mutagen and describe how chemicals and radiation act as mutagens.

18. State the classes of point mutation.

19. Define positive selection, negative selection, and auxotroph.

20. Describe the process of replica plating.

21. Discuss the importance of mutation in the evolution of new strains, genetic research, and cancer research.

22. Describe the Ames test.

23. Create the following table and fill in the information.

Process	How Is DNA Passed Between Cells?	Fate of Donor Cell?	Fate of Recipient Cell?
Transformation			
Transduction			
Conjugation			

24. Describe and state the significance of the Griffiths transformation experiment.

25. Define donor, recipient, competent, and recombinant cell.

26. Define F factor, F+, F–, and Hfr.

27. Discuss the importance of gene transfer to growth and virulence (ability to cause disease) of bacteria.

SELF-TEST

1. The process of transcription begins at

   A. origins of replication.
   B. start codons.
   C. promoters.
   D. initiation codons.
   E. inducers.

2. A mutation that results in a change in a single amino acid is an example of a

   A. nonsense mutation.
   B. silent mutation.
   C. missense mutation.
   D. frameshift mutation.
   E. None of the above is correct.

3. If these events that occur during replication were put in order, which would be third?

   A. Ligase forms covalent bonds in the sugar–phosphate backbone.
   B. Primase synthesizes short pieces of RNA complementary to DNA.
   C. DNA polymerase I removes RNA and replaces it with DNA.
   D. Helicase separates the parental DNA strands.
   E. DNA polymerase III synthesizes new DNA.

4. Where are anticodons found?

   A. mRNA
   B. tRNA
   C. Ribosome
   D. DNA
   E. Release factor

5. If *E. coli* is growing in the presence of glucose and lactose, which of the following is **not** true?

   A. The repressor is bound to the operator.
   B. Levels of cAMP are at low concentration.
   C. Transcription of the *lac* operon is occurring at very low levels.
   D. Glucose is being metabolized by the cell.
   E. cAMP is not bound to CAP.

6. Which of the following catalyzes peptide bond formation during translation?

   A. DNA polymerase III
   B. RNA polymerase
   C. tRNA
   D. Release factor
   E. An enzyme in the ribosome

7. Which process leads to the formation of recombinant cells?

   A. Transformation
   B. Conjugation
   C. Transduction
   D. All of the above are correct.
   E. None of the above is correct.

8. Use the genetic code (Figure 7.6) to answer the following: What is the anticodon of the tryptophan tRNA?

   A. 5'UGG3'
   B. 5'GGU3'
   C. 5'ACC3'
   D. 5'CCA3'
   E. None of the above is correct.

9. If the following steps in translation were put in order, which one would occur second?

   A. tRNA enters the A site.
   B. mRNA–ribosome shift.
   C. Peptide bond is formed.
   D. Met-tRNA binds to start codon.
   E. Release factor enters A site.

10. True or False: A mutation caused by exposure to the chemical 5-BU is an example of a spontaneous mutation.

11. AZT is an antiviral drug used to treat HIV infections. It is a thymine analog, which means it has a structure very similar to that of thymine and can fit into the active site of the viral DNA polymerase and then be incorporated into a DNA strand. Once incorporated, however, AZT stops further elongation of the growing chain. Compare the structure of thymidine and AZT shown in Figure 7.25. Based on this comparison and what you know about DNA replication, why does the inclusion of AZT in DNA block elongation of a growing strand?

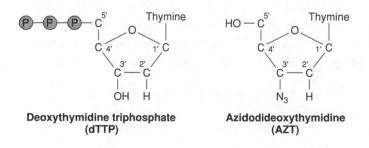

Deoxythymidine triphosphate (dTTP)   **Azidodideoxythymidine (AZT)**

FIGURE 7.25. Nucleotide analogs.

12. Erythromycin is an antibiotic that is effective against gram-positive bacteria, but not most gram-negative bacteria. It binds to the 50S subunit of the ribosome and prevents it from moving along a strand of mRNA. Why does erythromycin lead to cell death?

13. Given the following template strand of DNA, what is the sequence of the mRNA that would be made from this strand? What is the sequence of amino acids that would result from translation of the mRNA?

DNA   3'ATATTACACCACAGTAACT5'

Questions 14–17 refer to the data in the replica-plating experiment shown in Figure 7.26.

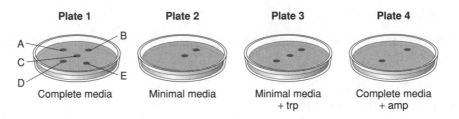

FIGURE 7.26. A replica plating experiment.

14. Which colonies represent tryptophan auxotrophs?

15. Which colonies represent auxotrophs that are resistant to ampicillin?

16. What type of selection is being used in plates 2 and 3?

17. What type of selection is being used in plate 4?

Questions 18 and 19 refer to the following scenario:

You are provided with cultures that have the following characteristics. A plus (+) indicates the organism has the trait; a minus (–) indicates the organism lacks the trait. S is sensitivity to an antibiotic; R is resistance to an antibiotic. Traits used are presence of F factor; ability to synthesize the amino acids tryptophan, serine, and lysine; and resistance to the antibiotic penicillin. The genes for amino acid synthesis are located on the main chromosome. The gene for antibiotic resistance is located on the F factor.

Culture 1: F+, Trp+, Ser+, Lys+, pen^R

Culture 2: F–, Trp–, Ser–, Lys–, pen^S

18. Indicate the possible genotypes of a recombinant cell resulting from the conjugation of cultures 1 and 2.

19. Indicate the possible genotypes of a recombinant cell resulting from conjugation of the two cultures after the F+ cell has become an Hfr cell.

Answers

Review Exercises

1. DNA replication is called semiconservative because each half of the double helix is used as a template for the synthesis of a new strand. From one double helix, two double helices are formed, each one half old DNA, half new. Because half of each new helix is the original DNA, the process is termed semiconservative.

2. See Figure 7.4.

3. DNA replication begins at origins of replication, which are specific sequences in the DNA. Enzymes called helicases unwind the DNA beginning at the origin. After the DNA is single-stranded, primase makes short RNA primers that are complementary to the DNA strand. DNA polymerase III starts making new DNA strands, beginning at the 3' end of the primers. It uses the parental DNA as a template. DNA polymerase I removes the RNA primers and replaces them with DNA. Finally, ligase forms covalent bonds between the fragments of DNA.

4. Nucleotides and nucleic acids are not symmetrical molecules. The chemical groups at one end of the molecule are different from those at the other end. A useful way of referring to the ends of these molecules is by referring to the carbons in the five-carbon sugar. The 5' carbon is the site of phosphate attachment in a nucleotide. A hydroxyl (OH) group is attached to the 3' carbon. When strands of nucleic acids attach to each other by hydrogen bonds between base pairs, the strands must always be lined up opposite to each other, with 5' ends pointing in the opposite directions. This is the proper orientation for the hydrogen bonding sites on the nitrogenous bases to line up.

5. DNA polymerase III makes new DNA that is complementary to an existing template strand. It cannot begin new strands because it must have a 3' OH group to which to add nucleotides. It always adds a new nucleotide to the 3' end of a growing chain (it cannot add nucleotides to 5' ends).

6. During transcription, information in a DNA molecule is copied into an RNA molecule. This is done by the enzyme RNA polymerase. RNA polymerase binds to promoters, the sequences that mark the beginnings of genes. It unwinds the DNA beginning at the promoter and then synthesizes RNA complementary to the DNA. It continues along the DNA until it reaches the transcription terminator, which causes it to release from the DNA.

7. During translation, information in mRNA is used to construct a protein. The mRNA is bound by the ribosome, which also contains binding sites for tRNAs. tRNAs enter the A site of the ribosome, carrying amino acids. Only the tRNA that

has the anticodon to the mRNA codon that is in the A site will be able to bond to the mRNA. When the correct tRNA is in place, an enzyme in the ribosome catalyzes bond formation between amino acids on adjacent tRNAs. The mRNA and ribosomes shift, placing a new codon in the A site. This process repeats itself until a stop codon is present in the A site. At that time, the release factor enters the ribosome and releases the polypeptide chain from the ribosome.

8.

Process	Template	Product	Starts at?	Ends at?
Replication	DNA	DNA	ORI	When forks meet
Transcription	DNA	RNA	Promoter	Transcription terminator
Translation	mRNA	Protein	Start codon	Stop codon

9. Translation of mRNA begins at the start codon (AUG) closest to the 5' end. Beginning with the start codon, the mRNA is broken up into units of 3 nucleotides (codons). The codons represent the amino acids, which can be looked up in the genetic code (Figure 7.6).

10. Feedback inhibition is when an end product in a metabolic pathway binds to an enzyme earlier in the pathway and inhibits it. Gene regulation is the process of controlling which genes are expressed (protein is available in the cell) at any given time. Constitutive genes are always being expressed or transcribed. Regulated genes may be turned on and off (gene product made, not made) in response to signals. Operons consist of sets of genes that share a common control region of DNA. Induction occurs when an environmental signal turns on transcription of a gene. Repression occurs when an environmental signal turns off transcription of a gene.

11. During induction, an environmental signal acts as an inducer to turn on transcription of the structural genes. The inducer binds to a repressor protein and inactivates it so that it cannot bind to the operator. This allows RNA polymerase to bind to the promoter. In the absence of the inducer, the repressor is bound to the operator, blocking transcription.

12. During the process of repression, an environmental signal acts as a co-repressor, binding to a repressor protein and activating it. The activated repressor protein binds to the operator and blocks transcription. In the absence of the co-repressor, the repressor is inactive, and RNA polymerase can bind to the promoter.

13. If glucose is available, its presence will affect the regulation of catabolic operons. When glucose is available to the cell, the concentration of cAMP within the cell is low. Thus, cAMP does not bind to its receptor protein, the cAMP receptor protein. cAMP-CAP is not available to help RNA polymerase bind to promoters, and tran-

scription of catabolic operons is low. As glucose is used up by the cell, the concentration of cAMP in the cell increases. cAMP binds to its receptor, the cAMP receptor protein. Together, they help RNA polymerase effectively bind promoters.

14. If lactose is the only carbon source, then lactose is bound to the repressor protein, and it is inactive. Thus, the promoter is available for RNA polymerase to bind. Also, glucose is not available, so cAMP levels are high within the cell. cAMP binds to CAP and together they bind near the promoter, helping RNA polymerase to bind. Transcription occurs at high levels.

    If glucose is the only carbon source, then the Lac repressor is active and is bound to the operator, blocking RNA polymerase. In addition, cAMP levels are low, so cAMP-CAP cannot bind near the promoter to help RNA polymerase. Transcription occurs at very low levels (because of the occasional release by the Lac repressor).

    If both lactose and glucose are available, lactose binds to the repressor protein and inactivates it. However, owing to the presence of glucose, cAMP levels are low, so cAMP-CAP cannot bind near the promoter to help RNA polymerase. Although the operator is not blocked by the repressor, RNA polymerase will only occasionally bind the promoter by itself. Transcription occurs at low levels.

15. In the presence of tryptophan, Trp binds to the repressor protein and activates it so that it binds to the operator. This blocks RNA polymerase from binding the promoter, and transcription does not occur. In the absence of Trp, the repressor is inactive and not bound to the operator. RNA polymerase can bind to the promoter, and transcription occurs, leading to the formation of the enzymes necessary for Trp synthesis.

16. Mutations may be spontaneous, occurring as a result of uncorrected mistakes by DNA polymerase III during replication, or it may be induced by environmental agents that increase the error rate of DNA polymerase III.

17. A mutagen is anything that increases the error rate of DNA polymerase III. Chemical mutagens may alter the binding properties of DNA nucleotides such that mismatches occur during base pairing by DNA polymerase III in replication. Radiation alters the structure of DNA by causing thymine dimers (UV radiation) or breaks in DNA molecules (ionizing radiation).

18. Silent mutations are mutations that do not result in a change in amino acids. Missense mutations are mutations that result in a change in amino acids. Nonsense mutations are mutations that result in the introduction of a stop codon. Frameshift mutations result from insertions or deletions that alter the reading frame (which codons should be read) of the mRNA.

19. Positive selection selects for mutants that have the ability to grow in certain circumstances. Negative selection screens for mutants that lack the ability to grow in certain circumstances. Auxotrophs are mutants that have a nutritional need not normally seen in wild-type (nonmutated) cells.

20. A common use of replica plating is to look for auxotrophs. To do this, bacteria that have been exposed to mutagens are first grown on complete media that contains amino acids. This will allow both wild-type (nonmutated) bacteria and auxotrophs to grow. Exact copies of these master plates are made by pressing sterile velvet onto the colonies and then transferring that velvet to new plates of minimal media. Wherever the velvet touched the colonies on the complete media, it will have picked up cells and deposited them in the same positions on the minimal media. The minimal media does not contain any supplements, and so the auxotrophs will not grow. The master plate is compared with the minimal media plate to determine which colonies grew on the master plate but not on the minimal media. These colonies represent colonies of auxotrophs.

21. Mutation is one of the driving forces in evolution. It is genetic change and leads to new traits in populations. Mutants are useful in genetic research because they can help us learn about the normal functions of genes and proteins. Mutants can also be used to look for things that cause cancer by testing whether or not chemicals can cause the mutants to mutate back to normal.

22. In the Ames test, auxotrophs are exposed to suspected mutagens. After exposure, the auxotrophs are plated onto minimal media. If the chemicals are mutagens, they will have induced mutations in the auxotrophs, some of which may have caused them to mutate back to normal. If colonies grow on the minimal media, it is evidence of this reversion. If more reversions occur in cultures that were exposed to the chemical than in cultures that were not exposed, the chemical is concluded to be a mutagen.

23.

Process	How Is DNA Passed Between Cells?	Fate of Donor Cell?	Fate of Recipient Cell?
Transformation	Naked DNA	Lysed	Recombinant
Transduction	Viruses	Destroyed by virus	Recombinant
Conjugation	Direct contact	Stays the same	Recombinant, may be F+

24. Griffiths experimented with two strains of *Streptococcus pneumoniae*, an encapsulated strain and an unencapsulated strain. When the encapsulated strain was injected into mice, the mice died and encapsulated bacteria were cultured from the blood. Injection with unencapsulated bacteria did not kill mice, and no bacteria were cultured from the blood. If the encapsulated bacteria were killed with heat, they no longer killed mice, and no bacteria were cultured from the blood. If heat-killed encapsulated bacteria were mixed with living unencapsulated bacteria, the mixture killed mice, and encapsulated bacteria were cultured from the blood. This

demonstrated that the living unencapsulated bacteria were able to acquire a trait from the dead encapsulated bacteria. Further investigations showed that the trait was passed when the living bacteria picked up DNA from the dead bacteria.

25. A donor cell gives DNA to another cell. A recipient cell receives DNA. A competent cell is capable of receiving DNA by transformation. A recombinant cell has DNA that has been combined from two sources (e.g., its own DNA plus DNA from another organism).

26. The F factor is a plasmid that contains the gene to make a sex pilus. A cell that has an F factor is F+ and can initiate conjugation. A cell that lacks an F factor is F−. An Hfr cell is a cell in which the F factor has integrated into the main chromosome.

27. When bacteria exchange genes, they may gain genes for useful traits. These might allow them to break down new food sources, thus enabling them to grow in new situations. They may also acquire genes for toxins or other proteins that might help them cause disease and thus make them more virulent.

Self-Test

  1. C   5. A
  2. C   6. E
  3. E   7. D
  4. B

8. D (Table of genetic code shows codons in 5' to 3' direction; nucleic acids always pair in an antiparallel fashion.)

9. A

10. F

11. AZT has an N_3 group at its 3' end. DNA polymerase must add new nucleotides to the 3' of the growing strand. After AZT is incorporated into the strand, DNA polymerase can't add anything to it. The chemical group at the 3' end is not the correct chemical group for DNA polymerase to use to form a bond with the 5' end of the new nucleotide.

12. Erythromycin blocks the elongation step of translation. The ribosome cannot shift along the mRNA to bring new codons into the A site. If translation is blocked, the cell cannot synthesize new proteins. Proteins are vital to cell function because they serve as enzymes, receptors, transporters, and the like. Without proteins, the cell dies.

13. mRNA: 5'UAUAAUGUGGUGUCAUUGA3'

Amino acids: Met Trp Cys His

14. Colony D. It cannot grow on the minimal media (plate 2), so it is not wild-type (has some mutation). When you add tryptophan to the media, it can grow (plate 3).

15. Colony D. It is an auxotroph (doesn't grow on plate 2), but it can grow on the complete media with ampicillin (Amp, plate 4). Although colony B also grows in Amp, it grew on plate 2, so it is not an auxotroph.

16. Negative or indirect

17. Positive or direct

18. Culture 2 is the recipient cell and would be recombinant. It's genotype after conjugation would be F+, Trp–, Ser–, Lys–, pen^R. During conjugation the recipient gets a copy of the F factor, which in this case also contained the penicillin-resistant gene.

19. If the donor cell (culture 1) becomes Hfr and then initiates conjugation, the recipient cell typically will receive a copy of part of the F factor and some of the main chromosome. Assuming the genes are on the main chromosome in the order written in the scenario, the possible genotypes are as follows, depending on how long the two cells stay together.

    F–, Trp+, Ser–, Lys–, pen^S

    F–, Trp+, Ser+, Lys–, pen^S

    F–, Trp+, Ser+, Lys+, pen^S

    F–, Trp+, Ser+, Lys+, pen^R

    F+, Trp+, Ser+, Lys+, pen^R (only if together for 90 min)

Molecular Biology and Recombinant DNA Technology

WHAT YOU WILL LEARN

This chapter introduces the molecular tools and techniques that are used in many areas of microbiology today. As you study this chapter, you will:

- discover some of the great moments in the history of molecular biology;
- explore the procedures used in genetic engineering;
- learn how scientists use recombinant DNA technology to study and manipulate DNA;
- investigate the practical applications of molecular biology.

SECTIONS IN THIS CHAPTER

- History of Molecular Biology and Recombinant DNA Technology
- Tools of Molecular Biology
- Applications of Recombinant DNA Technology

During the last few decades, there has been a revolution in biology and microbiology. The revolution has been fueled by our increasing understanding of the genetic code and by the development of tools that allow us to manipulate DNA. It has led to amazing discoveries, like the complete sequence of the entire human genome, and to significant advances in medicine, like the use of genetically engineered bacteria for the production of human hormones.

These innovations were possible because of advances in **molecular biology**, the study of how genetic information is contained in molecules of DNA and how this information determines the synthesis of proteins. Creative scientists took this knowledge and figured out ways to read the genetic code (**nucleic acid sequencing**), to make many copies of genes they were interested in studying (**polymerase chain reaction, plasmid replication**), and to put genes from one organism into the cells of another (**genetic engineering**). This has given us a greater ability than ever to change the behavior of cells and has raised the possibility of new cures for disease such as **gene therapy**, in which a normal copy of a gene is introduced into the cells of a patient with a defect in that gene. It has also raised concerns about genetically modified crop plants and other **genetically modified organisms (GMOs)**. As with all powerful tools, there is the potential for great good and great harm in these techniques. Understanding how these techniques work and the cellular consequences of genetic engineering is essential to informed participation in the current debate about recombinant DNA.

History of Molecular Biology and Recombinant DNA Technology

Genetic recombination occurs in nature in prokaryotes, eukaryotes, and viruses. Recombination in bacteria was first demonstrated by Frederick Griffiths' experiments with transformation in *Streptococcus pneumoniae* in 1928 (see Chapter 7), then again in 1946 when Joshua Lederberg and Edward L. Tatum showed that bacteria could exchange genes by conjugation, and in 1952 when Joshua Lederberg and Norton Zinder discovered transduction.

Griffiths' experiments not only showed that bacteria could acquire traits from each other but also contributed to the debate on the nature of the genetic information. At the time that Griffiths did his work, scientists were not sure whether the genetic information was contained in DNA or protein. In 1944, Oswald Avery, Colin MacLeod, and Maclyn McCarty followed up on Griffiths' experiments. Like Griffiths, they used two strains of *Streptococcus pneumoniae*, one that was encapsulated and one that was not. Griffiths had shown that if dead encapsulated bacteria were mixed with live unencapsulated bacteria, the unencapsulated bacteria were capable of acquiring the ability to make a capsule. Avery, MacLeod, and McCarty treated extracts from the dead encapsulated bacteria, with either proteases, which break down proteins, or nucleases, which break down nucleic acids like DNA. They found that if extracts treated with proteases were given to unencapsulated bacteria, the bacteria could still transform into encapsulated bacteria. However, if extracts treated with nucleases were given to unencapsulated bacteria, the bacteria could not transform. This demonstrated that the information that passed from the encapsulated bacteria to the unencapsulated bacteria was contained in DNA.

Shortly after this, in 1953, James Watson and Francis Crick proposed a model for the structure of DNA that explained how it could function as the genetic material. Their model of the double helix included information from X-ray pictures of DNA taken by Rosalind Franklin in the lab of Maurice Wilkins. Following the discovery of the structure of DNA were experiments that demonstrated how the genetic code was used in cells. In 1961, Marshall Nirenberg and J.H. Matthaei determined the first sequence of a codon UUU and showed that it represented the amino acid phenylalanine. In that same year, Sydney Brenner, François Jacob, and Matthew Meselson demonstrated that the ribosome is the site of protein synthesis. Taken together, these experiments are the foundation of molecular biology.

As our understanding of the nature of DNA grew, experiments were undertaken that increased our ability to manipulate DNA in cells. In 1970, Hamilton Smith showed that DNA could be cut into fragments using restriction enzymes. In 1973, Herbert Boyer and Stanley Cohen inserted pieces of DNA into bacterial plasmids in the first **gene cloning** experiments. In 1977, both Walter Gilbert and Fred Sanger developed methods to enable us to figure out the genetic code through **DNA sequencing**. Our ability to copy DNA and thus to work with very small amounts of DNA was given a tremendous boost by the invention of the **polymerase chain reaction** by Kary Mullis in 1983. These techniques enabled us to genetically engineer cells and rapidly read the genetic code, leading to events such as the creation of the first FDA-approved genetically engineered food in 1994 (the Flavr Savr tomato) and the first complete sequence of an entire organism's genome (*Haemophilus influenzae*) in 1995. One of the most dramatic outcomes of the advances in molecular biology and recombinant DNA technology was the completion of the Human Genome Project in 2003 by an international consortium of scientists.

Tools of Molecular Biology

RESTRICTION ENZYMES

Among the most useful tools for working with DNA are **restriction enzymes** (restriction endonucleases). These enzymes cut DNA into fragments and make it easier to manipulate DNA. The enzymes were originally isolated from bacteria that produce these enzymes in order to defend themselves

> **REMEMBER**
> Restriction enzymes act like molecular scissors, cutting DNA at specific sequences called restriction sites.

from viral attack by destroying the viral DNA (Figure 8.1). Scientists isolated these enzymes from bacteria, and now restriction enzymes are standard tools in every molecular biology lab.

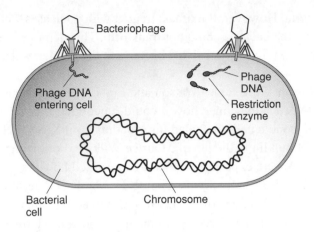

FIGURE 8.1. Restriction enzymes in nature.

Restriction endonucleases (restriction enzymes) are made by bacteria. They protect bacteria from viral attack by destroying viral DNA as it enters the bacterial cell.

Each restriction enzyme recognizes and cuts DNA at a specific sequence called a **restriction site**. For example, the restriction site for the enzyme *Eco*RI is shown in Figure 8.2. Many restriction enzymes make staggered cuts in the DNA backbone, creating fragments that have extensions of single-stranded DNA. These single-stranded ends are called *sticky ends* (Figure 8.2) because they are capable of base-pairing with other single-stranded ends that have the same sequence. Other restriction enzymes cut both DNA strands in the same place, producing blunt ends. Bacteria protect their chromosomal DNA from attack by their own restriction enzymes by attaching chemical groups to restriction sites in the DNA. Thus, a bacterium's restriction enzymes can't attack its own DNA.

DNA fragments that have been cut with restriction enzymes can be mixed together and used to create a **recombinant DNA molecule** (Figure 8.3) that contains DNA from two different sources. For example, the human gene for insulin has been combined with bacterial DNA. First, human DNA and bacterial DNA are cut with restriction enzymes. Then, the fragments are mixed together so they can attach to each other by hydrogen bonds between the complementary base pairs. Finally, **DNA ligase** is used to seal the breaks in the sugar phosphate backbone by forming covalent bonds between fragments. Recombinant DNA molecules can then be introduced into cells that will transcribe and translate the genes. For example, the recombinant DNA molecule containing the human gene for insulin has been introduced into bacterial cells that now produce the human protein for treatment of diabetes.

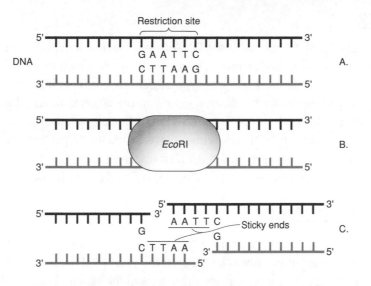

FIGURE 8.2. Restriction site for *Eco*RI.

EcoRI is a restriction enzyme, an enzyme that can cut the backbone of a DNA molecule at specific sequences. A. The restriction site is the sequence to which a restriction enzyme can bind. B. The EcoRI enzyme binds to the DNA at the restriction site and cuts the sugar-phosphate backbone. C. Many restriction enzymes make staggered cuts in the DNA, leaving short, single-stranded ends. Because these ends are complementary to each other and can bond by hydrogen bonds, they are called sticky ends.

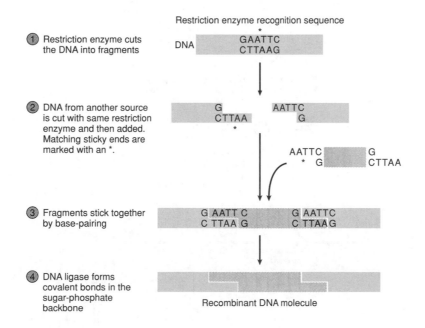

FIGURE 8.3. Recombinant DNA contains DNA from two sources.

To make recombinant DNA, DNA from two sources are cut with the same restriction enzyme and then combined.

VECTORS

Often, the goal of making recombinant DNA is to introduce a gene from one organism into a different organism. For example, genes for pest resistance might be introduced into crop plants. To do this type of **genetic engineering**, there must be a way to introduce the desired gene into the target cell. Anything that can carry DNA into a cell is called a **vector**. Bacterial plasmids and viruses are commonly used as vectors. Once inside the target cell, the vector must be capable of reproducing itself or inserting itself into the host genome so that the introduced gene will be copied as the cell divides.

> **REMEMBER**
> Vectors are used to introduce new genes into cells.

GENOMIC LIBRARIES

Another reason for making recombinant DNA is to store copies of genes for future use. Genes are stored in **genomic libraries**. A **genome** is all the DNA of an organism; for example, all forty-six chromosomes would represent the human genome from a human cell. Genomic libraries are not made of paper or books; they are actually populations of bacteria or viruses that have been genetically engineered to contain copies of the genes that are being stored. As the bacteria or viruses multiply, they make copies of the stored genes.

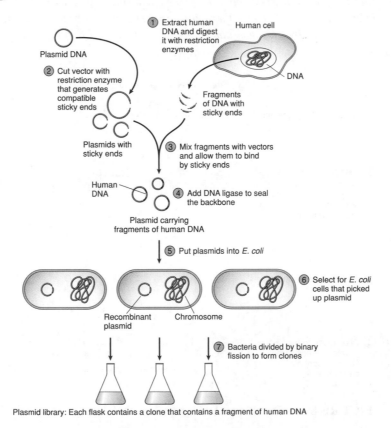

FIGURE 8.4. Making a genomic library.

A genomic library can be made by using bacteria or bacteriophage to house the information from another organism's genome.

The process of making a genomic library of human DNA is illustrated in Figure 8.4. First human DNA is extracted from human cells and digested with restriction enzymes. Vectors such as bacterial plasmids are also cut with the same restriction enzymes so that the human DNA and vector DNA have the same sticky ends. The human DNA and vector DNA are mixed, and DNA ligase is used to seal the fragments of human DNA into the vector DNA, forming recombinant plasmids that each contains different fragments of the human genome. These plasmids would then be introduced into bacterial cells by transformation. Each plasmid would be picked up by different bacterial cells. As each bacterial cell multiplies, it makes copies of the fragment of human DNA that it contains. The entire population of bacteria represents the library of the human genome.

Individual bacterial cells and their descendants that contain the same fragment of human DNA represent different **clones**. Because each of these clones can multiply and produce many copies of the human DNA fragment they carry, the genes they carry have been *cloned*. (This type of **gene cloning** is different from the cloning that produces genetically identical offspring from an animal cell.)

POLYMERASE CHAIN REACTION

The polymerase chain reaction (PCR) can be used to make many copies of one specific piece of a genome. This is very useful if scientists wish to work with a single gene, rather than an entire genome. PCR is also very useful if there are very few cells available from which to obtain DNA, as is often the case when evidence is collected from a crime scene. PCR targets a particular gene and then uses a special DNA polymerase to copy it. The copies are then copied over and over again until as many as a billion copies have been produced.

> **REMEMBER**
> PCR allows scientists to make billions of copies of specific pieces of DNA.

To target a particular gene of interest, special **primers** need to be made. Recall that DNA polymerase cannot start copying DNA unless it has a primer with a 3' end available for the attachment of new nucleotides (see Chapter 7). To start the process of copying the gene of interest, two single-stranded DNA primers that are complementary to the DNA are made on either side of the target DNA (Figure 8.5). When the DNA polymerase begins making new DNA, it will begin at these primers and move toward the gene of interest.

To begin the process of PCR (Figure 8.6), the DNA to be copied, lots of primers, nucleotides, and a special DNA polymerase are combined. The DNA polymerase that is often used is called *Taq* polymerase. It is a DNA polymerase that was isolated from the thermophilic bacterium *Thermophilus aquaticus*. It is used in PCR because it is stable at the high temperatures necessary to run the reactions in a machine called a thermocycler, which uses heat to separate the strands of double-stranded DNA.

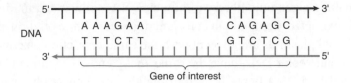

DNA

Gene of interest

① Make primers that are complementary to DNA on either side of the gene of interest.

② For PCR, combine DNA to be copied, lots of primer, nucleotides, and Taq DNA Polymerase.

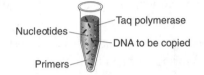

Taq polymerase

Nucleotides

DNA to be copied

Primers

③ Heat tube to 95°C for 1 minute. This will separate the double-stranded DNA.

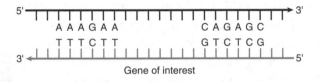

Gene of interest

④ Cool to 65°C for 1 min. This allows hydrogen bonds to form between DNA strands. Because there is so much primer, it is likely that primer will stick to the gene of interest.

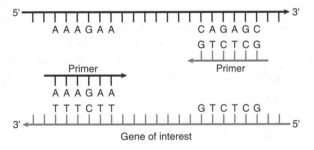

Primer

Primer

Gene of interest

⑤ Raise the temperature to 75°C for 1 min. TaQ Polymerase makes DNA beginning at the primers.

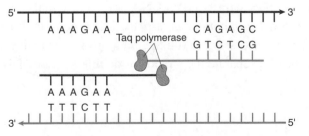

Taq polymerase

FIGURE 8.5. Using primers to target a gene of interest.

The thermocycler heats the tube at approximately 95°C for 1 minute (Figure 8.6). This disrupts the hydrogen bonds holding the double-stranded DNA together. The thermocycler then cools to an annealing temperature, for example, to 65°C, which is cool enough to allow hydrogen bonds to reform but high enough to avoid nonspecific hydrogen bonding. Because there is so much primer in the solution, it is very likely that primers will attach to the DNA that is being copied rather than the original DNA strands getting back together. The temperature is then raised to 72°C, which is the optimal temperature for *Taq* polymerase. The enzyme makes new DNA complementary to the original DNA strands beginning at the 3' ends of the primers. For every original strand of DNA that we wanted to copy, we now have two strands. This entire cycle is repeated, and the two strands become four strands. When it is repeated again, the four become eight. The thermocycler is capable of automatically cycling multiple times until the desired number of copies is produced. Only thirty cycles are required to produce a billion copies of a single piece of DNA!

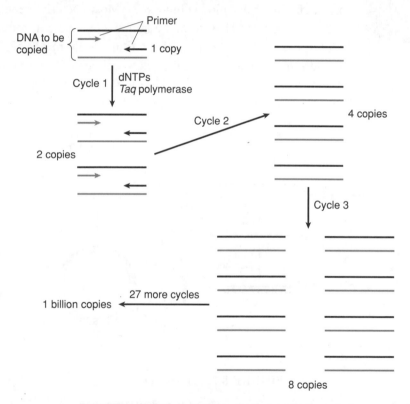

FIGURE 8.6. The polymerase chain reaction.

During PCR, the DNA to be copied is combined with primers, Taq *polymerase (a DNA polymerase), and nucleotides (dNTPs). The primers point the* Taq *polymerase to the DNA that needs to be copied. Every time the DNA is copied, all the copies are used as templates in the next round. Because of this, PCR quickly generates many copies of the target DNA.*

cDNA

One reason for making recombinant DNA is to insert human genes into bacteria, enabling bacteria to make the human gene product. For example, bacteria that contain the human gene for insulin produce human insulin for the treatment of diabetes. This is possible because the structure of DNA and the processes of transcription and translation are essentially the same in all cells on planet Earth. However, there is one difference between the organization of genes in bacteria and in eukaryotes that complicates the production of eukaryotic gene products in bacteria.

Like bacterial genes, eukaryotic genes contain the code that will be transcribed into mRNA and then translated into protein. In bacteria, the code consists of one uninterrupted strand of DNA. In eukaryotes, however, the code for the protein is broken up into sections called **exons**. The exons are separated by pieces of DNA called **introns** that will not be used to make the protein. These introns are removed from the mRNA after transcription and before translation (Figure 8.7). Most bacterial genes do not contain introns. Therefore, if a human gene were introduced directly into a bacterial cell, the bacterium would make mRNA that contained both exons and introns then translate the mRNA directly. The protein produced would not be the normal human protein because it would have been made from this abnormal mRNA.

> **REMEMBER**
> cDNA molecules are reverse transcribed from mRNA, so they don't contain any introns.

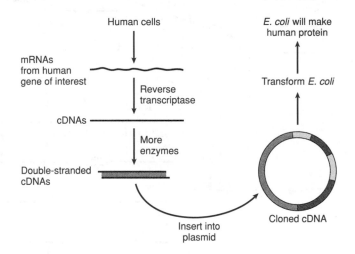

FIGURE 8.7. Making cDNA.

The viral enzyme, reverse transcriptase, is used to make DNA from mRNA. This technique creates cDNA genes that can be transcribed and translated into useful proteins by bacteria.

To solve this problem, human genes are not directly inserted into bacterial cells. Instead, copies of the genes that do not contain the intron information are made. To make these copies, mRNA that has already had the introns removed is used as a template. A viral enzyme, **reverse transcriptase**, that can make DNA from RNA is used to produce single-stranded **complementary DNA (cDNA)** molecules to the mRNA (Figure 8.7). After reverse transcriptase produces a single-stranded DNA molecule that is complementary to the mRNA, enzymes remove the mRNA. DNA polymerase is then used to synthesize a second strand of DNA complementary to the first, creating a double-stranded cDNA molecule. These cDNA molecules contain all the necessary coding information for building the human protein, without the extra information that would normally be found in the human gene. Thus, when cDNA molecules are introduced into bacteria, the bacteria can correctly make the human protein.

DNA PROBES

When working with DNA or genomic libraries, it is often necessary to identify which pieces of DNA, or which clones, contain a particular gene. Genes of interest can be located with **probes**, single-stranded pieces of DNA that are complementary to the gene you are trying to locate (Figure 8.8). Basically, this is like going fishing using a piece of DNA as bait. First, the library is treated with heat or chemicals to separate the DNA strands. This creates single-stranded DNA molecules that are ready to base-pair with complementary molecules of DNA. The probe, which is marked with an identifiable tag such as a fluorescent chemical, is allowed to mix with the single-stranded DNA from the library. The probe will stick to its complementary sequence, thus revealing the location of the gene of interest.

> **REMEMBER**
> Scientists use probes to locate specific pieces of DNA that are complementary to the probe.

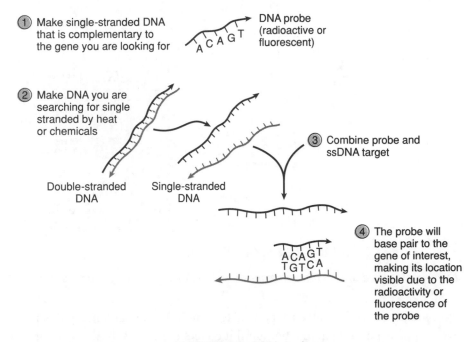

FIGURE 8.8. DNA probes.

Single-stranded DNA probes are used to locate specific DNA sequences within larger amounts of DNA. For example, the location of a gene on a chromosome can be identified with DNA probes.

GEL ELECTROPHORESIS

Gel electrophoresis can be used to separate molecules on the basis of their size and electrical charge. This technique has many uses, including DNA fingerprinting, DNA sequencing, and comparison and identification of molecules. During gel electrophoresis, molecules are moved through a gel. Smaller molecules move more quickly through the gel; larger molecules move more slowly. Thus, over a fixed interval of time, small and large molecules will separate from each other. A molecule's electrical charge also affects the speed of its movement through the gel. Charged molecules will be attracted more strongly to the positive and negative electrodes of the gel apparatus than uncharged molecules, increasing their speed through the gel.

> **REMEMBER**
> Gel electrophoresis is used to separate proteins and nucleic acids based on their size, shape, and charge.

When working with DNA samples, the gel is made of agarose, a polysaccharide, or a polyacrylamide. Liquid gel is poured into a tray called a *mold* or *platform*. Before the gel hardens, a plastic *comb* is inserted in one end. After the gel hardens, the comb is removed. Wherever the teeth of the comb stuck into the gel, there will be a small depression called a *well*. The samples to be separated will be placed into the wells.

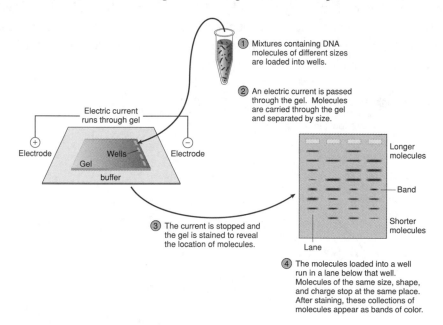

FIGURE 8.9. Gel electrophoresis.

Electricity is used to separate molecules based on their size, shape, and charge.

To begin the process of gel electrophoresis (Figure 8.9), the hardened gel is placed into a box. *Buffer* is poured into the box until it covers the gel. Buffers are solutions that maintain pH at a desired point. In this case, the buffer will also act to conduct electricity. The samples to be separated are loaded into the wells. The gel box is connected to a power supply, and an electrical current is run through the gel. Because DNA molecules are negatively charged, they are attracted to the positive electrode and will move through the gel toward that electrode. The smaller DNA molecules can slide more easily through the gel, whereas larger molecules get tangled up more easily. All molecules of the same size, shape, and charge move at the same pace. When the power supply is shut off, the molecules stop moving through the gel. For each well, there will be a *lane* that contains *bands* of molecules that were the same size and therefore stopped in the same place. Bands that contain larger molecules will be closer to the wells; bands that contain smaller molecules will be farther from the wells.

DNA SEQUENCING

DNA sequencing determines the exact code of a molecule of DNA. Like PCR, this process is now automated and very rapid. Many copies of the DNA to be sequenced are combined with *Taq* polymerase, primers that target the DNA of interest, normal deoxyribonucleotides, and special dideoxyribonucleotides. Deoxyribonucleotides are the normal building blocks of DNA. They have a hydroxyl group at their 3' ends, which is necessary for the attachment of new nucleotides by DNA polymerase (Figure 8.10). Dideoxyribonucleotides are the same as deoxyribonucleotides, except that they do not have a hydroxyl group at their 3' ends (Figure 8.10). Thus, if dideoxyribonucleotides are incorporated into a growing strand of DNA, DNA polymerase cannot add any more nucleotides to that strand. The four different dideoxyribonucleotides (ddA, ddT, ddC, and ddG) are each marked with a differently colored fluorescent tag.

> **REMEMBER**
> DNA sequencing can be used to determine the entire genome of an organism.

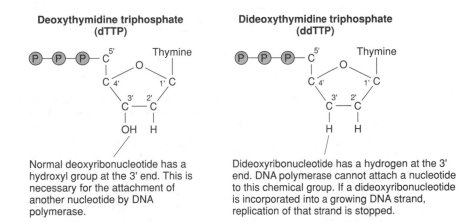

FIGURE 8.10. A comparison between deoxyribonucleotides and dideoxyribonucleotides.

After the necessary materials are loaded into the sequencer, DNA sequencing can begin (Figure 8.11). First, the DNA is heated to separate the strands. Then, the temperature is cooled to allow the primers to bond to the DNA. The temperature is raised to the optimal temperature of *Taq* polymerase, and DNA synthesis begins at the 3' ends of the primers. *Taq* polymerase randomly grabs complementary nucleotides to add to the growing strands. Whenever it grabs a deoxyribonucleotide, synthesis of that strand can continue. However, whenever it grabs a dideoxyribonucleotide, further elongation of that strand is blocked.

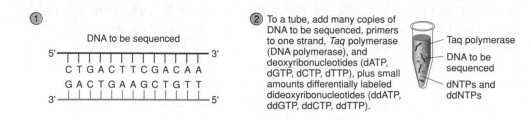

① DNA to be sequenced

② To a tube, add many copies of DNA to be sequenced, primers to one strand, *Taq* polymerase (DNA polymerase), and deoxyribonucleotides (dATP, dGTP, dCTP, dTTP), plus small amounts differentially labeled dideoxyribonucleotides (ddATP, ddGTP, ddCTP, ddTTP).

③ Heat tube to separate the DNA strands, then cool to allow primers to stick.

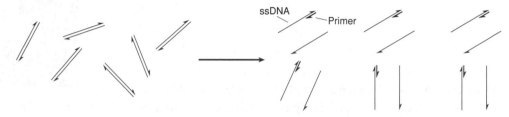

④ Raise temperature to optimal temperature for *Taq* polymerase. Beginning at the primer, the enzyme begins DNA replication. DNA ploymerase usually inserts deoxyribonucleotides into the growing chain because these nucleotides are more common than the dideoribonucleotides. However, dideoribonucleotides are occasionally inserted. Whenever a dideoxynucleotide is inserted, further synthesis of that strand is stopped. This creates a collection of partial sequences, each one stopped at a difference place.

⑤ Gel electrophoresis is used to separate the partial sequences by size. A computer examines the gel and uses a laser to determine the color of the fluorescent tag on the dideoxynucleotides. Each dideoxynucleotide has a differently colored tag, allowing the computer to read the code.

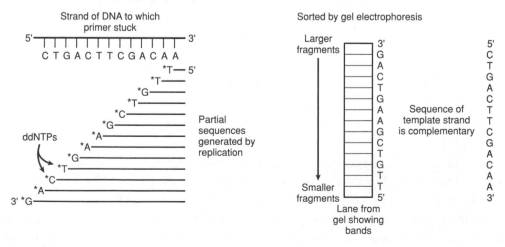

FIGURE 8.11. DNA sequencing.

At the end of DNA replication, *Taq* polymerase has generated many copies of the original DNA molecules, each one blocked in a different place. This is because the reaction started with many copies of the DNA to be sequenced, and each copy randomly got blocked at a different place. Thus, each of these copy molecules is a different length, and each one ends with a dideoxyribonucleotide. Gel electrophoresis is used to separate the fragments by size. As the fragments pass through the gel, a computer uses a laser to read the colored fluorescent tags on the dideoxyribonucleotides to determine which nucleotide belongs in that place. Because the fragments of DNA are lined up by size, the fluorescent tags get read in order and reveal the sequence of the original DNA molecule.

Applications of Recombinant DNA Technology

GENETIC ENGINEERING

Genetic engineering is used to put genes of interest into target cells. Human genes have been placed into bacteria to produce a variety of therapeutic human proteins, including insulin, human growth hormone, and interferon. Crop plants have been engineered to have the ability to resist pathogens, survive certain pesticides, and produce vitamins. Bacteria have been engineered to help clean up pollution, produce viral proteins for vaccines, and make antibiotics. Genetic engineering is being used to develop human gene therapy, in which people with a genetic disease could receive a copy of the normal gene. Cells or organisms that contain DNA from more than one course are said to be **recombinant**.

> **REMEMBER**
> Recombinant DNA technology is used to create DNA molecules that contain DNA from more than one source.

As an example of the process of genetic engineering, we will look at putting a human gene into a bacterium to get the bacterium to produce the human gene product (Figure 8.12). First, a cDNA copy of the human gene is made. Next, the cDNA gene and a bacterial plasmid are both cut with the same restriction enzyme so they will have the same sticky ends. In order to prevent DNA ligase from re-sealing the plasmid, the phosphate group at the 5' end of the plasmid is removed by dephosphorylation. The open plasmids and cDNA genes are combined, and some plasmids will bind to the cDNA fragments. DNA ligase is used to seal the breaks in the sugar–phosphate backbone. At this point, some of the plasmids will have the human sequence recombinant. Other plasmids may have resealed themselves without picking up the human sequence.

For bacteria to make the human gene product, the recombinant plasmids must be transformed into bacterial cells. The bacteria and plasmid mixture are combined, and the bacteria are given a brief heat shock to encourage them to take up plasmids from the plasmid mixture. Some bacterial cells will take up plasmids and others won't. Of those that take up plasmids, some will have taken up recombinant plasmid that contains the human sequence, others will have plasmid that is missing the human sequence.

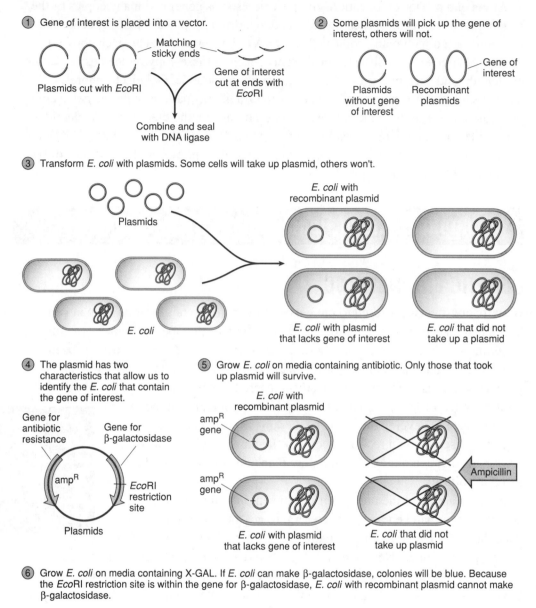

① Gene of interest is placed into a vector.

Matching sticky ends

Plasmids cut with *Eco*RI

Gene of interest cut at ends with *Eco*RI

Combine and seal with DNA ligase

② Some plasmids will pick up the gene of interest, others will not.

Plasmids without gene of interest

Recombinant plasmids

Gene of interest

③ Transform *E. coli* with plasmids. Some cells will take up plasmid, others won't.

Plasmids

E. coli

E. coli with recombinant plasmid

E. coli with plasmid that lacks gene of interest

E. coli that did not take up a plasmid

④ The plasmid has two characteristics that allow us to identify the *E. coli* that contain the gene of interest.

Gene for antibiotic resistance

Gene for β-galactosidase

amp^R

*Eco*RI restriction site

Plasmids

⑤ Grow *E. coli* on media containing antibiotic. Only those that took up plasmid will survive.

E. coli with recombinant plasmid

amp^R gene

amp^R gene

E. coli with plasmid that lacks gene of interest

E. coli that did not take up plasmid

Ampicillin

⑥ Grow *E. coli* on media containing X-GAL. If *E. coli* can make β-galactosidase, colonies will be blue. Because the *Eco*RI restriction site is within the gene for β-galactosidase, *E. coli* with recombinant plasmid cannot make β-galactosidase.

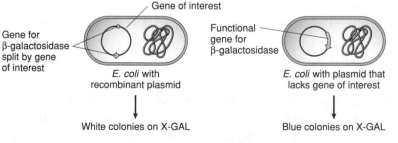

Gene of interest

Gene for β-galactosidase split by gene of interest

E. coli with recombinant plasmid

Functional gene for β-galactosidase

E. coli with plasmid that lacks gene of interest

White colonies on X-GAL

Blue colonies on X-GAL

⑦ Select white colonies from X-GAL plate and screen to confirm presence of human gene.

FIGURE 8.12. Genetic engineering.

In this particular case, two features of the plasmid vector will enable us to identify the bacterial cells that have taken up the recombinant plasmid. First, the plasmid vector contains a gene for antibiotic resistance. Second, it contains the gene for β-galactosidase, an enzyme that can break a particular type of bond like the one found in the sugar lactose. Because the restriction site for the enzyme used to cut the plasmid is located within the gene for β-galactosidase, any plasmid that contains the human gene sequence will have a nonfunctional β-galactosidase gene. Any plasmid that did not pick up the human gene sequence will have a functional β-galactosidase gene.

To identify the bacteria that have the recombinant plasmid, the transformed bacteria are plated onto media that contains an antibiotic and a reporter compound. The presence of the antibiotic in the medium allows us to use positive selection to identify the bacterial cells that took up any plasmid. Because all plasmids contain the gene for antibiotic resistance, only those cells that took up a plasmid will be able to grow in the presence of the antibiotic. The reporter compound in the medium is a substrate for bacterial enzymes and produces a colored product that allows us to identify which cells also have the human gene sequence. One example of a reporter compound is X-Gal. If X-Gal is broken down by β-galactosidase, one of the products will be blue. Thus, any bacteria that contain plasmids that lack the human gene and thus still have a functional gene for β-galactosidase will produce blue colonies. Bacteria that picked up plasmids containing the human gene will form white colonies because their cells will not be able to make β-galactosidase. Because some white colonies may also form from bacteria that picked up nonrecombinant vectors that had mutations in the gene for β-galactosidase, cells from white colonies are screened to confirm that they actually contain the human gene. Bacterial cells that do contain the human gene are now able to make a human protein. These bacteria can be cultured in large quantities, and the human gene product can be purified and used for therapeutic purposes.

> **REMEMBER**
> Genetic engineering is the process of inserting a gene from one organism into a cell from another organism.

IDENTIFICATION OF PATHOGENS

Increasingly, the identification of pathogens has incorporated more molecular techniques than were previously used. Molecular techniques are useful for the identification of viral pathogens that cannot be identified on the basis of metabolic characteristics, and they also enable more rapid identification of bacterial pathogens, even those that cannot be cultured in the lab. If samples of DNA from the pathogen are available, these may be sequenced and compared to sequences of known pathogens. Another method of identification uses nucleic acid probes specific to known pathogens. Cells to be identified are screened with the probes. If a probe sticks to the nucleic acid of the test culture, it reveals the identity of the bacteria.

Scientists are working on the development of DNA chips that will contain probes for many known pathogens. In the near future, it may be possible to determine the cause of a disease by placing a sample from the patient onto a DNA chip and loading it into a computer. The computer would very quickly reveal which probes, if any, were capable of binding to the sample.

GENOMICS

The science of **genomics** is a rapidly developing field dedicated to the study of entire genomes. The goal of genomics is to understand both the coding regions of DNA that are responsible for protein production and the noncoding regions of DNA whose functions are still being determined. By studying and comparing the genomes of organisms, we are improving our understanding of evolution, animal development, and disease.

By sequencing the genomes of pathogenic bacteria, it is possible to identify all of the regions that contain code for proteins. The sequences from the coding regions can be used to identify virulence factors that the pathogen might possess. The first bacterial genome to be completely sequenced was that of *Haemophilus influenzae* in 1995. *H. influenzae* causes ear and respiratory infections as well as meningitis in children. Since then, the genomes of *E. coli* O157:H7 (bloody diarrhea, hemolytic uremic syndrome), *Helicobacter pylori* (ulcers), *Mycobacterium tuberculosis* (tuberculosis), *Treponema pallidum* (syphilis), and many others have followed.

REVIEW EXERCISES FOR CHAPTER 8

1. Describe how restriction enzymes work.

2. Define vector and genomic library.

3. State the purpose of the polymerase chain reaction and describe how it works.

4. Describe the significance of cDNA and its use in genetic engineering.

5. Define DNA probe.

6. State the purpose of gel electrophoresis and describe how it works.

7. Describe the significance of dideoxynucleotides and their use in nucleic acid sequencing.

8. State three beneficial uses of genetically modified organisms.

9. Describe how molecular biology is changing the way we identify pathogens.

10. State the significance of sequencing the entire genome of organisms.

SELF-TEST

1. To make cDNA, you would use

   A. RNA polymerase.
   B. DNA ligase.
   C. *Taq* polymerase.
   D. restriction enzyme.
   E. reverse transcriptase.

2. To make many copies of a single gene, you would use

   A. nucleic acid sequencing.
   B. polymerase chain reaction.
   C. gel electrophoresis.
   D. DNA probes.
   E. None of the above is correct.

3. All of the following are true of restriction enzymes **except**:

   A. They are made by bacteria.
   B. They cut DNA at specific sequences.
   C. They are used in the lab to fight viruses.
   D. They may create sticky ends of double-stranded DNA.
   E. All of the above are true.

4. True or False. After gel electrophoresis, the smallest molecules will be closest to the wells, while the larger molecules will be further away from the wells.

5. True or False. To locate a specific DNA sequence within a larger amount of DNA, you would use a DNA probe.

6. True or False. A genomic library is a computer database that contains information about the genome of an organism.

7. True or False. If a human gene were taken directly from a human chromosome, placed into a plasmid, and then transformed into *E. coli, E. coli* would be able to make the human protein.

8. If the following steps in the process of inserting a human gene into a plasmid were put into order from start to finish, which step would be fourth?

   A. Cut plasmid with restriction enzyme.
   B. Use DNA ligase to seal the sugar–phosphate backbone.
   C. Cut human gene with restriction enzyme.
   D. Mix plasmids and human genes together.
   E. Create cDNA of human gene.

9. True or False: Heat can be used to break the hydrogen bonds that hold DNA strands together.

10. To identify bacteria that have taken up plasmids,

   A. bacteria are grown on X-Gal.
   B. bacteria are grown on antibiotics.
   C. DNA primers are used.
   D. Both A and B are correct.
   E. All (A–C) are correct.

11. Describe the process you would use to engineer *E. coli* to make the human insulin protein. For your source of DNA, you have a very small sample of cells from a person who can make the insulin protein.

Answers

Review Exercises

1. Restriction enzymes can bind to specific DNA sequences called restriction sites. They bind to the DNA backbone at these sites and catalyze the breaking of bonds in the sugar–phosphate backbone. This creates fragments of DNA.

2. A vector carries DNA; two examples are plasmids and phage. A genomic library is a population of bacteria or phage that contain fragments of DNA from the genome of another organism.

3. PCR makes many copies of a specific sequence of DNA. The DNA to be copied is combined with primers, short pieces of single-stranded DNA that will target the gene of interest, nucleotides, and a heat-tolerant DNA polymerase (*Taq* polymerase). The mixture is heated, which separates the strands in the DNA to be copied. It is then cooled to allow the primers to bind. It is heated again to allow Taq polymerase to synthesize new DNA. This entire process is repeated until the desired number of copies is generated.

4. cDNA is DNA that was made by reverse transcriptase from mRNA. It is useful because it does not contain any of the noncoding information (exons) normally found in eukaryotic genes. Thus, cDNA can be introduced into bacteria so that they will produce human proteins.

5. A DNA probe is a short piece of single-stranded DNA that is complementary to a sequence of DNA that you wish to locate. The probe is labeled with a tag such as a fluorescent molecule.

6. Gel elecrophoresis separates molecules on the basis of size and shape. The molecules to be separated are placed into a gel that is submerged in a solution of buffer. An electric current is passed through the gel. The current causes the molecules to move through the gel. Smaller molecules move more quickly and farther

in a given time period than larger molecules. This separates the larger and smaller molecules from each other.

7. Dideoxynucleotides have a hydrogen atom attached to their 3' carbons rather than the hydroxyl group of deoxyribonucleotides. Whenever a dideoxyribonucleotide is inserted into a growing DNA strand, DNA polymerase is blocked from adding any more nucleotides to that strand (recall that DNA polymerase requires a 3' OH). In nucleic acid sequencing, dideoxyribonucleotides are used to randomly stop synthesis of a specific DNA sequence at different locations. The dideoxyribonucleotides can be identified, and their position in the fragments allows the code to be determined.

8. GMOs are currently used to produce medicines such as insulin and human growth hormone, to produce proteins for vaccines, and to increase the nutritional value of certain crop plants.

9. Traditional techniques for identifying bacteria involved staining and metabolic testing. Our increasing ability to manipulate DNA is enabling new approaches like direct sequencing of DNA from pathogens. This is especially useful for viral pathogens that cannot be identified by the methods traditionally used for bacteria. It is also useful for the identification of bacteria and archaea that cannot yet be grown in the lab.

10. By determining the sequence of a genome, it is possible to identify all of the regions in the DNA that code for protein. Previously unidentified proteins may thus be revealed. Also, noncoding regions of DNA are revealed and can be studied.

Self-Test

1. E

2. B

3. C

4. F, the largest molecules move slowly and will be closer to the wells.

5. T

6. F, genomic libraries are copies of genomes maintained in populations of bacteria or viruses.

7. F, the gene would contain exons. A cDNA copy must first be made.

8. D

9. T

10. D

11. First, use PCR to make many copies of the insulin gene. Then, use reverse tran-scriptase to make cDNA copies of the gene. Cut a plasmid that contains both an antibiotic resistance gene and a reporter gene with a restriction enzyme. Cut the ends of the cDNA gene with the same restriction enzyme. Mix the genes with the plasmids and seal with DNA ligase. Transform the plasmid into *E. coli*. Plate the *E. coli* on antibiotics to kill any cells that did not take up the plasmid. The media should also contain the substrate for the reporter gene. For example, the media could contain X-Gal, and the reporter gene could be the gene for β-galactosidase. If the restriction site on the plasmid is within the gene for β-galactosidase, then any cells that have a recombinant plasmid would appear white. Cells with a plas-mid that lacked the gene would appear blue. Pick white colonies, and screen them to confirm the presence of the gene for insulin. Culture recombinant bacteria, and then extract and purify the insulin.

9

Classification of Microorganisms

WHAT YOU WILL LEARN

This chapter introduces the systems that scientists use to organize the kinds of life on Earth. As you study this chapter, you will:

- discover how scientists build phylogenetic trees;
- explore the history of classification systems;
- learn the standard protocols scientists use to name organisms;
- consider the difficulties in assigning bacteria to a particular species;
- be introduced to some of the modern molecular methods used to identify bacteria.

SECTIONS IN THIS CHAPTER

- Phylogeny
- History of Classification Systems
- Taxonomy
- Modern Methods Used to Classify and Identify Bacteria

When you get sick, probably one of the first things you want to know is "What is it?" Of course, this is usually quickly followed by "How do I get rid of it?" When you visit a health professional to get answers to these questions, they will observe your vital signs and may collect samples of your body fluids. If they suspect a bacterial infection, these samples are sent to a microbiology laboratory where they are incubated on media. Tests are run on any bacteria that grow from the cultures. In many cases, the results of these tests lead to the **identification** of the bacteria. This information helps decide the best course of treatment for your infection.

In many parts of the world today, identification and elimination of a bacterial pathogen is routine for many common infections. In more serious cases of very invasive infectious diseases, rapid identification of the pathogen can mean the difference between life and death. In these cases, correct identification of the pathogen is absolutely critical so that effective treatment can begin immediately. So, what does it mean to identify a bacterium? How do we separate one bacterium from another?

Just as you might distinguish between a cat and a dog based on their physical characteristics, microbiologists distinguish among different kinds of bacteria based on their biochemical and genetic characteristics. Physical characteristics of bacteria are also used, but these are not as useful as they are for larger organisms because there aren't as many physical differences that can readily be distinguished among bacteria. Just as we have names for larger organisms, microbiologists have studied many types of bacteria, determined their characteristics, and agreed upon names for bacteria. An unknown bacterium can be identified if its characteristics match those of a known bacterium.

Long before anyone tried to name or identify bacteria, probably as long as people have existed, we have made up names for organisms in the world around us. We have also recognized the similarities between organisms and tried to organize them based on their relationships. For example, most people would agree that somehow dogs and cats are more like each other than they are like trees. We make these decisions automatically based on similarities in characteristics. Scientists make these same types of observations and decisions, although their observations often go beyond what can be seen by the naked eye. The science of naming organisms is called **taxonomy**, and the organization of organisms into categories based on their similarities is called **classification**.

Phylogeny

If you were asked to group four animals, a yellow cat, a yellow dog, a black cat, and a black dog, into categories, how would you do it? Would you put the animals together by color or by type? When designing classification systems, scientists have to make these types of decisions all the time. The best classification system is considered to be the one that most accurately reflects the relatedness, or **phylogeny**, of the organisms. By this rule, it would be better to group the animals by type because the cats have many more characteristics in common with each other than they do with the dogs. Characteristics that two organisms have in common are called **shared characteristics**.

When scientists are deciding how to categorize organisms, they compare their characteristics. The more characteristics that are shared, the more closely related they are thought to be. For example, you probably have more characteristics in common with the members of your immediate family than you do with your next-door neighbors. This is because you have DNA sequences in common with your family, and DNA determines our characteristics. Thus, the more related two organisms are, the more similar they generally appear.

Scientists who are interested in phylogeny use computer programs to make family trees, or **phylogenetic trees**, based on similarities in characteristics (Figure 9.1). The types of characteristics that are used may be structural, chemical, or genetic. Because bacteria do not have a great deal of structural complexity, genetic characteristics have been very important in trying to sort out relationships among the bacteria. When making phylogenetic trees, scientists sometimes have to make decisions about which characteristics are most important in determining relationships. For example, in the dog and cat example, they might decide that color is not as important as the presence of whiskers. This would lead to the correct assortment of showing cats as more related to each other than they are to dogs, despite color differences. In addition to showing relationships among organisms, phylogenetic trees also illustrate their evolutionary history (Figure 9.1).

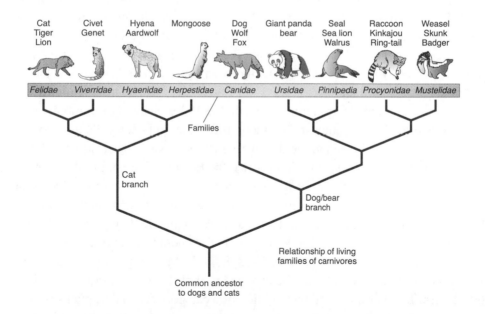

FIGURE 9.1. A phylogenetic tree of carnivores, including cats and dogs.

This tree shows the relationships among living carnivores. The closer the two branches are to each other, the more closely related are the groups. For example, the closest relatives to the cats (Felidae) are the civets (Viverridae). This tree also indicates that dogs and bears evolved from a common ancestor.

History of Classification Systems

Carolus Linnaeus established the foundation for our modern system of taxonomy and classification in 1735. Linnaeus proposed that each organism should have a unique scientific name. Today, scientists and medical professionals around the world regardless of culture or language use these names, such as *Escherichia coli* or *Homo sapiens*. Linnaeus also began to organize all of the named organisms into categories that reflected their similarities. Based on what he knew at the time, this first **classification system** had just two categories, plant and animal. In very simple terms, things that were green and didn't move were considered plants, whereas things that ate other things and moved were animals.

The plant versus animal distinction is a very old one, based on differences that are easily observed with the naked eye. Since the time of Linnaeus, however, scientific methods of observation have become increasingly more sophisticated. As we have the gained the ability to look more closely at the organisms around us, our understanding of how they should be organized has changed and continues to change.

One of the first advances in our ability to observe came with the invention of the light microscope. One person who used a microscope to make many detailed observations of the microbial world was Ernst Haeckel. He observed, drew, and described many fantastic creatures that did not seem to fit into the category of either plant or animal. There were little green creatures that swam and small bizarre creatures that moved yet had no recognizable characteristics of animals. Based on Haeckel's observations, a new category of organism, called Protista, was added to those of Plantae and Animalia in 1866. Within the Protista, he included a group called Monera, which consisted of cells without nuclei.

The invention of the electron microscope in the 1930s enabled even more detailed observations of cells and demonstrated more differences between cells that have nuclei and cells that do not. This led to the discovery that there was a basic structural difference between all cells. These observations led the French scientist, Edouard Chatton, to propose in 1937 that the term prokaryote (*procariotique*) be used for cells without nuclei and the term eukaryote (*eucariotique*) be used for cells with nuclei. In 1938, H.F. Copeland suggested that the prokaryotes were so unique they deserved their own kingdom, resulting in his four kingdom system of Monera (prokaryotes), Protista (Protoctista), Plantae, and Animalia. In 1968, R. G. E. Murray proposed that the name of the prokaryote kingdom Monera be changed to Prokaryotae. Shortly after this, in 1969, R. H. Whittaker proposed the addition of a category for Fungi, plant-like organisms that are not photosynthetic, resulting in a classification system that consisted of five categories called kingdoms. The distinctions between the five kingdoms were based upon structural, metabolic, and developmental differences between the organisms (Table 9.1).

TABLE 9.1. The Five Kingdom System of Classification

Kingdom	Defining Characteristics
Prokaryotae (Monera)	Cells lack nuclei
Protista	Eukaryotic cells that do not fit into the Fungi, Plantae, or Animalia kingdoms
Fungi	Heterotrophic, absorptive nutrition Form spores No flagellated cells at any point in development
Plantae	Multicellular Autotrophic (photosynthetic) Develop from embryos supported by maternal tissue
Animalia	Multicellular Heterotrophic, ingestive nutrition Form from sperm and egg Early in development a blastula (hollow ball of cells) is formed

The next major breakthrough in our understanding of the organisms around us came after the discovery of a new type of cell. Although microscopes help us to visualize structures of cells, another branch of science gives us the ability to compare the chemicals of cells. In the 1970s, Carl Woese began to focus on one particular molecule, **ribosomal RNA (rRNA)**, and to make comparisons between the rRNA of different types of organisms. Ultimately, this led to a whole new way of categorizing life on Earth.

Ribosomal RNA is thought to have evolved a long time ago. It is part of ribosomes, which are found in all cells. It is a nucleic acid and is made up of four building blocks—A, C, G, and U (see Chapter 2). Woese extracted the ribosomes from a wide variety of cells, isolated the rRNA molecules, and determined the sequence of the building blocks. For example, a partial sequence might read "ACCUUACGGAU." He then compared the sequences among the different organisms to see who was the most alike. As expected, when he compared sequences between eukaryotes, he found they were more similar to each other than they were to sequences from prokaryotes. The big surprise came when he compared the sequences among the prokaryotes and found that there were two distinct groups. He also found that the differences in rRNA sequence between these two groups of prokaryotes were equal to the differences between the sequences of prokaryotes and eukaryotes. In other words, even though all prokaryotes looked very similar, genetically they were very different. This led to the proposal, in 1978, that all life on Earth should be divided into three main categories called Domains (Figure 9.2), the Bacteria, Archaea, and Eukarya. The three domains are based on fundamental biochemical and genetic (rRNA) differences among the types of cells on Earth (Table 9.2).

> **REMEMBER**
> The currently accepted phylogeny of life on Earth is represented by the 3 Domain System.

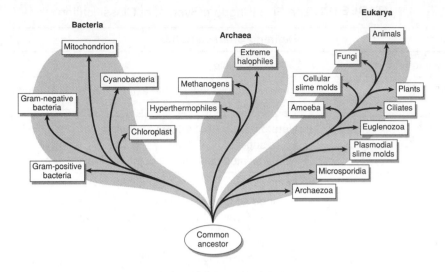

FIGURE 9.2. The three domains.

Based on comparisons of rRNA sequences, living things can be organized into three categories called domains. These domains are the Bacteria, Archaea, and Eukarya. Cells in the Bacteria and Archaea are prokaryotic, whereas those in the Eukarya are eukaryotic.

TABLE 9.2. Distinguishing Biochemical Characteristics of the Three Domains

Characteristic	Archaea	Bacteria	Eukarya
Cell wall	No peptidoglycan	Peptidoglycan	No peptidoglycan
Lipids in membrane	Branched carbon chains ether-linked (a type of bond) to glycerol	Straight carbon chains ester-linked (a type of bond) to glycerol	Straight carbon chains ester-linked (a type of bond) to glycerol
First amino acid in translation	Methionine	Formylmethionine	Methionine
Presence of loop in rRNA	No	Yes	No
Presence of common sequence in tRNA	No	Yes	Yes

Taxonomy

Woese's three-domain system of classification is the one that is generally accepted by most biologists today. The recognition of the three domains, however, does not wipe out the earlier observations and resulting classification systems. Instead, Woese's observations have been added to what was already known about the characteristics of life on Earth. The domain level represents the most fundamental difference between organisms. Within domains, organisms are further sorted into kingdoms. Within kingdoms, there are phyla (singular is phylum), then classes, then orders, then families, then genera (singular is genus), then species (Table 9.3). Placement into each of these categories is based on certain defining characteristics. For example, humans belong to the class Mammalia. All mammals, whether humans or dogs or elephants, have hair and nurse their young. There are many tricks for remembering the order of the taxonomic hierarchy, one of them is **D**arn **K**ids **P**icking **C**acti **O**n **F**ridays **G**et **S**tuck. As you move down the taxonomic hierarchy from domain to species, the categories become more specific and the organisms within them become more similar.

TABLE 9.3. A Comparison of the Taxonomy of the Domestic Dog, *Canis familiaris*, and the Intestinal Bacterium, *E. coli*

Taxon	Domestic Dog	Intestinal Bacterium
Domain	Eukarya	Bacteria
Kingdom	Animalia	Eubacteria
Phylum	Chordata	Proteobacteria
Class	Mammalia	Gamma-proteobacteria
Order	Carnivora	Enterobacteriales
Family	Canidae	Enterobacteriaceae
Genus	*Canis*	*Escherichia*
Species	*C. familiaris*	*E. coli*

Species is the most narrowly defined category. If two organisms are the same species, they are considered essentially the same in some way. For most eukaryotic species, the deciding factor is whether or not two organisms can breed with each other and produce living offspring that can also breed. Thus, all humans, regardless of superficial differences, belong to the same species. Because bacteria reproduce asexually by binary fission, the species definition that works for eukaryotes does not work for bacteria. In addition, gene transfer can occur between bacteria that are considered unrelated. Thus, the definition of bacterial species is difficult. Essentially, a bacterium is considered to belong to a particular species if it has certain characteristics that are used to define that species and if its genome is at least 70 percent identical to the genomes of other members of the species.

REMEMBER
Bacterial species are defined by the degree of similarity in traits and DNA sequences between bacterial cells.

Species are referred to by the precise naming system developed by Linnaeus. This system is called **binomial nomenclature** because each species is referred to by two names (*bi* = two; *nom* = name). The first name is the name of the genus to which the organism belongs; the second name is a describing word called the **specific epithet**. Each species gets a unique combination of genus and specific epithet. Thus, all humans are *Homo sapiens*. The tiger is *Panthera tigris*, whereas the lion is *Panthera leo*. To indicate that a proper scientific name is being used, the genus is capitalized and the entire name is either italicized or underlined. Scientists and medical personnel around the world use the proper name, making communication much easier. Proper notation is particularly important in microbiology because some words have dual meanings. For example, staphylococcus refers to any clump of round bacteria. *Staphylococcus aureus*, however, refers to a specific opportunistic pathogen with known characteristics. Likewise, a bacillus is any rod-shaped bacterium, whereas *Bacillus* refers to a genus of bacteria that contains *Bacillus anthracis*, the causative agent of anthrax.

Even among bacteria that share certain key characteristics and are therefore considered to belong to the same species, there can be considerable variation in other characteristics. This is the result of both spontaneous mutations and transfer of genes between cells. For example, a bacterium that belongs to one species might pick up an unusual characteristic by acquiring a plasmid from a bacterium belonging to a different species. The variants within a species are called **strains** and are designated by letters and numbers after the specific epithet. These letters and numbers refer to unique traits that are used to identify the strain. For example, all *E. coli* are gram-negative, lactose-fermenting, rod-shaped bacteria that break down the amino acid tryptophan and produce mixed acids during fermentation. However, within the species of *E. coli* there are many variants from harmless bacteria such as *E. coli* K12 to dangerous pathogens such as *E. coli* O157:H7. The difference between the strains is that the pathogenic *E. coli* have acquired genes from other species of bacteria that enable them to cause disease.

Much of our current knowledge of the taxonomy of prokaryotes is compiled into one resource, *Bergey's Manual of Systematic Bacteriology*. Up until very recently, most of the information we have had on the classification of prokaryotes has come from structural and biochemical characteristics. Because bacteria have fairly simple morphologies, our understanding of the relationships among bacteria has been limited. Since the work of Carl Woese opened the door to genetic analysis, however, our understanding of the relationships among prokaryotes has changed greatly. The recently released *Bergey's Manual* reflects our current understanding of prokaryote phylogeny.

Modern Methods Used to Classify and Identify Bacteria

The classical methods of identifying bacteria based on structural and biochemical methods are still valuable today. One of the most important initial distinctions is to determine the cellular morphology and characterize the cell wall using the Gram stain (see Chapter 3). The result of the Gram stain then determines which biochemical tests should be run or if any additional structural information is needed. Certain biochemical tests are important for the identification of particular groups of bacteria. Microbiology labs that identify bacteria usually follow a certain sequence of tests designed to distinguish rapidly among the types of bacteria they are most likely to encounter depending on the source of the bacterium.

In recent years, molecular approaches have increasingly been used for both identification and classification (Table 9.4). These methods directly examine cellular components such as proteins or DNA. The size of particular proteins can be examined using **gel electrophoresis** (see Chapter 8). In addition, the presence or absence of particular proteins can be detected using **serological techniques.** In these tests, antibodies that are known to react with specific molecules are tested against an unknown bacterium (Figure 9.3). For example, the pathogen *E. coli* O157:H7 is characterized in part by the presence of a particular O polysaccharide, type 157, in its cell wall. If an unknown sample of *E. coli* is tested with antibodies that are specific to the O157 antigen and a reaction occurs, then the unknown *E. coli* has the antigen and is identified as being that particular strain.

> **REMEMBER**
> Serological techniques use highly specific antibody proteins to screen cells for the presence of certain identifying molecules (antigens).

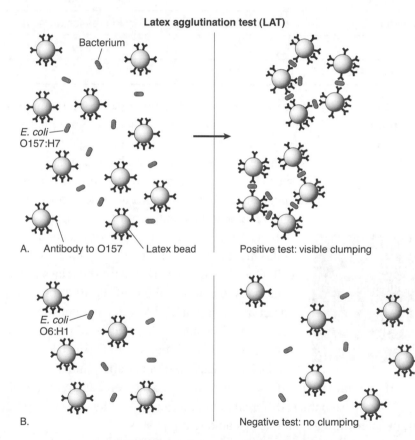

FIGURE 9.3. A latex agglutination test.

First, latex beads that are coated with antibody specific to a particular antigen are mixed with a sample of the bacterium to be identified. If the bacterium has the antigen, the antibodies will attach, and agglutination will occur. This is visible as clumping. For example, if antibody to the O157 polysaccharide is on the bead, and the bacterium is E. coli *O157:H7, then the antibody will bind the polysaccharide, and clumping will occur. If the bacterium is not* E. coli *O157:H7, no clumping will occur.*

TABLE 9.4. Selected Methods Used for
Classification and Identification of Bacteria

Method	Used for Classification?	Used for Identification?
Differential staining such as the Gram stain	Yes	Yes
Morphology (rod, coccus, spiral)	No	Yes
Biochemical testing	Yes	Yes
Serology	Yes	Yes
Phage typing	No	Yes
DNA base composition	Yes	No
Nucleic acid sequencing	Yes	No (Yes for viruses)
PCR	Yes	Yes
Nucleic acid hybridization	Yes	Yes

Another method for detecting whether a bacterium possesses certain antigens is to use **phage typing**. Bacteriophage are viruses that attack bacteria. For a phage to attack a particular bacterium successfully, the bacterium must have the right molecules on its surface to which the phage can bind. Thus, the ability of a phage to attack and destroy bacteria is very specific to the species, or even strain, of the bacterium. Unknown bacteria can be tested with various phage to see which phage can successfully attack them, thus revealing the identity of the bacterium.

REMEMBER
Nucleic acid hybridization measures the similarities between two sources of DNA by determining how many DNA sequences are complementary to each other.

In addition to examining antigens, scientists use several other methods that directly examine the DNA of the cell. One characteristic that is used in classification of bacteria is to compare the **G+C content** of different bacteria. Recall that DNA is made up of four building blocks, A, T, G, and C, and that within the double helix, A pairs with T and G pairs with C (see Chapter 2). Thus, the DNA is made up of a certain number of A–T base pairs and a certain number of G–C base pairs. Because the percentage of the DNA that is made up of G–C base pairs is different among different species, this characteristic is often useful in determining relationships among organisms.

The degree of relationship among organisms can also be determined by directly comparing the similarities in their DNA by using **nucleic acid sequencing**. In this case, the actual sequence of a gene or molecule of rRNA is determined and compared among organisms. An alternative method for comparing sequence similarity is to use **nucleic acid hybridization**. In this case, the DNA from two different species can be compared (Figure 9.4). First, the DNA is extracted from each species, and the strands of the double helix are separated from each other. This creates single-stranded DNA molecules that can base-pair with other single-stranded DNA molecules. The DNA from both sources is combined and allowed to base-pair. If the sequences are very similar, they will pair well. If the sequences are not similar, base pairing will not occur.

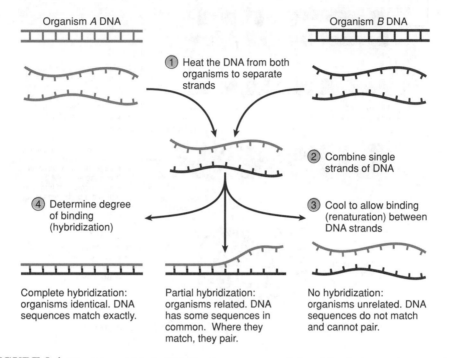

FIGURE 9.4. Nucleic acid hybridization to compare relationships between two species.

Another use of nucleic acid hybridization is to test the DNA of unknown bacteria to see if it matches that of a known pathogen. First, DNA from the unknown bacterium is extracted, and the strands of the double helix are separated from each other. Next, a single-stranded copy of a gene from the pathogen, called a **probe**, is allowed to contact the separated DNA. The more similar the two DNA sequences are, the better they will be able to pair with each other (Figure 9.5). This is like going fishing with a piece of DNA as bait: Only the unknown bacteria that have complementary sequences to the probe DNA will be able to stick to it, or take the bait.

To examine the DNA of bacteria using these techniques, it is necessary to have enough DNA to work with. Sometimes, it may be possible to grow the bacteria in culture to obtain sufficient numbers of cells to provide the DNA. In many instances, however, this is not possible. In these cases, when only a very small sample of bacteria or even viruses are available, copies of their DNA can be made using the **polymerase chain reaction** (see Chapter 8).

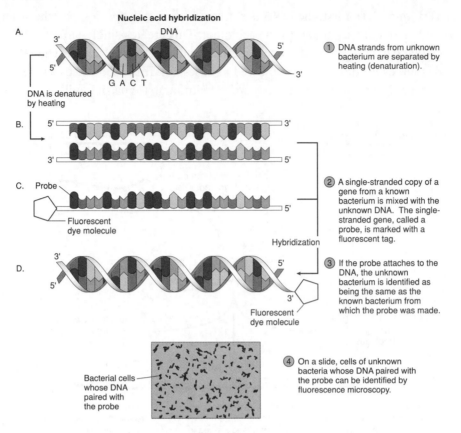

Nucleic acid hybridization

A. DNA is denatured by heating

① DNA strands from unknown bacterium are separated by heating (denaturation).

② A single-stranded copy of a gene from a known bacterium is mixed with the unknown DNA. The single-stranded gene, called a probe, is marked with a fluorescent tag.

③ If the probe attaches to the DNA, the unknown bacterium is identified as being the same as the known bacterium from which the probe was made.

④ On a slide, cells of unknown bacteria whose DNA paired with the probe can be identified by fluorescence microscopy.

FIGURE 9.5. Using a DNA probe to identify a bacterium.

REVIEW EXERCISES FOR CHAPTER 9

1. Name the three domains and state the evidence that supports this system of classification.

2. Describe proper binomial nomenclature.

3. State the basis for the definition of species in bacteria and explain why bacterial species cannot be defined using the same criteria that are used for defining eukaryotic species.

4. Why does classification of bacteria rely heavily on genetic testing?

5. Give a brief description of how the following types of tests are used in identification and/or classification of microorganisms:

   a. Morphology

   b. Differential staining such as the Gram stain

   c. Biochemical tests

   d. Serology

   e. Phage typing

   f. Nucleic acid sequencing

   g. PCR

   h. Nucleic acid hybridization

6. Discuss the significance of rRNA in determining phylogeny and explain how and why rRNA is used.

SELF-TEST

1. Which of the following represents a proper scientific name?

   A. Anthrax
   B. Bacillus
   C. *Bacillus anthracis*
   D. Rod-shaped bacterium
   E. Anthrax bacillus

2. Which characteristic is unique to organisms in Kingdom Animalia?

   A. Motility
   B. Heterotrophic nutrition
   C. Spore production
   D. Develop from embryo
   E. Develop from hollow ball of cells

3. True or False. Based on comparisons of rRNA sequences, humans and plants are much more closely related to each other than they are to bacteria.

4. The G–C content of *E. coli* is 51 percent and the G–C content of *Shigella flexneri* is 51 percent. From this information, what would you conclude about the relationship between *E. coli* and *Shigella flexneri*?

   A. They are very closely related.
   B. They are not closely related.
   C. There is not enough information to tell how related they are.

5. Which of the following is found in all three domains?

   A. Nucleus
   B. Peptidoglycan
   C. Prokaryotic (70S) ribosome
   D. rRNA
   E. None of the above is correct.

6. Which of the following represents the correct order of the taxonomic hierarchy from most inclusive group to least inclusive group?

   A. Species Genus Family Order Class Phylum Kingdom Domain
   B. Domain Kingdom Phylum Class Order Family Genus Species
   C. Genus species
   D. Kingdom Phylum Class Order Family Genus Species
   E. Phylum Class Order Kingdom Domain Genus Species Family

7. Which of the following characteristics is **least** useful for the classification of bacteria?

   A. Morphology
   B. G–C content
   C. rRNA sequencing
   D. Nucleic acid hybridization
   E. Differential staining

8. The relationship, or evolutionary history, of organisms is called

   A. taxonomy.
   B. classification.
   C. phylogeny.
   D. identification.
   E. domain.

9. True or False. Bacterial species are defined based on their ability to conjugate with each other.

10. Which of the following is not a eukaryotic kingdom?

    A. Fungi
    B. Plantae
    C. Animalia
    D. Protista
    E. All of the above are eukaryotic kingdoms.

11. Use the information in the following table to construct a phylogenetic tree for the organisms listed. Give the rationale for your tree, indicating the relative importance of each characteristic.

Characteristic	Species 1	Species 2	Species 3	Species 4
Morphology	Rod	Rod	Coccus	Rod
Gram stain	Negative	Negative	Positive	Positive
Utilization of glucose	Fermentative	Fermentative	Oxidative	Oxidative
G–C content (mol %)	51	52	75	73

Answers

Review Exercises

1. The three domains are the Eukarya, the Bacteria, and the Archaea. The separation of life into three domains was originally based on sequence comparisons of rRNA molecules. These comparisons showed that the prokaryotes actually contained two different groups of cells. It also showed that eukaryotes were much more closely related to each other than they were to either of the two groups of prokaryotes. Since the initial rRNA sequence comparisons, other fundamental differences between the three types of cells have given support to the three-domain system. These include differences in the cell walls, membrane lipids, tRNA, and the first amino acid used in protein synthesis.

2. Proper binomial nomenclature describes each organism by a combination of two names: the genus followed by the specific epithet. The genus is capitalized and the specific epithet is lowercase. The name is italicized or underlined to indicate that it is a proper scientific name.

3. Species in bacteria are defined based on similarities in characteristics. It is not possible to define bacterial species based on the ability to sexually reproduce, as it is in eukaryotes, because sexual reproduction does not exist in bacteria. Gene exchange in bacteria is not linked to reproduction. Reproduction occurs by binary fission, which is an asexual process.

4. The classification of bacteria relies heavily upon genetic testing because bacteria do not have a great deal of morphological complexity, so morphology is not very useful.

5. a. Morphology is determined by staining and is used in identification to narrow the list of possible bacteria.

    b. Differential staining, such as the Gram stain, reveals unique characteristics about the bacteria. The Gram stain indicates the cell wall structure of the bacterium and is used in both classification and identification.

    c. Biochemical tests determine whether particular enzymes are present. They are used primarily in identification.

    d. Serological tests are used to test for the presence of specific molecules. Antibodies that are known to bind specifically to certain molecules (antigens) are tested for their ability to bind to an unknown bacterium. If the antibodies can bind, it indicates that the antigen is present. These tests are used in identification of bacteria.

    e. Phage typing tests for the presence of specific molecules by determining whether bacteriophage can infect unknown bacteria. Because bacteriophage are often very specific to the bacteria they can infect, the ability of bacterio-phage to infect a particular cell can reveal its identity. This is used for identification of bacteria.

f. For nucleic acid sequencing, nucleic acids such as RNA and DNA are isolated, and their specific sequence is determined. These sequences can be compared to establish relationships, as in the example of rRNA sequence comparisons that led to the identification of the three domains. DNA sequences of unknown viruses can be compared to those of known viruses in a computer database in order to identify the unknown virus.

g. The polymerase chain reaction makes many copies of a targeted piece of DNA. This technique is used to generate enough DNA for sequence comparisons, evaluation of G–C content, and nucleic acid hybridization. It is thus important in both the identification and classification of organisms.

h. Nucleic acid hybridization tests the ability of DNA from different sources to base-pair with each other. The more similar the DNA sequence is, the greater the amount of pairing, or hybridization. The more hybridization that occurs, the greater the degree of relationship. This technique is used for both classification and identification of bacteria.

6. One of the reasons that ribosomal RNA is a useful molecule for comparisons between cells is because it is found in all cells. Also, even though there has been enough change in rRNA sequences over evolutionary time to show differences between cells, there are still enough similarities to make comparisons possible. When rRNA sequences are compared, cells fall into three groups, called domains. These are the Bacteria, Archaea, and Eukarya.

Self-Test

1. C	5. D	9. F
2. E	6. B	10. E
3. T	7. A	
4. A	8. C	

11. The G+C content is the most important characteristic because it is a genetic trait. However, in this case, the G+C content, Gram stain, and utilization of glucose all agree that species 1 and 2 are closely related and species 3 and 4 are closely related. A possible tree would show two major branches, each of which split into two smaller branches. The two smaller branches on one of the major branches would lead to species 1 and species 2. The two smaller branches on the other major branch would lead to species 3 and species 4.

The Bacteria

WHAT YOU WILL LEARN

This chapter provides a brief introduction to the many diverse groups of bacteria that have been identified and described by microbiologists. As you study this chapter, you will:

- review the current view of the phylogeny of the bacteria;
- discover the diversity of bacteria;
- explore the impact of bacteria on human health and the environment.

SECTIONS IN THIS CHAPTER

- Phylum Cyanobacteria
- Phylum Proteobacteria
- Phylum Firmicutes
- Phylum Actinobacteria
- Phylum Chlamydiae
- Phylum Spirochaetes

Most of the prokaryotes you have ever heard of are probably part of the Domain Bacteria, which includes most of the familiar bacteria along with thousands of species you may never have heard of or thought about. Only a very small fraction of bacteria are either human pathogens or used in human industry. Most bacteria are part of the complex ecosystem of planet Earth and perform vital roles in the cycling of nutrients and energy.

According to the preliminary taxonomic outline of *Bergey's Manual of Systematic Bacteriology, 2nd edition*, there are twenty-three phyla within Domain Bacteria. The bacteria within these phyla exhibit an incredible diversity of metabolism, structure, and role in the environment. Here, we will consider a representative sample of those genera that demonstrate the medical, economic, and environmental importance of the bacteria. In addition, we will look at examples of bacteria (Table 10.1) that illustrate the structural and metabolic diversity of bacteria.

TABLE 10.1. Significant Features of Selected Bacteria from *Bergey's Manual of Systematic Bacteriology, 2nd edition*

Phylum	Class	Important Genera	Significant Features
Cyanobacteria	Cyanobacteria	*Anabaena, Oscillatoria, Synechococcus*	Perform oxygenic photosynthesis (O_2 is released) and nitrogen fixation
Proteobacteria	Alphaproteo-bacteria	*Acetobacter, Gluconobacter*	Production of acetic acid (vinegar) from ethanol
		Rhodospirillum, Rhodobacter, Rhodopseudomonas	Perform anoxygenic photosynthesis (no O_2 is released)
		Erhlichia, Rickettsia	Obligate intracellular parasites of humans
		Wolbachia	Symbionts of insects
		Caulobacter	Produces a stalk for attachment to surfaces and absorption of nutrients
		Hyphomicrobium	Divides by budding
		Rhizobium	Symbionts in plant roots, performs nitrogen fixation

TABLE 10.1. (continued)

Phylum	Class	Important Genera	Significant Features
Proteobacteria	Alphaproteo-bacteria	*Azospirillum, Beijerinkcia*	Free-living soil bacteria, performs nitrogen fixation
		Agrobacterium	Pathogens of plants, used to introduce genetic material to plants
		Nitrobacter	Performs nitrification
		Alcaligenes	Oxidizes hydrogen and carbon monoxide
	Betaproteo-bacteria	*Nitrosomonas*	Performs nitrification
		Thiobacillus	Oxidizes sulfur compounds
		Gallionella	Produces a stalk of iron-containing compounds for attachment to surfaces
		Sphaerotilus	Sheathed
		Spirillum	Large, spiral-shaped bacteria, motile by polar flagella
		Neisseria	Pathogens of humans
		Zoogloea	Forms flocs when growing, used in sewage treatment
	Gammaproteo-bacteria	*Beggiatoa*	Oxidizes sulfur compounds
		Thiomargarita	Oxidizes sulfur compounds, giant bacterium
		Azomonas, Azotobacter	Free-living bacteria, performs nitrogen fixation
		Chromatium, Ectothiorhodospira	Performs anoxygenic photosynthesis (no O_2 is released)
		Legionella	Pathogen of humans

TABLE 10.1. (continued)

Phylum	Class	Important Genera	Significant Features
Proteobacteria	Gammaproteo-bacteria	*Coxiella*	Obligate intracellular parasites of humans
		Pseudomonas	Opportunistic pathogens of humans
		Vibrio	Pathogens of humans
		Escherichia	Normal intestinal bacteria of humans, some are pathogens of humans
		Proteus	Normal intestinal bacteria of humans, some are pathogens of humans, exhibits swarming motility
		Salmonella, Shigella Enterobacter, Klebsiella, Serratia	Pathogens of humans Opportunistic pathogens of humans
		Haemophilus	Pathogens of humans
	Deltaproteo-bacteria	*Bdellovibrio*	Predators of gram-negative bacteria
		Desulfovibrio	Sulfate reducers
		Myxococcus	Gliding, fruiting
	Epsilonproteo-bacteria	*Campylobacter, Helicobacter*	Pathogens of humans
Firmicutes	Clostridia	*Clostridium*	Forms endospores, pathogens of humans
	Mollicutes	*Mycoplasma, Spiroplasma*	Very small bacteria, lacks a cell wall
	Bacilli	*Bacillus*	Forms endospores
		Staphylococcus	Forms clusters of cocci, opportunistic pathogens of humans
		Lactobacillus	Rod-shaped bacteria that produce lactic acid during fermentation, important in production of fermented milk products

TABLE 10.1. (continued)

Phylum	Class	Important Genera	Significant Features
Firmicutes	Bacilli	*Streptococcus*	Chains of coccoid bacteria, pathogens of humans
		Enterococcus	Chains of coccoid bacteria, opportunistic pathogens that cause nosocomial infections
		Listeria	Coccobacilli, causes food poisoning from contaminated deli products
Actinobacteria	Actinobacteria	*Corynebacterium*	Club-shaped rods, some are human pathogens
		Mycobacterium	Acid-fast cell wall, pathogens of humans
		Propionibacterium	Produces propionic acid during fermentation, important in manufacture of Swiss cheese
		Streptomyces	Filamentous growth habit, produces many antibiotics
Chlamydiae	Chlamydiae	*Chlamydia*	Obligate intracellular parasites, pathogens of humans
Spirochaetes	Spirochaetes	*Spirochaete, Cristispira, Treponema, Leptospira, Borrelia*	Corkscrew-shaped bacteria that are motile by axial filaments, some are human pathogens

Phylum Cyanobacteria

Because they are photosynthetic and aquatic, **cyanobacteria** are sometimes referred to as blue-green algae. Their unique color is due to their photosynthetic pigments, which are a combination of chlorophyll a (green), phycocyanin (blue), and phycoerythrin (red). Cyanobacteria grow in freshwater, saltwater, and terrestrial environments and can survive in harsh conditions such as salt lakes, hot springs, and desert soils. They range from small unicellular organisms such as the marine plankton *Synechococcus* to large filamentous species such as *Oscillotoria*. Because some species produce toxins, a bloom of cyanobacteria can result in the closure of swimming areas.

Cyanobacteria perform photosynthesis in the same way that plants do, using water as a source of electrons and releasing oxygen as waste (**oxygenic photosynthesis**, Chapter 5). Fossil evidence suggests that cyanobacteria were among the first oxygen-producing organisms and were responsible for the introduction of oxygen into the atmosphere of the early Earth. As photosynthetic organisms, they also play an important role in the **carbon cycle** (Chapter 14), taking up carbon dioxide (CO_2) and producing sugar ($C_6H_{12}O_6$).

> **REMEMBER**
> Cyanobacteria do photosynthesis in much the same way as do plant chloroplasts. As food-makers (autotrophs), these blue-green bacteria help support many aquatic food webs.

In addition to photosynthesis, many species of cyanobacteria are capable of **nitrogen fixation** (Chapter 14). These species produce special cells called *heterocysts* in which enzymes capture atmospheric nitrogen (N_2) and convert it into ammonium (NH_4^+), which is more easily used by cells as a source of nitrogen. In fact, some rice farmers use cyanobacteria such as *Anabaena* as a natural nitrogen fertilizer, growing it with their rice in the rice paddies. Because of their ability to perform nitrogen fixation, cyanobacteria perform a very important role in the **nitrogen cycle** (Chapter 14).

Phylum Proteobacteria

The phylum Proteobacteria is the most diverse group of the known bacteria, with over 1600 identified species. Within these species are found most of the gram-negative bacteria that are of medical, veteterinary, and agricultural importance. There are also many ecologically important species, including nitrogen-fixing soil bacteria. Proteobacteria come in many shapes, from the more typical rods and cocci to ring-shaped, filamentous, and sheathed. One of the few types of bacteria that have a multi-cellular stage, the myxobacteria, is also found in the Proteobacteria. In addition to morphological variability, the group also shows a tremendous metabolic diversity, from photolithoautotrophy to chemoorganoheterotrophy to chemolithoautotrophy. One amazing bacterium, *Rhodopseudomonas palustris*, can use all of these metabolic strategies plus photoorganoheterotrophy! It is no wonder that this group was named after the Greek god Proteus, who could assume many different shapes. The phylum is organized into five classes, each of which is designated by a Greek letter (alpha, beta, gamma, delta, or epsilon).

CLASS ALPHAPROTEOBACTERIA

The Alphaproteobacteria come in many different shapes and perform many different types of metabolism. They include phototrophic bacteria, nitrogen-fixing bacteria, chemolithoautotrophs, and chemoorganoheterotrophs. Many species form associations with eukaryotes, either as symbionts or pathogens. Indeed, an ancient symbiosis between a relative of this group and a eukaryote appears to have led to the evolution of the mitochondrion (see Chapter 4). Finally, alphaproteobacteria thrive in a wide variety of environments and are important ecologically.

ACETIC ACID BACTERIA

Acetobacter and *Gluconobacter* are examples of acetic acid bacteria, aerobic bacteria that produce large amounts of organic acids such as acetic acid (vinegar). These bacteria are acid-tolerant and can often be found growing in wine or beer. When ethanol is available as a carbon source, they oxidize it, producing acetic acid (vinegar) as waste. For the commercial production of vinegar, the bacteria are added to wine, cider, or distilled alcohol, producing wine vinegar, cider vinegar, or distilled (white) vinegar as a result.

PURPLE NONSULFUR BACTERIA

The purple nonsulfur bacteria such as *Rhodospirillum*, *Rhodopseudomonas*, and *Rhodobacter* are photosynthetic bacteria that do not produce oxygen during photosynthesis. This **anoxygenic photosynthesis** (Chapter 5) occurs because the bacteria do not use water (H_2O) as an electron source. The bacteria employ cyclic photophosphorylation (Chapter 5) to make ATP, then obtain electrons by oxidizing organic molecules (**photoorganoheterotrophy**) or by oxidizing inorganic molecules other than water (**photolithoautotrophy**). The electrons are transferred to electron carriers like NADPH for use in biosynthesis. The dramatic colors of these bacteria, from red to purple, result from their combinations of **bacteriochlorophyll** and various **carotenoid** pigments. Some purple nonsulfur bacteria fall within the class Betaproteobacteria.

> **REMEMBER**
> Rickettsias are obligate intracellular parasites that cause diseases such as Rocky Mountain spotted fever and epidemic typhus.

RICKETTSIAS

Rickettsias are one of the unusual groups of bacteria that are **obligate intracellular parasites**. They are small coccoid or rod-shaped bacteria that enter the host cell by phagocytosis and then escape into the cytoplasm where they multiply by binary fission. Members of this group cause several diseases and may be transmitted between animal hosts by arthropod vectors. In humans, epidemic typhus is caused by *Rickettsia prowazekii*, which is transmitted by lice. Another species of *Rickettsia*, *R. rickettsii*, is transmitted by ticks and causes Rocky Mountain spotted fever. The genus *Ehrlichia* is also transmitted by ticks to humans and causes the disease ehrlichiosis. Another genus, *Wolbachia*, infects over a million species of insects and other animals. The bacterium lives within the host cells and can be passed from one generation to the next in infected egg cells. *Wolbachia* can have dramatic effects on its insect hosts such as causing males to become females, killing male offspring, or causing females to produce offspring in the absence of male fertilization (parthenogenesis).

BUDDING AND STALKED BACTERIA

Some of the morphological diversity of the bacteria is demonstrated by species such as *Caulobacter*, which forms a **stalk** that it uses to attach to surfaces (Figure 10.1). In addition to attachment, the stalk increases the surface area of the cell that can be used for absorption of nutrients. This may be an advantage to stalked bacteria like *Caulobacter* that are often found in nutrient-poor conditions. Another genus of stalked bacteria, *Gallionella*, forms a stalk made of iron-containing compounds and is a member of the Betaproteobacteria.

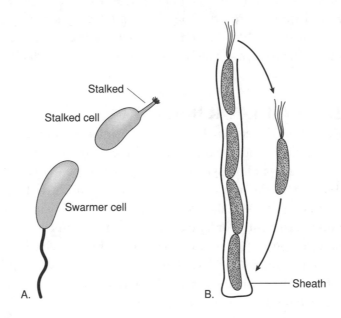

FIGURE 10.1. Unusual morphologies in bacteria.

A. Caulobacter, *a stalked bacterium. B.* Sphaerotilus, *a sheathed bacterium.*

Another unusual feature of some Alphaproteobacteria is the ability to divide by **budding**. Recall that most bacteria divide by binary fission, in which the cell elongates and then divides equally into two daughter cells. By contrast, in budding, elongation occurs from one point on the cell, producing an unequal daughter cell (Figure 10.2). In the genus *Hyphomicrobium*, buds form at the tip of a cellular extension called a **hypha** (Figure 10.2).

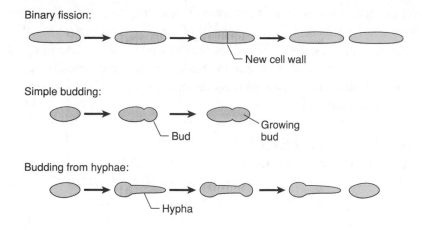

Binary fission:

New cell wall

Simple budding:

Bud

Growing bud

Budding from hyphae:

Hypha

FIGURE 10.2 Budding in bacteria.

*Most bacteria divide by binary fission, in which the products of cell division are equal.
Some bacteria divide by budding, in which the products of cell division are initially not equal.*

NITROGEN-FIXING BACTERIA

Members of the genus *Rhizobium* are of great agricultural
importance because of the association they form with the roots
of leguminous plants such as peas, beans, and alfalfa. These
bacteria enter the plant roots from the soil and cause the formation of special structures called **root nodules** (Chapter 14). The
bacteria colonize the nodules and establish a symbiosis with
the plants. The bacteria perform **nitrogen fixation** (Chapter 14),
converting atmospheric nitrogen (N_2) into nitrogen-containing compounds that are
usable by both the bacteria and the plants. Because of the nitrogen-fixing bacteria
associated with the roots of leguminous plants like alfalfa, farmers will often use these
plants as a cover crop to increase the nitrogen content of the soil. In addition to the
genus *Rhizobium*, the Alphaproteobacteria also includes several genera of free-living
soil bacteria that are capable of nitrogen fixation, including *Azospirillum* and
Beijerinckia. Other free-living nitrogen-fixing bacteria are the large rods *Azotobacter*
and *Azomonas*, which are members of the Gammaproteobacteria.

REMEMBER
Nitrogen-fixing bacteria are
incredibly important to all life
on Earth because they can
capture nitrogen gas from
the atmosphere (N_2) and
convert it into forms that
other living things can use.

AGROBACTERIUM

Another genus of bacteria that can colonize plant cells is
Agrobacterium. *A. tumefaciens* enters plant cells and causes the
disease crown gall, producing a large abnormal growth at the
junction of the roots and stem. This gall is produced as a result
of the transfer of a plasmid containing bacterial genes, called the
Ti plasmid, into the plant chromosomes. Because of its ability

REMEMBER
*Agrobacterium
tumefaciens*
is very useful in the
bioengineering of
plant cells.

to transfer genetic material, it has been used by scientists to genetically engineer plant cells. The scientists first modify the bacterial plasmid to contain the gene they want to add to the plant cell; then they reintroduce the plasmid into the bacterium and allow the bacterium to infect the plant. Using methods such as these, scientists have successfully engineered plants to introduce characteristics such as herbicide resistance or increased vitamin content.

NITRIFYING BACTERIA

Just as nitrogen-fixing bacteria are important for their ability to capture atmospheric nitrogen (N_2) and convert it to ammonia (NH_3), **nitrifying bacteria** are essential because they convert the ammonia (NH_3) to nitrate (NO_3^-), an important source of nitrogen for plants. There are actually two groups of bacteria that participate in the conversion of ammonia (NH_3) to nitrate (NO_3^-). One group, which includes the betaproteobacteria *Nitrosomonas*, converts the ammonia (NH_3) to nitrite (NO_2^-). The second group, which includes the alphaproteobacteria *Nitrobacter*, converts the nitrite (NO_2^-) to nitrate (NO_3^-). Both groups of bacteria are **chemolithoautotrophs** (Chapter 5) that oxidize the nitrogen-containing compounds and use the electrons to generate ATP by chemiosmosis. Their conversion of the nitrogen-containing compounds from one form to another is an essential part of the **nitrogen cycle** (Chapter 14).

HYDROGEN-OXIDIZING BACTERIA AND CARBOXYDOTROPHIC BACTERIA

Another type of chemolithoautotrophy is seen in the hydrogen-oxidizing bacteria. Most of these bacteria are also chemoheteroorganotrophs that have the ability to obtain energy and carbon by breaking down organic matter. Under certain circumstances, however, they can begin to oxidize hydrogen (H_2), producing water (H_2O) as waste. Several genera in the Betaproteobacteria include species that can act as hydrogen oxidizers, including the genus *Alcaligenes*. Several other genera of bacteria from diverse phylogenetic groups also have this ability.

In addition to the ability to oxidize hydrogen (H_2) to obtain electrons, some hydrogen-oxidizing bacteria, including *Alcaligenes*, can also oxidize carbon monoxide (CO), producing carbon dioxide (CO_2) as waste. These bacteria, called carboxydotrophic bacteria, probably help to maintain the relatively constant levels of carbon monoxide (CO) in the atmosphere.

CLASS BETAPROTEOBACTERIA

Like the Alphaproteobacteria, the class Betaproteobacteria contains bacteria that are diverse metabolically and morphologically. Photoautotrophs, chemolithoautotrophs, and chemoorganoheterotrophs are all found in this group, as are nitrogen-fixing bacteria. Some betaproteobacteria have interesting morphologies, such as the sheathed bacteria that produce and grow within a straw-like sleeve. Other betaproteobacteria are important ecologically and as pathogens.

SULFUR- AND IRON-OXIDIZING BACTERIA

Just as we have seen for nitrifying and hydrogen-oxidizing bacteria, some chemolithoautotrophic bacteria obtain their electrons and energy by oxidizing reduced sulfur- or iron-containing compounds. Members of the genus *Thiobacillus* can use sulfur-containing compounds such as hydrogen sulfide (H_2S) and elemental sulfur (S^0) as electron donors, producing sulfate (SO_4^{2-}) as waste. Thus, they are important contributors to the global **sulfur cycle** (Chapter 14). One species, *T. ferrooxidans*, can also oxidize reduced iron compounds and is often found growing near coal mines. The acids produced by the bacterial oxidation of the reduced sulfur compounds in coal can lead to the formation of acid rivers that damage natural ecosystems. Other species of sulfur-oxidizing bacteria include the large filamentous bacteria in the genus *Beggiatoa*, which is a member of the Gammaproteobacteriaceae.

SHEATHED BACTERIA

Sheathed bacteria like those in the genus *Sphaerotilus* are usually found in freshwater environments that are rich in organic matter and in sewage. The bacteria produce and live inside a hollow tube, as if they were jellybeans slipped into a soda straw (Figure 10.1). The bacteria divide within the sheath by binary fission, and the cells on the ends produce new sheath as the line of cells gets longer. Under certain conditions, such as low nutrient availability, the cells develop flagella and burst out of the sheath swimming away to establish themselves elsewhere.

SPIRILLUM

Members of the genus *Spirillum* are large spiral-shaped bacteria. Some of the larger species were seen and drawn by the early microscopists, including Anton van Leeuwenhoek. Although some spirilla resemble spirochetes, they are distinguished by the presence of polar flagella, or flagella that are located at the ends of the cells.

NEISSERIA

The genus *Neisseria* contains aerobic cocci that are commonly isolated from animals. In humans, *N. gonorrhoeae* causes gonorrhea, while *N. meningitidis* causes meningococcal meningitis.

REMEMBER
Members of the genus *Neisseria* are gram-negative cocci that cause significant human diseases such as gonorrhea and meningitis.

ZOOGLOEA

As members of the genus *Zoogloea* grow, they form **flocs**, slimy masses of bacteria that can float in water. This characteristic makes them very useful for the treatment of sewage (Chapter 14). The bacteria are added to sewage in large tanks that are continuously aerated. As the bacteria digest the organic matter in the sewage, they form large floating flocs, which eventually settle and form tight masses of organic matter that can be dried and then used as fertilizer or burned.

CLASS GAMMAPROTEOBACTERIA

The Gammaproteobacteria is the largest group within the Proteobacteria and includes at least 750 species. These range from phototrophic bacteria to many familiar chemoheterotrophs such as the enteric bacteria, pseudomonads, and vibrios. Many members of these groups are important human pathogens. Chemolithoautotrophs, such as the iron- and sulfur-oxidizing bacteria, are also found in the Gammaproteobacteria.

PURPLE SULFUR BACTERIA

Like the purple nonsulfur bacteria, purple sulfur bacteria like *Chromatium* and *Ectothiorhodospira* perform **anoxygenic photosynthesis** (Chapter 5). These bacteria use cyclic photophosphorylation (Chapter 5) to generate ATP, then obtain electrons by oxidizing sulfur-containing compounds such as hydrogen sulfide (H_2S). They transfer the electrons to NAPH for use in biosynthesis. As a result of H_2S oxidation, they produce elemental sulfur (S^0), which is often deposited as visible granules within the cell. By oxidizing sulfur-containing compounds, these bacteria contribute to the global **sulfur cycle** (Chapter 14).

THIOMARGARITA

In recent years, giant bacteria have been isolated from nature, challenging our ideas about how big bacterial cells can be. These bacteria are so large they can be seen with the naked eye. *Thiomargarita namibiensis* is as large as 750 micrometers (μm), or as large as a period on this page! Most bacteria are fairly small, ranging in size from 1 to 10 μm. This size limitation was thought to be because bacteria rely upon diffusion to obtain nutrients from their environment. Giant bacteria like *Thiomargarita* have provided interesting examples of how bacterial cells can solve the diffusion problem. *Thiomargarita* relies upon the diffusion of hydrogen sulfide (H_2S) from the environment into the cell so that it can oxidize the hydrogen sulfide (H_2S) for energy. However, cells of *Thiomargarita* have a large internal vacuole, much like that of a plant cell, so that the cytoplasm is restricted to the outer perimeter of the cell, and thus the hydrogen sulfide (H_2S) does not have to diffuse very far.

> **REMEMBER**
> *Thiomargarita* is one of the largest bacteria ever discovered and is visible to the naked eye.

LEGIONELLA

In 1976, a mysterious outbreak of a respiratory illness occurred during a convention of Legionnaires in Philadelphia. Attempts to culture and identify the pathogen that caused Legionnaire's disease were initially unsuccessful. Finally, special media was developed that allowed the causative agent, which was named *Legionella*, to be cultured. Since then, *Legionella* has been found to be common in aquatic environments. It also routinely colonizes air conditioning systems and water supply lines, from which it can infect humans as it did in the case of the Legionnaire's convention.

COXIELLA

Like the rickettsias, members of the genus *Coxiella* are obligate intracellular parasites of mammalian cells. *C. burnetii*, which causes Q fever, forms a resistant spore that enables it to survive outside the host, thus allowing transmission of the pathogen through the air.

PSEUDOMONAS

Species of *Pseudomonas* are well known for their ability to utilize a wide variety of organic molecules as carbon and energy sources, including things like adhesives used on bottle cap liners, soap residues, and even some disinfectants. This ability enables them to colonize many environments in hospitals where they are a significant cause of hospital-acquired, or **nosocomial infections** (Chapter 16). They are **opportunistic pathogens** that can colonize patients whose defenses are compromised by burns or invasive procedures such as catheterization. They are also a significant problem for patients with cystic fibrosis. Because of their resistance to antibiotics and disinfectants, it is difficult to control these organisms. In nature, *Pseudomonas* is common in the soil and water.

> **REMEMBER**
> *Pseudomonas* are well known for their ability to use a wide variety of carbon sources and their ability to form biofilms. These two abilities contribute to the importance of *Pseudomonas aeruginosa* as an opportunistic pathogen in hospital settings.

VIBRIO

Most members of the genus *Vibrio* are slightly curved rods that are common in aquatic systems. The most famous of the *Vibrio* is undoubtedly *V. cholerae*, the causative agent of cholera. *V. cholerae* is transmitted in contaminated water and causes severe watery diarrhea. Regular outbreaks of cholera occur in countries that do not have good water sanitation systems and can result in death due to dehydration, particularly in children.

ENTERIC BACTERIA

The **enteric bacteria** are facultatively anaerobic, rod-shaped bacteria that are common in the intestines of humans and other animals. Because there are many significant human pathogens within this group, they have been well studied, and there are many tests available for their rapid identification. One important test, which is used to separate the enteric bacteria from *Vibrio* or *Pseudomonas,* is the *oxidase test*, which tests for the presence of the enzyme cytochrome c oxidase. The enteric bacteria, which use different cytochrome oxidases, give a negative result for the oxidase test, whereas *Vibrio* and *Pseudomonas* give a positive result. Within the enteric bacteria, species are often identified by differences in biochemistry or antigens.

The most well-known enteric bacterium is probably *Escherichia coli*. Although *E. coli* is infamous for causing deadly outbreaks of severe diarrhea, these outbreaks are caused by a few particular pathogenic strains of *E. coli*. Most strains of *E. coli* are harmless to humans. In fact, *E. coli* is a normal inhabitant of the human intestinal tract where it actually benefits us by synthesizing vitamin K. *E. coli* is also used in microbiology labs around the world and has taught us a great deal about genetics. Two genera that are closely related to *Escherichia* are *Salmonella* and *Shigella*, both of which are usually pathogenic to humans. *Salmonella* species can cause salmonellosis, which causes diarrhea, and *Salmonella typhi* is the causative agent of typhoid fever.

> **REMEMBER**
> The *Enterobacteriaceae* includes many human pathogens such as *Salmonella typhi*, *Shigella dysenteriae*, and pathogenic strains of *Escherichia coli*.

Shigella dysenteriae also causes a diarrheal illness called bacillary dysentery or shigellosis.

Members of the genus *Proteus* can cause infections of wounds and the urinary tract. In culture, they exhibit an unusual colony morphology. The edges of the colonies produce **swarmer** cells that have many flagella and move rapidly away from the edge of the colony. These cells then settle down, reduce their motility, and divide. Periodically, another group of swarmer cells is produced, which creates another wave of movement away from the edge of the colony. These periodic waves of movement create a distinctive pattern of concentric rings in the colony.

Enterobacter, Klebsiella, and *Serratia* are three closely related genera that are occasionally pathogenic in humans. Species of *Enterobacter* are commonly found in water, sewage, and the intestinal tracts of animals. *E. aerogenes* causes urinary tract infections in humans. *Klebsiella* species are common in soil and water, and some are capable of **nitrogen fixation**. One species, *K. pneumoniae*, causes pneumonia in humans. *Serratia* species are found in water, soil, and the guts of various insects. *S. marcescens* makes a distinctive red pigment and can be a problem in hospitals when it contaminates catheters and saline solutions.

HAEMOPHILUS

Species of *Haemophilus* are found in the mucous membranes of the mouth, upper respiratory tract, gastrointestinal tract, and genitourinary tract. *Haemophilus influenza* is a significant human pathogen, causing earaches and meningitis in young children, as well as epiglotitis, septic arthritis, pneumonia, and bronchitis.

CLASS DELTAPROTEOBACTERIA

The Deltaproteobacteria contain some of the most unusual bacteria known to man. One of these, *Bdellovibrio*, is the vampire of the bacterial world. This bacterium burrows into the cells of other bacteria and then remains there, absorbing its victim's nutrients. After killing its victim, *Bdellovibrio* reproduces and then moves on to more prey. Another highly unusual type of bacteria found within the Deltaproteobacteria are the gliding and fruiting bacteria, such as *Myxococcus*, which have multicellular stages during their life cycle. The other major group within the Deltaproteobacteria are the sulfate-reducing bacteria, which are chemoheterotrophs that use sulfate as a final electron acceptor during anaerobic respiration.

BDELLOVIBRIO

The lifestyle of the *Bdellovibrio* is among the most unusual in the Bacteria. These small, curved rods are bacterial predators that attack and destroy other gram-negative bacteria. After cells of *Bdellovibrio* attach to their gram-negative prey, they cross the outer membrane and take up residence in the periplasmic space. Once in the periplasmic space, *Bdellovibrio* absorbs host molecules (*bdello* = leech) and then reproduces (Figure 10.3), creating a new crop of bacterial predators.

> **REMEMBER**
> *Bdellovibrio* is the vampire of the bacterial world, inserting itself inside other gram-negative bacteria and then absorbing their molecules.

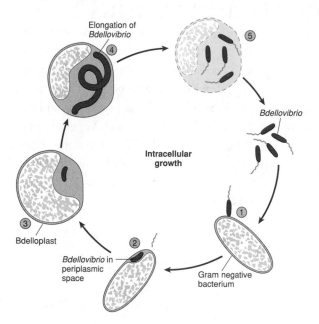

FIGURE 10.3. *Bdellovibrio*, a bacterial predator.

1. Bdellovibrio *attaches to a bacterium. 2.* Bdellovibrio *enters periplasmic space of bacterium. 3. The host bacterium forms a circular shape called a bdelloplast. 4.* Bdellovibrio *elongates as it digests its host's cytoplasm. 5. The host cell is lysed, and motile* Bdellovibrio *are released.*

SULFATE- AND SULFUR-REDUCING BACTERIA

Sulfate-reducing bacteria such as *Desulfovibrio* are common in anaerobic aquatic and terrestrial systems, such as the black, stinky mud you might find at the beach. They have the ability to use sulfate (SO_4^{2-}) instead of oxygen (O_2) as the final electron acceptor in their electron transport chains, and thus perform **anaerobic respiration** (Chapter 5). When sulfate (SO_4^{2-}), thiosulfate ($S_2O_3^{2-}$) or sulfite (SO_3^{2-}) accept electrons, they are reduced to hydrogen sulfide (H_2S). Some genera of bacteria in these environments can use elemental sulfur (S^0) instead of sulfate (SO_4^{2-}) as their electron acceptor and are therefore referred to as **sulfur-reducing bacteria**. Both sulfate- and sulfur-reducing bacteria play an important role in the global **sulfur cycle** (Chapter 14), and they may be valuable tools in biotechnology and for cleaning up pollution (**bioremediation**), like that from the mining industry. These bacteria are also part of a fascinating microbial community that lives within the crust of the Earth, where there is no light. Sulfate-reducing bacteria can obtain energy and electrons by oxidizing inorganic molecules like hydrogen, enabling them to live in the **deep biosphere** where photosynthesis can't occur.

GLIDING AND FRUITING BACTERIA

Two characteristics that are most commonly associated with bacteria are that they are single-celled and too small to see with the naked eye. However, a few species of bacteria, such as those in the genus *Myxococcus*, dare to contradict both of those trends! These bacteria have a very complex life cycle (Figure 10.4), which results in the production of a multicellular **fruiting body** that can be seen with a hand lens on the surface of decaying wood. Vegetative cells of *Myxococcus* move by **gliding** along a trail of slime that they secrete. When nutrients become low, the vegetative cells glide toward each other and join together to form a fruiting body that can produce resting structures called **myxospores**.

> **REMEMBER**
> Some bacteria, like *Myxococcus*, have multicellular stages.

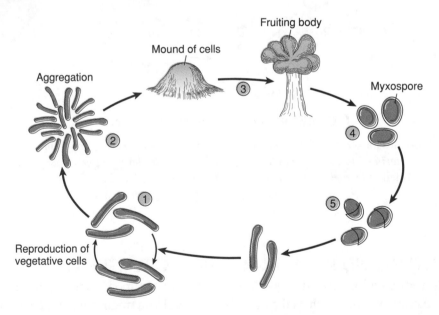

FIGURE 10.4. *Myxococcus*, a gliding and fruiting bacterium.

1. Vegetative cells reproduce. 2. Cells glide along slime trails and aggregate into a mound of cells. 3. The mound of cells differentiates into a fruiting body, and myxospores are produced. 4. Myxospores, which are resistant to drying out, are released. 5. When conditions are favorable, myxospores germinate to produce vegetative cells.

CLASS EPSILONPROTEOBACTERIA

Most members of the Epsilonproteobacteria are microaerophilic chemoheteroorgan-otrophs. They have curved or spiral morphologies. The most well-known genera, *Campylobacter* and *Helicobacter*, are both enteric pathogens.

CAMPYLOBACTER

The genus *Campylobacter* contains several species of motile, curved rods, most of which are pathogenic to humans and other animals. *C. jejuni* is frequently involved in outbreaks of foodborne intestinal disease.

HELICOBACTER

Members of the genus *Helicobacter* are motile, curved rods that have several flagella. *Helicobacter pylori* can withstand the acidic conditions of the human stomach and causes peptic ulcers.

Phylum Firmicutes (Low G+C Gram-Positive Bacteria)

Phylum Firmicutes is the largest group of known bacteria, with over 2400 identified species. These bacteria are gram-positive and include familiar human pathogens such as *Staphylococcus*, *Streptococcus*, and *Clostridium*. The gram-positive bacteria are separated into phyla based on their G+C ratios (Chapter 9), the percentage of the DNA that consists of G–C base pairs as opposed to A–T base pairs. The genera in phylum Firmicutes have low G+C ratios; for example, *Streptococcus* has a G+C ratio of 33 to 44 percent and *Clostridium* has a G+C ratio of 21 to 54 percent. These are in comparison to gram-positive bacteria in the phylum Actinobacteria, which have higher G+C ratios.

CLOSTRIDIA

The clostridia are obligately anaerobic, endospore-forming, rod-shaped bacteria that are common in soil. They lack electron transport systems and make ATP by complicated and diverse fermentative pathways. In the past, fermentation by clostridia was the major source of industrial products like acetone and butanol. Because they are endospore formers, clostridia can resist killing by heat and disinfectants. If foods are not canned properly, *Clostridium botulinum* can survive and produce botulinum toxin, resulting in botulism if the food is eaten. If *C. tetani* is introduced into deep wounds, it can cause tetanus as a result of its production of the potent toxin, tetanospasmin. Other species of *Clostridium* that are introduced into wounds, such as *C. perfringens*, cause gas gangrene. *C. perfringens* can also cause foodborne diarrhea. *C. difficile* causes diarrhea associated with antibiotic use and can spread rapidly among compromised individuals, such as elderly people in long-term care facilities.

MYCOPLASMAS

Mycoplasmas are the smallest cells that are capable of being free-living. They lack a cell wall, and as a result they are **pleomorphic**, or have variable shapes. The mycoplasmas are more resistant to osmotic lysis, or bursting, than other wall-less cells because their plasma membranes are reinforced with special molecules like sterols and lipoglycans. Morphology of the mycoplasmas ranges from coccoid and filamentous forms like those of *Mycoplasma* to those of the helical *Spiroplasma*. *M. pneumoniae* causes pneumonia in humans, and species of *Spiroplasma* are plant pathogens.

> **REMEMBER**
> The gram-positive rods in the genera *Clostridium* and *Bacillus* can form endospores, long-lived and extremely protective structures that help ensure the survival of these bacteria.

BACILLUS

Like the clostridia, members of the genus *Bacillus* are endospore-forming rods that are very common in soils. They differ from the clostridia, however, in that they are not obligate anaerobes. Several species of *Bacillus* are of medical or agricultural importance to humans. *Bacillus anthracis* causes the disease anthrax, which can affect both cattle and humans. Because *B. thuriengensis (B.T.)* produces a toxin that is effective against lepidopterans like moths, *B.T.* is used in agriculture to control crop pests. Several species of *Bacillus* produce useful antibiotics, including bacitracin and polymixin.

STAPHYLOCOCCUS

Members of the genus *Staphylococcus* typically form grape-like clusters of cocci. They can be distinguished from species of *Streptococcus* by their positive reaction to the catalase test. Many are facultative halophiles that are capable of growing in fairly salty conditions. Because of this, they can successfully colonize environments such as the human nose, human skin, and foods such as ham that have been preserved with salt. The most significant pathogen in this genus is *Staphylococcus aureus*. *S. aureus* is an opportunistic pathogen that can cause serious problems when introduced into surgical wounds. It produces toxins that are responsible for toxic shock syndrome and some types of food poisoning. Some strains of *S. aureus* are especially problematic because of their resistance to antibiotics. Because some of these strains, such as methicillin-resistant *Staphylococcus aureus* (MRSA), are very difficult to control, they can prove fatal if a serious infection is contracted.

LACTOBACILLUS

Lactobacilli are rod-shaped bacteria that produce lactic acid as a by-product of fermentation. They are normal inhabitants of the human mouth, vagina, and intestinal tract. Commercially, lactobacilli are important in the production of dairy products and sourdough bread. The next time you eat a container of yogurt, check to see whether it contains live cultures of *Lactobacillus acidophilus*. Or if you are ever in San Francisco, try some of that city's famous sourdough bread, which owes its unique tang to the fermentation by *L. sanfrancisco*.

STREPTOCOCCUS

Many members of the genus *Streptococcus* form distinctive chains of coccoid cells and can be distinguished from members of the genus *Staphylococcus* on the basis of their negative reaction to the catalase test. This genus contains some human pathogens, the most significant of which is *Streptococcus pyogenes* also known as Group A streptococcus. Depending on the strain and the site of colonization, *S. pyogenes* causes several diseases, including strep throat, scarlet fever, impetigo, rheumatic fever, and necrotizing fasciitis (flesh-eating disease). Another species, *S. mutans*, is an inhabitant of the human mouth and contributes to tooth decay.

REMEMBER
The gram-positive cocci in the genera *Staphylococcus* and *Streptococcus* have made a name for themselves as human pathogens, but many species are normal inhabitants of the human body. These two groups of bacteria cause some similar diseases and can look similar in appearance on the microscope, but they can be distinguished by the presence of the enzyme catalase; staphylococci produce the enzyme, whereas streptococci do not.

One way to distinguish species of *Streptococcus* from each other is on the basis of their reactions on blood agar. Some species, such as *S. pyogenes*, produce proteins called **hemolysins** that allow them to break down blood cells. These species will produce a clear zone around their colonies when grown on blood agar. This reaction is called **beta hemolysis**. Other species cause a discoloration in the blood cells that appears as a greenish or brownish zone around their colonies and is referred to as **alpha hemolysis**. Finally, some species produce no reaction on blood agar, which is called **gamma hemolysis**.

Species of *Streptococcus* and related cocci are also identified based on **serological tests** (Chapter 9) that look for the presence of specific carbohydrate antigens. These tests divide the streptococci into antigenic groups called Lancefield groups that are designated by the letters A through O. *Streptococcus pyogenes* is in Lancefield group A, which is why it is also referred to as Group A *Streptococcus*.

ENTEROCOCCUS

Like the streptococci, enterococci also form chains of coccoid cells and give a negative response to the catalase test. These organisms are found in the human mouth, intestinal tract, and vagina. Two species, *Enterococcus faecalis* and *Enterococcus faecium*, are responsible for many **nosocomial infections**. They are opportunistic pathogens that cause problems when they enter the body through surgical wounds or during catheterization. Because so many strains are resistant to antibiotics, treatment and control of these organisms are significant health concerns.

LISTERIA

Listeria are coccobacilli that form short chains. *L. monocytogenes* can contaminate foods like cheeses, hot dogs, and deli meats and is responsible for a type of food poisoning called **listeriosis**. Listeriosis can range in severity from a mild illness to fatal meningitis.

Phylum Actinobacteria (High G+C Gram-Positive Bacteria)

The Actinobacteria are gram-positive bacteria that have high G+C ratios. This group includes many genera that are **pleomorphic**, or have variable morphology, ranging from irregular rods to filamentous, fungal-like growth forms. Many Actinobacteria are members of soil and freshwater communities and are important decomposers. Some genera, like *Mycobacterium* and *Corynebacterium,* include pathogenic species. Representative G+C ratios for this phylum include 69 to 73 percent for *Streptomyces* and 51 to 63 percent for *Corynebacterium*.

CORYNEBACTERIUM

Cells of *Corynebacterium* are pleomorphic, often appearing as club-shaped cells that are fatter on one end than on the other. The cells may also be arranged in characteristic groups of two cells angled toward each other to form a V-shaped structure. The V-shaped forms result from a type of cell division called **snapping division**. During snapping division, the inner layer of the cell wall separates the dividing cell into two cells, but the outer layer of the cell wall remains around the cells. When the outer layer ruptures, the two cells snap apart into the V-shaped form. Some species of *Corynebacterium* are plant and animal pathogens, including *C. diphtheriae*, which causes diphtheria in humans.

> **REMEMBER**
> *Mycobacterium tuberculosis* causes tuberculosis, which spreads easily and is hard to treat, making it one of the most significant human pathogens in the world.

MYCOBACTERIUM

The mycobacteria are rod-shaped bacteria that are distinguished by their unique cell wall, the **acid-fast cell wall**. Like the cell wall of gram-negative bacteria, the acid-fast cell wall has an extra layer on the outside of the peptidoglycan (Chapter 4). In the mycobacteria, this layer is composed of a very waxy substance called **mycolic acid**. Because the waxy outer layer resists disinfectants and prevents many antibiotics from entering the cell, mycobacteria are particularly difficult to control. Two important human pathogens in this group are *Mycobacterium tuberculosis*, which causes tuberculosis, and *M. leprae*, which causes leprosy.

PROPIONIBACTERIUM

Species of *Propionibacterium* perform a unique form of fermentation that leads to the production of propionic acid. They are important in the production of Swiss cheese. The characteristic holes in the cheese are made by carbon dioxide released by the bacteria during fermentation of the cheese. The characteristic flavor of Swiss cheese is also attributable in part to the propionic acid. *Propionibacterium acnes* are found on human skin and are involved in the development of acne.

STREPTOMYCES

Streptomyces is the most famous genus in the group of bacteria called actinomycetes. These bacteria grow in long branching filaments called **hyphae** that intertwine to form a tangled mass called a **mycelium**. Because of this filamentous growth, colonies of actinomycetes often have a fuzzy or hairy appearance. Members of the genus *Streptomyces* are very common in the soil and produce a chemical called geosmin, which gives freshly turned soil its characteristic odor. They are also experts at chemical warfare and produce a variety of compounds that inhibit the growth of other bacteria. Many of these compounds, including tetracycline, streptomycin, and chloramphenicol, have proven to be very useful antibiotics. *Streptomyces* reproduces by forming new cells called **conidiospores** at the tips of its hyphae (Figure 10.5). The conidiospores break off of the hyphae and can form new colonies.

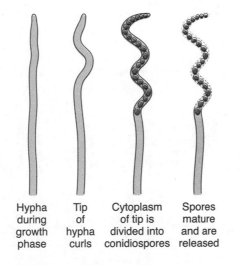

Hypha during growth phase | Tip of hypha curls | Cytoplasm of tip is divided into conidiospores | Spores mature and are released

FIGURE 10.5. Conidiospores in *Streptomyces*.

Phylum Chlamydiae

Chlamydias are gram-negative coccoid bacteria that are obligate intracellular parasites. They have a unique life cycle (Figure 10.6) that forms two different types of cells, **elementary bodies** and **reticulate bodies**. Elementary bodies are resistant to drying and allow species of *Chlamydia* to be transmitted through the air. Reticulate bodies form and multiply when the bacterium is inside a host cell.

C. psittaci causes psittacosis, a respiratory illness that is epidemic in birds and is occasionally transmitted to humans. Pneumonia in humans can also be caused by another species, *C. pneumoniae*. *C. trachomatis* causes trachoma, which leads to blindness, and can also be transmitted sexually, causing nongonococcal urethritis and lymphogranuloma venereum.

> **REMEMBER**
> *Chlamydia* is the most common sexually transmitted bacterium in the United States.

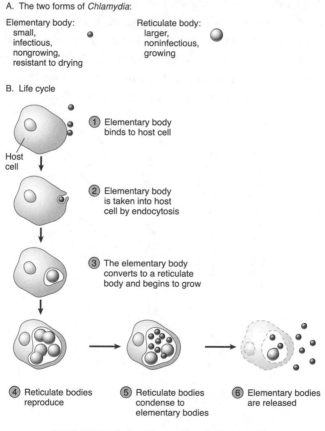

A. The two forms of *Chlamydia*:

Elementary body:
  small,
  infectious,
  nongrowing,
  resistant to drying

Reticulate body:
  larger,
  noninfectious,
  growing

B. Life cycle

Host cell

1. Elementary body binds to host cell
2. Elementary body is taken into host cell by endocytosis
3. The elementary body converts to a reticulate body and begins to grow
4. Reticulate bodies reproduce
5. Reticulate bodies condense to elementary bodies
6. Elementary bodies are released

FIGURE 10.6. Life cycle of *Chlamydia*.

A. The two forms of Chlamydia. *B. Life cycle.*

Phylum Spirochaetes

Spirochetes are tightly coiled bacteria that look like little wavy snakes. They are common inhabitants of aquatic systems and animals. They have a unique form of motility that results from an additional layer, called the **outer sheath**, which surrounds the cell, encasing both the cell and its flagella. Because the flagella, which emerge from each end of the cell, are trapped underneath the outer sheath, they are called endoflagella or **axial filaments**. When the flagella rotate underneath the sheath, they cause the cell to move like a corkscrew.

The genera *Spirochaeta* and *Cristispira* include many free-living species that are found in marine and freshwater. The genus *Treponema* includes many parasites of humans and other animals, including *T. pallidum*, the causative agent of **syphilis**. *Leptospira* includes both free-living and parasitic species. Species of *Leptospira* can spread to humans by the infected urine of cats and dogs and cause **leptospirosis**, which can lead to kidney failure. The genus *Borrelia* includes several pathogenic bacteria, such as *B. burgdorferi*, which is transmitted by the bite of ticks and causes **Lyme disease**.

REVIEW EXERCISES FOR CHAPTER 10

1. Name two important contributions of cyanobacteria to natural ecosystems.

2. Describe and give examples of the unusual morphologies found in the Proteobacteria.

3. What types of metabolism are found in the Proteobacteria?

4. Name five human pathogens that are members of the Proteobacteria. Give an example of a disease caused by each.

5. Name two genera of spore-forming bacteria found in the phylum Firmicutes. What is the significance of each genus?

6. Name three human pathogens that are members of the phylum Firmicutes. Give an example of a disease caused by each.

7. Name two genera in the Actinobacteria that are of economic significance. What is their economic value to humans?

8. Name three human pathogens that are members of the phylum Actinobacteria. Give an example of a disease caused by each.

9. Describe how chlamydias meet the definition of *obligate intracellular parasites*.

10. Describe motility in spirochetes.

SELF-TEST

1. True or False. All photosynthetic bacteria produce oxygen, just like photosynthetic plants.

2. Which of the following genera does **not** contain species that are human pathogens?

   A. *Streptococcus*
   B. *Staphylococcus*
   C. *Escherichia*
   D. *Streptomyces*
   E. *Salmonella*

3. True or False. *Legionella* was hard to culture because it is an obligate intracellular parasite.

4. Which of the following does not belong with the others?

   A. *Escherichia*
   B. *Proteus*
   C. *Shigella*
   D. *Salmonella*
   E. *Rhizobium*

5. True or False. Some species of bacteria can obtain energy by oxidizing hydrogen gas.

6. True or False. All bacteria are one of the following three shapes: rods, cocci, or spiral.

7. True or False. All bacteria have cell walls.

8. True or False. The Bacteria that are human pathogens are all very closely related to each other.

9. True or False. All motile bacteria have flagella.

10. Which of the following does not belong with the others?

    A. *Ehrlichia*
    B. *Zoogloea*
    C. *Rickettsia*
    D. *Chlamydia*
    E. *Wohlbachia*

Answers

Review Exercises

1. Cyanobacteria perform photosynthesis, fixing carbon dioxide into organic carbon that acts as food for other organisms and releasing oxygen into the atmosphere. They are also capable of fixing nitrogen from the atmosphere and converting it to forms of nitrogen that are usable by other organisms.

2. *Caulobacter* produces a stalk that it uses for attachment. *Hyphomicrobium* reproduces by budding, forming elongated cells that produce buds at their tips. Cells of *Sphaerotilus* grow inside of a sheath. *Myxococcus* cells come together to form a multicellular fruiting body that looks like a flower.

3. Members of the Proteobacteria perform fermentation, aerobic respiration, anaerobic respiration, and anoxygenic photosynthesis.

4. *Rickettsia*, Rocky Mountain spotted fever; *Neisseria*, gonorrhea; *Legionella*, legionnaire's disease; *Vibrio*, cholera; *Salmonella*, salmonellosis.

5. The two spore-forming genera in Firmicutes are *Clostridium*, which includes the causative agents of gangrene and tetanus, and *Bacillus*, which includes the causative agents of botulism and anthrax.

6. *Clostridium*, gangrene; *Staphylococcus*, toxic shock syndrome; *Streptococcus*, strep throat.

7. *Propionibacterium* is used to manufacture Swiss cheese. *Streptomyces* is used to manufacture antibiotics.

8. *Corynebacterium*, diphtheria; *Mycobacterium*, tuberculosis; *Propionibacterium*, acne.

9. Chlamydias invade human cells and reproduce inside of them. Chlamydia cannot reproduce on their own.

10. Spirochetes have flagella that are surrounded by an outer membrane. These flagella are called axial filaments. When they beat, they cause the spirochetes to wiggle like corkscrews.

Self-Test

1. F	5. T	9. F
2. D	6. F	10. B
3. F	7. F	
4. E	8. F	

The Archaea

WHAT YOU WILL LEARN

This chapter introduces you to the most recently discovered Domain of life on Earth—the Archaea. As you study this chapter, you will:

- learn how the Archaea was discovered;
- compare the Archaea with the Bacteria;
- get to know the types of organisms found in the Archaea.

SECTIONS IN THIS CHAPTER

- Phylum Crenarchaeota
- Phylum Euryarchaeota

If you stuck your hand in boiling water or a bucket of acid, the resulting damage to your hand would not feel good. Your cells cannot survive such environmental extremes. Thus, when scientists first started exploring extreme environments on the Earth like hot springs, salt lakes, or acid rivers, some of them did not expect to find any living things. Imagine their surprise when they discovered that these environments not only contained living prokaryotes, but that the prokaryotes that lived there actually *preferred* their extreme habitat and would not grow without it! Because these newly discovered prokaryotes love their extreme environments, they were nicknamed **extremophiles**. The discovery and subsequent study of these extremophiles revealed unique strategies for survival (Chapter 6) and made people who are interested in life on other planets reconsider their definition of habitable environments.

In addition to adding to our understanding of the metabolic diversity of life on Earth, the extremophiles also taught us something profound about the evolution of life. In the 1970s, a scientist named Carl Woese constructed a family tree of life on Earth by comparing the sequences of ribosomal RNA of different organisms (Chapter 9). At the time, it was thought that all prokaryotes were closely related and represented one branch on the family tree. When Woese included rRNA sequences of extremophiles into his comparisons, however, he made a startling discovery. The rRNA sequences of most extremophiles were very different from those of other prokaryotes and from those of eukaryotes. His discovery led to the creation of an entirely new branch on the tree of life, the domain Archaea.

> **REMEMBER**
> Although archaea were first discovered in extreme environments, they've since been found in all types of environments on Earth.

Since the extremophiles led us to the discovery of the Archaea, other members of the domain have been found in all types of environments, not just extreme ones. All archaea are prokaryotic cells that lack peptidoglycan in cell walls. They have fundamental chemical differences from both the Bacteria and the Eukarya (Chapter 9). Comparison of rRNA sequences suggests that there are three phyla within the Archaea, the Crenarchaeota, the Euryarchaeota, and the Korarchaeota. Very little is known about the Korarchaeota, however, so only the first two phyla will be discussed here. Table 11.1 summarizes the significant features of the genera presented.

PHYLUM CRENARCHAEOTA

The archaea that have been identified in the Crenarchaeota so far are mostly **hyperthermophiles**, bacteria that require temperatures above 80°C to grow (Chapter 6). These hyperthermophiles have been isolated from hot sulfur-rich environments like the hot springs found in Yellowstone National Park in Wyoming (United States) and from **hydrothermal vents** deep on the ocean's floor. Some members of the Crenarchaeota are **psychrophiles**, or cold-loving bacteria, that have been identified in cold ocean waters such as those of the Antarctic. Not much is known yet about the psychrophilic crenarchaeotes, so they will not be considered further here.

TABLE 11.1. *Significant Features of Selected Archaea from* Bergey's Manual of Systematic Bacteriology, 2nd edition

Phylum	Class	Important Genera	Significant Features
Crenarchaeota	Thermoprotei	*Sulfolobus*	Hyperthermophile from hot springs, oxidizes reduced sulfur and iron compounds
		Thermoproteus	Hyperthermophile from hot springs, oxidizes hydrogen
		Pyrolobus	Hyperthermophile from hydrothermal vents, grows at temperatures up to 113°C
Euryarchaeota	Halobacteria	*Halobacterium*	Extreme halophile, can produce chemical energy by absorbing light with bacteriorhodopsin
	Methanobacteria	*Methanobacterium*	Methanogen (produces methane)
	Thermoplasmata	*Thermoplasma*	Thermophile and acidophile, lacks a cell wall

HYPERTHERMOPHILES

The hot sulfur-rich environments favored by the hyperthermophiles of the Crenarchaeota are often associated with volcanoes. The archaea use the sulfur-containing compounds that are plentiful in these environments to produce chemical energy. Some **chemolithoautotrophic** hyperthermophiles oxidize reduced sulfur compounds such as hydrogen sulfide (H_2S) to supply an electron transport chain and produce ATP by chemiosmosis. Other hyperthermophiles use oxidized sulfur compounds such as sulfate (SO_4^{2-}) instead of oxygen (O_2) as an electron acceptor during the **anaerobic respiration** of organic molecules. The conversion of sulfur from one form to another by hyperthermophiles is part of the global **sulfur cycle** (Chapter 14).

Hyperthermophiles that live in terrestrial environments like hot springs may experience temperatures up to 100°C, and those that live in hydrothermal vents, where pressure keeps the water from boiling, may experience even higher temperatures! To survive in these hot environments, all hyperthermophiles must have strategies to avoid being destroyed by the heat (Chapter 6). Their proteins are sturdier, with more internal bonds, so that they don't unfold in the heat. Special molecules protect their DNA from coming apart, and their membranes are designed so that they don't melt in the heat. Some of the heat-resistant molecules produced by hyperthermophiles are being tested for their use in industrial applications.

> **REMEMBER**
> The proteins and DNA of hyperthermophiles don't denature at high temperatures, enabling hyperthermophilic archaea and bacteria to live at temperatures close to that of boiling water (100°C).

The first hyperthermophile to be discovered was *Sulfolobus,* a lobed coccus that was isolated from a sulfur-rich and acidic hot spring in Yellowstone National Park. *Sulfolobus* can function as a chemo-lithoautotroph by obtaining electrons either from reduced sulfur compounds (H_2S, S^0) or from reduced iron (Fe^{2+}) and by obtaining carbon from carbon dioxide (CO_2). It can also function as a chemo-organoheterotroph, breaking down organic molecules by aerobic respiration. The ability of *Sulfolobus* to oxidize reduced iron has made it useful in the process of **leaching**, a type of mining in which useful metals are extracted from low-grade ore.

Another hyperthermophile that grows in hot springs is *Thermoproteus*, a strictly anaerobic, rod-shaped archaeon. *Thermoproteus* can function as a chemoorganoheterotroph by using elemental sulfur (S^0) as an electron acceptor in anaerobic respiration. It can also act as a chemolithoau-totroph using hydrogen (H_2) as an electron source. The metabolic diversity exhibited by both *Thermoproteus* and *Sulfolobus* is typical of many of the hyperthermophiles.

The ability of hyperthermophiles to function as chemolithoautotrophs is especially critical to the lightless deep-sea ecosystems in which some of them are found because they form the basis for food webs. Chemolithoautotrophic archaea and bacteria utilize the sulfur-rich gases that spew out of cracks in the ocean floor to obtain electrons and produce ATP by chemiosmosis. Dissolved carbon dioxide is converted to sugar by a variety of pathways, including the Calvin–Benson cycle (Chapter 5). The chemolitho-autotrophs provide food for other organisms, enabling the existence of vent communi-ties that include worms, clams, and crabs.

Although both bacteria and archaea are present in vent communities, an archaeon, *Pyrolobus fumarii,* holds the record for growth at the highest known temperature. *P. fumarii* can grow in water up to 113°C and can survive at 121°C. It is found in the walls of *black smokers*, chimneys that form from the metals that spew out of the hot hydrothermal vents. Coccoid in shape, *P. fumarii* is an obligate chemolithoautotroph that oxidizes hydrogen gas (H_2) for ATP production.

PHYLUM EURYARCHAEOTA

Like the crenarchaeotes, many euryarchaeotes live in extreme environments. Some eur-yarchaeotes are **hyperthermophiles**, whereas others thrive in high salt environments and are called **halophiles**. Members of a third group of euryarchaeotes are called **methanogens** because they produce methane gas as a by-product of their metabolism.

EXTREME HALOPHILES

Extreme halophiles not only tolerate high salt, they require it in order to grow. These organisms are found in environments like salt lakes, soda lakes, and salt evaporation ponds. One of the main challenges to living in these environments is that the high salt concentration outside the cell could cause water to leave the cell by osmosis (Chapter 6). To counteract this, halophilic archaea stock their cytoplasm with inorganic molecules that balance the concentration of the salt outside the cell. For example, some halophiles increase the amount of potassium (K^+) inside the cell. The proteins of halophiles must also be able to withstand the high salt environment. In fact, the proteins of extreme halophiles are constructed such that they actually require postassium ions (K^+) to be stable!

> **REMEMBER**
> Extreme halophiles can protect themselves from plasmolysis by increasing the amount of potassium ions (K^+) in their cytoplasm.

Most extremely halophilic euryarchaeotes are chemoorganoheterotrophs that break down organic molecules by aerobic respiration, although a few strains can perform fermentation or anaerobic respiration. Some members of the genus *Halobacterium*, rod-shaped bacteria found in salt lakes, also have the ability to make ATP using energy from the sun. These bacteria have a special pigment, **bacteriorhodopsin**, which is similar to the rhodopsin found in the human eye. When bacteriorhodopsin absorbs light energy, it changes shape, causing hydrogen ions (H^+) to be pumped to the outside of the plasma membrane of the *Halobacterium*. The hydrogen ions then pass back across the membrane through ATP synthase, generating ATP (Figure 11.1). This process is not considered photosynthesis because it does not involve chlorophyll or an electron transport chain, and, like all extremely halophilic euryarchaeotes, *Halobacterium* obtains its carbon by oxidizing organic molecules. It is thus considered a **photoheterotroph**.

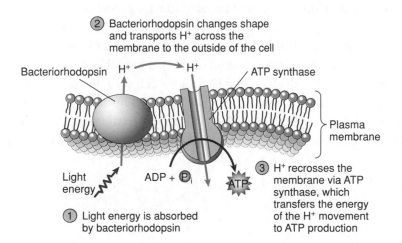

FIGURE 11.1. Chemiosmosis in Haloarchaea.

METHANOGENS

The methanogens are a diverse group of anaerobic euryarchaeotes that produce methane (CH_4) as a by-product of their metabolism. They are found in sediments, the intestinal tracts of animals, sewage treatment facilities, bogs, and deep soils. Also, many methanogens are thermophiles that are found in environments like hydrothermal vents. These archaea exhibit different morphologies, including rods, cocci, spiral, and irregular forms, and have chemical differences in their cell wall construction. The reactions that produce methane differ between the methanogens, but all are coupled to ATP production by chemiosmosis. In the genus *Methanobacterium*, electrons from hydrogen (H_2) are used to generate a proton and sodium motive force during the reduction of carbon dioxide (CO_2), resulting in methane production (CH_4) by the following reaction:

$$CO_2 + 4H_2 \rightarrow CH_4 + 2H_2O$$

> **REMEMBER**
> Methanogens produce methane as a byproduct of their method for making ATP. Biologically produced methane is a component of biogas, which is produced during the decomposition of human wastes and agricultural materials. Biogas is a useful alternative energy source for some populations.

The proton and sodium motive force is also used to generate ATP. Methane production by methanogens in the past produced the natural gas that is used for energy by humans. Some current sources of methane, such as that produced by methanogens at sewage treatment plants, is trapped and used for energy on a small scale. Methane production by the methanogens that live in the rumen of cows contributes significantly to the greenhouse gases in our atmosphere.

THERMOPHILES AND EXTREME ACIDOPHILES

In addition to the thermophilic methanogens, the Euryarchaeota contains other thermophilic archaea that are also extreme acidophiles (Chapter 6). The genus *Thermoplasma*, which has been isolated from coal refuse, grows best at temperatures around 55°C and at a pH of 2. A unique feature of this genus is that it lacks a cell wall. Like the mycoplasmas in the Bacteria that also lack a cell wall, the plasma membrane of *Thermoplasma* is chemically reinforced so that it can resist lysis. *Thermoplasma* and other thermophilic euryarchaeotes also have characteristics similar to those of the hyperthermophilic crenarchaeotes that enable them to survive high temperatures.

REVIEW EXERCISES FOR CHAPTER 11

1. Describe some of the characteristics that the hyperthermophilic members of the Crenarchaeota have in common.

2. What is bacteriorhodopsin? What is its importance to *Halobacterium*?

3. How do methanogens produce methane?

4. Describe the characteristics of *Thermoplasma*.

SELF-TEST

1. Which of the following is not a thermophile?

   A. *Sulfolobus*
   B. *Pyrolobus*
   C. *Halobacterium*
   D. *Thermoproteus*
   E. *Thermoplasma*

2. True or False. *Halobacterium* performs photosynthesis using bacteriorhodopsin instead of bacteriochlorophyll.

3. True or False. All archaeons are found in extreme environments.

4. True or False. Methanogens oxidize methane for energy.

5. True or False. Some archaeans lack a cell wall.

6. True or False. Extreme halophiles cannot survive outside of high salt environments.

7. True or False. Some archaeans are psychrophiles.

8. Which of the following cannot obtain energy by oxidizing inorganic molecules?

   A. *Sulfolobus*
   B. *Thermoproteus*
   C. *Pyrolobus*
   D. *Halobacterium*
   E. All of the above are correct.

9. Which of the following requires organic carbon?

    A. *Sulfolobus*
    B. *Thermoproteus*
    C. *Pyrolobus*
    D. *Halobacterium*
    E. All of the above are correct.

10. True or False. Methanogens can be found in the rumen of cows.

Answers

Review Exercises

1. The hyperthermophiles in the phylum Crenarchaeota are resistant to high temperatures, have sturdy heat-resistant proteins, and protect their DNA with special binding proteins. Many of them can switch between chemolithoautotrophy and chemoorganoheterotrophy.

2. Bacteriorhodopsin is a bacterial protein that is very similar to rhodopsin, a light-absorbing pigment in the eye. When it absorbs light, it changes shape. It is found in the plasma membrane of *Halobacterium*, where it absorbs light energy and uses that energy to transport hydrogen ions (H^+) outside the plasma membrane. The hydrogen ion gradient is then used to synthesize ATP.

3. Methanogens oxidize hydrogen gas and transfer the electrons to carbon dioxide, reducing it to methane.

4. *Thermoplasma* lacks a cell wall. It is a thermophile whose plasma membrane is reinforced to help it resist lysis. It is also an acidophile and can tolerate acidic pH.

Self-Test

1. C	5. T	9. D
2. F	6. T	10. T
3. F	7. T	
4. F	8. D	

Eukaryotic Microorganisms

WHAT YOU WILL LEARN

This chapter explores the world of eukaryotic microorganisms. As you study this chapter, you will:

- be introduced to the groups of eukaryotic microorganisms;
- learn about the diversity of the Protista;
- discover the differences among the algae;
- examine the life cycles of fungi.

SECTIONS IN THIS CHAPTER

- Protista
- Fungi

If you have ever looked at a drop of pond water under a microscope and seen little creatures zooming around, you were probably looking at eukaryotic microorganisms. Although prokaryotes would certainly be present in a drop of pond water, their small size makes them more difficult to see without staining the sample. Because they are larger, eukaryotes such as protozoa, algae, and even small crustaceans would be much easier to see. If you were to investigate the organisms in soil, you would also find protozoa along with fungi and small worms. Eukaryotic microorganisms are as diverse as they are plentiful, and they are an important part of the communities to which they belong.

Protozoa, algae, crustaceans, worms, and fungi all belong to the domain Eukarya, which is currently divided up into four kingdoms (Chapter 9): Plantae, Animalia, Fungi, and Protista. The crustaceans and worms are members of the kingdom Animalia. Even though they are small, these organisms are not typically considered to be microorganisms and so will not be considered here. Kingdom Fungi includes molds, yeasts, and mushrooms. Kingdom Protista is made up of a large and very diverse group of organisms that do not fit into any of the other three eukaryotic kingdoms. As we examine the major groups of eukaryotic microorganisms (Table 12.1), the emphasis will be placed upon representative organisms and those that have medical or economic importance.

TABLE 12.1. Significant Features of Selected Groups of Eukaryotic Microorganisms.

Kingdom	Common Name*	Important Genera	Significant Features
Protista	Amoebomastigotes (amoeboflagellates)	*Naegleria*	Transform between flagellated cells and amoebae; some are parasitic
	Pelobiontids	*Pelomyxa*	Amoebae; lack mitochondria and other organelles, symbiotic bacteria live within them
	Amastigote amoebae	*Amoeba, Entamoeba*	Amoebae; *Entamoeba* causes amebic dysentery
	Diplomonads	*Giardia*	Lack mitochondria; form cysts; cause hiker's diarrhea
	Parabasalids	*Trichonympha, Trichomonas*	Lack mitochondria; have up to hundreds of flagella; form symbiotic associations with animals
	Kinetoplastids	*Trypanosoma*	Have an enlarged mitochondrion and one to two flagella; some are parasitic
	Euglenids	*Euglena*	Photosynthetic; have a protein coat called a pellicle; can move by a wriggling movement called metaboly

TABLE 12.1. (continued)

Kingdom	Common Name*	Important Genera	Significant Features
Protista	Ciliates	*Paramecium*	Have cilia and two types of nuclei
	Foraminifera	*Rotaliella*	Amoebae that produce a shell called a test; shells are present in fossil record and sediments associated with oil deposits
	Dinoflagellates	*Gonyaulax, Pfiesteria*	Photosynthetic, have a test and two flagella; cause red tide and paralytic shellfish poisoning
	Apicomplexans	*Plasmodium, Toxoplasma, Cryptosporidium*	Obligate parasites; complicated life cycle sometimes involving more than one host; some are parasitic in humans
	Oomycetes	*Phytophthora*	Fungal-like decomposers; produce motile cells called zoospores; some are parasitic
	Diatoms	*Thalassiosira*	Photosynthetic; contain chlorophylls a and c, carotene, and xanthophylls; produce a shell called a frustule that is made of silica
	Brown algae	*Laminaria*	Photosynthetic; contain chlorophylls a and c, as well as xanthophylls; among the largest of the algae
	Red algae	*Porphyra*	Photosynthetic; contain chlorophylls a and d, as well as phycobilins; important food source
	Green algae	*Chlamydomonas, Ulva*	Photosynthetic; contain chlorophylls a and b; gave rise to the land plants
	Cellular slime molds	*Dictyostelium*	Individual amoebae come together to form a giant slug that migrates and then develops into a sporangium

TABLE 12.1. (continued)

Kingdom	Common Name*	Important Genera	Significant Features
Protista	Acellular slime molds	*Physarum*	Feeding stage is a giant multinucleate amoeba called a plasmodium; sporangia are formed when food is scarce; useful for studies on muscle function
Fungi	Zygomycetes	*Rhizopus*	Asexual sporangiospores produced in a sporangium; sexual reproduction produces zygospores
	Ascomycetes	*Penicillium, Aspergillus, Saccharomyces, Morchella, Cryphonectria, Ceratocystis*	Asexual conidiospores produced in conidia; sexual ascospores produced in an ascus
	Basidiomycetes	*Agaricus, Lentinus, Amanita, Psilocybe*	Asexual reproduction by fragmentation; sexual basidiospores produced in a basidium

*Because the classification of protists is still being debated, common names rather than taxa are given.

Protista

You are probably aware of the many remarkable animals on planet Earth, from the bald eagle to bizarre deep-sea fish. As strange and wondrous as the animals, the **protozoa** have just as much power to fascinate on a microscopic level. The term "protozoa" refers rather loosely to those members of the kingdom **Protista** that are unicellular, lack a cell wall, and are generally not photosynthetic. This includes the fast-swimming ciliates, the blob-like amoebae, the spinning dinoflagellates, and many other unique creatures. Traditionally, the protozoa were seen as distinct from the **algae**, photosynthetic organisms that range in size from single cells to giant kelp.

Protozoa and algae are not taxonomic terms; that is, they do not refer to organisms that are closely related to each other. They date back to the time when all organisms were

> **REMEMBER**
> Protozoa are eukaryotic microbes that are predatory and not photosynthetic. Algae are eukaryotes that are photosynthetic, may be single-celled or multicellular (like seaweeds, for example), but aren't considered true plants.

divided up into two kingdoms: the animals, which moved and ate others, and plants, which did not move and performed photosynthesis. Because these terms are a convenient way to describe the overall lifestyle of an organism, they are still in use.

The protozoa are such a diverse group of organisms that it is difficult to make generalizations about them. Most are motile, some by cilia or eukaryotic flagella (Chapter 4), some by a crawling type of movement called **amoeboid movement**. Most live in aquatic environments where they feed upon bacteria or dissolved particulate matter. Food is either transported across the membrane, or taken in by some form of **phagocytosis** (Chapter 18), in which they wrap their membrane around the food and drag it into the cell for digestion. Many protozoa are symbiotic in the guts of animals, and some are parasitic. Most have a means of **asexual reproduction** in which the parent cell makes two exact copies of itself, and some have the ability to perform **sexual reproduction** during which genes are exchanged between two individuals. Asexual or sexual reproduction can lead to the production of many **spores**, sturdy cells that are easy to disperse. Some protozoa have the ability to make a resting cell called a **cyst**, which may enable survival of the individual in adverse conditions.

In contrast to the motile protozoa, the algae, many of which are nonmotile or have flat leaf-life structures, seem more like plants. The one characteristic that all of the algae share is the presence of a chloroplast. Even these chloroplasts show differences among the algae: although all algae contain the pigment chlorophyll a, the different types of algae contain various other pigments that give them their unique colors. Because they perform photosynthesis (Chapter 5), algae are incredibly important in marine and freshwater ecosystems. They use energy from sunlight and carbon dioxide to make the organic molecules that other organisms can eat, thus forming the basis for most food chains. In addition, more than half of the oxygen in the Earth's atmosphere is a by-product of photosynthesis. Because the various types of algae are not all closely related, the various groups of algae will be discussed along with those of the protozoa according to current ideas about protist phylogeny.

The classification of the protists has changed many times, most recently as a result of comparisons of genes for rRNA and proteins (Chapter 9). The relationships among many of the protists are still not well understood, and there isn't an agreed upon classification system for this group. This text will follow one of the most recent organizations, which is presented by the University of California at Berkeley's Museum of Paleontology (*www.ucmp.berkeley.edu*). Current information on the status of protist phylogeny is also available on the Tree of Life web (*www.tolweb.org*).

PRIMITIVE FLAGELLATES, AMOEBAE, AND PARASITIC TAXA

When the phylogenetic tree of life is drawn based on rRNA comparisons, several groups of eukaryotes appear very close to the bottom of the tree. Thus, these groups may represent the descendants of our very earliest eukaryotic ancestors. They are sometimes referred to as *primitive eukaryotes* to reflect their ancestry.

AMOEBAE

Amoebae may be the most famous protists. They star in many *The Far Side* cartoons, may have inspired movies like *The Blob,* and even have a line in the song "Thing Called Love" which says, "We can choose, you know we ain't no amoebas." From these sources, you may already know that amoebae are irregularly shaped cells (Figure 12.1) that seem to change shape and crawl around. When an amoeba moves, parts of the cell project outward forming **pseudopods**, and then the rest of the cell flows toward the pseudopods. This crawling type of movement is called **amoeboid movement**.

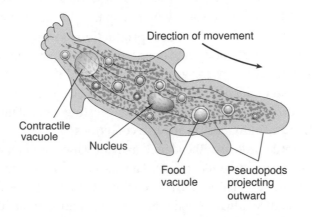

FIGURE 12.1. *Amoeba* and amoeboid movement.

> **REMEMBER**
> Amoebae move by amoeboid movement, extending their pseudopods, then pulling the rest of the cell along to catch up.

Several types of protozoa from different lineages have amoeboid stages, so the presence of amoebae does not indicate a close relationship. Within the primitive eukaryotes, amoeboid stages are found within the **amoebomastigotes**, which transform between small flagellated cells and amoebae depending on environmental conditions. When in a liquid, nutrient-poor environment, they grow flagella and elongate into a swimming form. Once they encounter food such as bacteria, they retract their flagella, return to the amoeboid form, and begin feeding. One species, *Naegleria fowleri,* can cause serious brain infections and death in humans.

The **pelobiontids** are also amoebae that appear to form an early branch in the eukaryotic lineage. These giant amoebae can be up to 1 millimeter in length, making them visible to the naked eye. They are different from all other known eukaryotes in that they lack mitochondria, endoplasmic reticulum, Golgi bodies, or mitotic spindle (Chapter 4). The most intensely studied species is *Pelomyxa palustris*, which was isolated from the mud at the bottom of a pond in England. *P. palustris* contains three different types of symbiotic bacteria that appear to participate in its metabolism almost like mitochondria. If antibiotics are used to kill its symbionts, *P. palustris* dies also.

The phylogeny of another group of amoebae, the **amastigote amoebae**, is not yet clear, but they will be included here because of their superficial resemblance to other amoebae. This group includes free-living species such as *Amoeba proteus* which are common in freshwater and marine environments, as well as parasitic species such as *Entamoeba histolytica*. *E. histolytica* can colonize the human intestine, where it binds to intestinal cells and causes ulcers, resulting in the diarrheal disease **amoebic dysentery**. The amoeba forms cysts that are transmitted to new hosts by fecal contamination of food or water.

DIPLOMONADS

Diplomonads typically have two eukaryotic flagella that are attached to the front end of the cell. They are one of the few types of eukaryotic cells that lack mitochondria. At one time, it was thought that these cells might be descendents of very ancient eukaryotes that existed before endosymbiosis created mitochondria (Chapter 4), but recent evidence suggests that these eukaryotes may be descended from cells that lost their mitochondria. The diplomonad *Giardia lamblia* causes **hiker's diarrhea** when it colonizes the human intestine. Cysts that exit in the feces of the host and survive in the environment transmit the parasite. The disease is often acquired when people drink untreated water from streams or lakes.

PARABASALIDS

> **REMEMBER**
> Diplomonads and parabasalids lack mitochondria.

The flagellated cells of **parabasalids** (Figure 12.2) are very similar to those of the diplomonads, including the lack of mitochondria. However, parabasalids may have anywhere from four to hundreds of flagella and are not capable of forming cysts. They are always found in symbiotic association with animals. Symbiotic species such as *Trichonympha ampla* are part of the thriving microbial community found in the hindgut of termites. In one of the most fascinating examples of symbiosis known in the natural world, parabasalids like *Trichonympha* live inside of termites, while symbiotic bacteria live inside *Trichonympha*! In fact, it is the symbiotic bacteria inside *Trichonympha* that produce enzymes to break down cellulose, making it possible for the termites to digest wood. To add another layer to this complex web of relationships, some parabasalids also have spirochetes that attach to the outside of their cells and act like extra flagella! Another species of parabasalid, *Trichomonas vaginalis*, is a human parasite that can be transmitted by sexual intercourse.

KINETOPLASTIDS

The **kinetoplastids** include free-living and parasitic cells that have one or two flagella and a large mitochondrion called the kinetoplast. *Trypanosoma brucei gambiense* is parasitic in humans, causing African sleeping sickness, a chronic and usually fatal disease. Like other trypanosomes, *T. brucei gambiense* is an elongated, crescent-shaped cell that has one flagellum. The flagellum folds back along the cell and is enclosed under a flap of membrane. When the flagellum beats, the **undulating membrane** ruffles along with it, helping to propel the organism through thick solutions like human blood, where the parasite lives and grows. In the later stages of infection, the parasite also invades the central nervous system. It is passed from host to host by the bite of the tsetse fly.

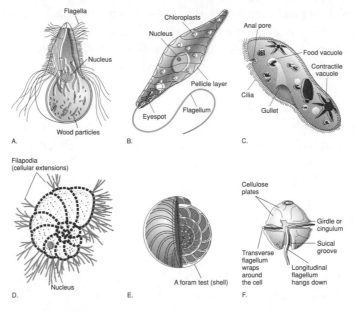

FIGURE 12.2. Eukaryotic microorganisms.

A. Trichonympha, *a parabasalid. B.* Euglena, *a euglenoid. C.* Paramecium, *a ciliate. D. A foraminiferan. E. A foram test (shell). F.* Gonyaulax, *a dinoflagellate.*

EUGLENIDS

Because some **euglenids** contain chloroplasts, they are sometimes included along with the algae. However, based on comparisons of their rRNA and disk-shaped mitochondria, they are actually more closely related to the kinetoplastids. In fact, one species, *Euglena gracilis*, can survive as a heterotrophic organism after its chloroplast has been destroyed. Euglenids are most commonly found in freshwater, but some live in saltwater or are parasites of animals. Most contain two flagella, one that projects out from the top of the cell and one, called the preemergent flagellum, that is contained within the cell (Figure 12.2). The preemergent flagellum is associated with a cluster of pigments called the **eyespot**.

> **REMEMBER**
> Euglenids can move by swimming with their flagella or by a unique wriggling motion called metaboly.

Light signals received by the eyespot direct euglenids toward the light. In addition to swimming with their flagella, euglenids can also move by a strange wriggling motion called **metaboly**. Metaboly is possible because euglenids have a flexible protein coat called a **pellicle** that compresses and changes the shape of the cell.

ALVEOLATES

The **alveolates** are a recently proposed group of protists that are united, in part, because they all have a system of sacs, called **alveoli**, under their cell membranes. This group includes the ciliates, foraminifera, dinoflagellates, and apicomplexans.

CILIATES

The **ciliates** are an enormous group of protists, with over 7000 identified species. They are found in marine, freshwater, and terrestrial environments, and as symbionts in the guts of animals. Although some species have unusual morphologies, many ciliates are similar to those in the genus *Paramecium* (Figure 12.2), which are generally oval cells that are covered with **cilia** (Chapter 4). They are unusual because they contain two types of nuclei: the **micronuclei**, which function in reproduction, and the **macronuclei**, which function in the protein synthesis (Chapter 8) needed for the day-to-day maintenance of the cell. They feed upon bacteria or organic matter that enters through a groove-like mouth called a **cytostome**. At the base of the mouth, the food is taken into the cell in food vacuoles formed by **phagocytosis**. Some ciliates form **cysts**, and many contain endosymbiotic bacteria.

> **REMEMBER**
> Foraminifera live in shells of calcium carbonate, called tests. Foram tests can be found in the fossil record and make up the famous White Cliffs of Dover.

FORAMINIFERA

The **foraminifera**, or forams, are amoeboid cells that live within a shell called a **test**. The shells of forams are made of minerals like calcium carbonate, which resists decay and accumulate in the environment. The famous White Cliffs of Dover in England are made of foram shells, and there are many examples in the fossil record. If you were to examine some beach sand under a microscope, you would be able to see them too, looking like teeny, tiny, seashells. They are often present in sediments that cover oil deposits and so are used to locate new sources of oil. When forams like *Rotaliella* are alive, extensions of the cell called filopodia extend out through holes in the test (Figure 12.2) to gather food or to swim.

DINOFLAGELLATES

Dinoflagellates are probably best known as the organisms that cause **red tide**. They are photosynthetic and part of the marine **plankton**, small organisms that float in the sea. They produce a distinctive **test** and have two flagella, one that wraps around the cell and one that hangs off the cell (Figure 12.2). When the flagella beat, dinoflagellates swim in a distinctive spinning fashion. In addition to chlorophyll, they contain the accessory pigment **xanthophyll**, which gives the cells a red color that can actually be seen in the water when there is a bloom, or increase, in the dinoflagellate population.

> **REMEMBER**
> Some dinoflagellates, including those that cause red tide, produce toxins that can kill people and fish.

The number of dangerous blooms of dinoflagellates has increased in recent years, perhaps in response to pollution of marine waters or global warming: Dinoflagellates grow well in warm, nutrient-rich water. Red tide is dangerous because dinoflagellates such as *Gonyaulax* produce powerful neurotoxins. These neurotoxins can kill fish or marine mammals that come into contact with the dinoflagellates. If people eat fish or shellfish that is contaminated with toxin-producing dinoflagellates, it causes **paralytic shellfish poisoning**. One fish-killing dinoflagellate, *Pfiesteria piscicida*, produces a neurotoxin that is so potent it can cause symptoms in people who simply touch contaminated fish.

APICOMPLEXANS

The Apicomplexa is a large group of obligate parasites that have very complex life cycles. By far, the most infamous apicomplexans are *Plasmodium vivax* and *Plasmodium falciparum*, two of the species of *Plasmodium*, which cause malaria. These organisms require two hosts, humans and the *Anopheles* mosquito, to complete their life cycle (Figure 12.3). They undergo sexual reproduction in the mosquito, producing infective cells called **sporozoites** that travel to the salivary gland of the mosquito. When the mosquito bites a human, it injects sporozoites into the blood. The liver removes sporozoites from the blood, and the sporozoites infect the liver cells, where they reproduce asexually to produce another type of cell called a **merozoite**. Merozoites leave the liver and infect red blood cells. Inside the red blood cells, the merozoites enlarge and form a **trophozoite**, or **ring cell**, in which the cytoplasm looks like a ring around the nucleus of the cell. The trophozoite fragments into more merozoites, which can infect more blood cells. This cycle of merozoite production and release repeats at regular intervals, causing the distinctive fever cycle of malaria. Some of the cells produced by the trophozoite are not merozoites, but rather **gametocytes**. These gametocytes cannot attack new red blood cells. Instead, they are picked up when a mosquito bites an infected person. Inside the mosquito, the gametocytes participate in sexual reproduction, producing more sporozoites, and the entire life cycle can begin again. Other apicomplexans that cause significant disease are *Toxoplasma gondii*, which causes **toxoplasmosis**, and *Cryptosporidium*, which causes respiratory and gallbladder infections in people with suppressed immune systems.

> **REMEMBER**
> Malaria is caused by members of the genus *Plasmodium*, which is within the Apicomplexan group.

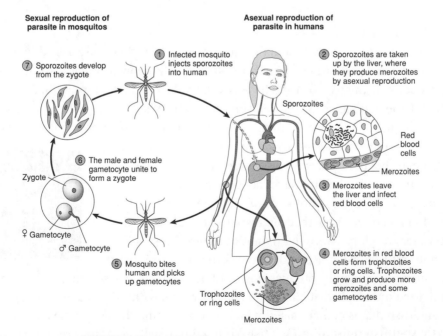

FIGURE 12.3. The life cycle of a malarian parasite.

CHROMISTA

The **Chromista**, or Stramenopiles, is a large group of protists that contains protozoa and algae, as well as the water molds, which were once thought to be closely related to the fungi. These organisms, once considered very different from each other, have been united by both rRNA sequence comparisons and also by the presence of very distinctive hairs on the flagella.

OOMYCETES

The oomycetes are also called water molds, white rusts, and downy mildews. Many of them live in freshwater or soil, where they decompose dead matter. Some species are parasites, such as *Phytophthora infestans,* which caused the Irish potato famine in the 1800s. Oomycetes look and act like fungi, growing as masses of white filaments, or **hyphae**, that release enzymes into organic matter and then absorb the nutrients. However, oomycetes reproduce asexually by producing flagellated **zoospores** in the tips of the hyphae, whereas fungi never produce flagellated cells. Sexual reproduction can also occur in oomycetes when hyphae become specialized to produce sperm or eggs. The fusion of sperm and egg results in formation of **oospores,** which can survive low nutrients and the lack of water.

DIATOMS

Diatoms are photosynthetic cells that live in little glass boxes that come in a beautiful array of shapes. The boxes are their shells, or **frustules**, which are composed of silica and formed by two halves that come together just like a box. Some diatoms, like *Thalassiosira*, are colonial, with many individual cells held together in chains. In addition to chlorophyll a, the chloroplasts of diatoms have chlorophyll c, carotene, and xanthophylls. These pigments, particularly the yellow xanthophylls, combine to color diatoms in a range of pigments from green to golden-yellow. Because their frustules are resistant to decay, diatoms have an excellent fossil record extending back 200 million years. **Diatomaceous earth**, which is composed of diatom frustules, is used in pool filters, as abrasives, and in the production of chalk. Diatoms produce a toxin called **domoic acid**, which can cause **domoic acid intoxication** in people who eat shellfish contaminated with diatoms.

BROWN ALGAE

The **brown algae**, which are found mostly in marine intertidal environments, are the largest of all the algae. They include the giant multicellular kelps, which can be up to 100 meters long and form vast underwater forests that are home to many marine animals. Brown algae are brown owing to their combination of the green chlorophylls a and c, as well as the yellowish xanthophylls. Many brown algae have structures that appear similar to those of plants, such as the leaf-like **blades**, stem-like **stipe**, and root-like **holdfast** (Figure 12.4). The holdfast secretes a sticky substance that allows the algae to anchor themselves to rocks and other substrates. In water, the leaf-like blades spread out to maximize their area for photosynthesis. In giant kelp, like *Laminaria*, an air bladder is located next to the blades so they will float near the surface of the water where there is plenty of sunlight (Figure 12.4). The brown algae are a source of the chemical **algin**, which is used as a thickener in ice cream, toothpaste, and other products.

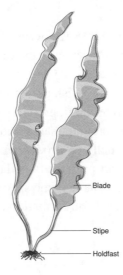

Blade

Stipe

Holdfast

FIGURE 12.4. *Laminaria*, a brown alga.

RED ALGAE

The **red algae** form a distinct branch of the eukaryotic tree of life. In addition to the pigments chlorophylls a and d, they contain **phycobilins**, red and blue pigments that are also found in the cyanobacteria (Chapter 10). This combination of pigments color the various red algae in a range from pink to almost black. Because the phycobilins absorb the wavelengths of light that readily penetrate water, red algae can live deeper in the oceans than any other type of algae. Red algae are also distinct in that, unlike the other algae, they lack any flagellated stages. Red algae can be single-celled or multicellular, and they range in appearance from small filaments to large plant-like organisms. One type of red algae, the **coralline red algae**, secretes a calcium carbonate shell. These algae are very important members of coral reefs.

Red algae are of economic importance to humans in several ways. *Porphyra*, also called **nori**, is an important food source for many cultures. Because of this, it is farmed in many parts of the world. **Agar**, the thick-

> **REMEMBER**
> Brown, red, and green algae all contain the green pigment chlorophyll a. The different colors in the algae are due to their additional pigments. Brown algae contain chlorophyll c and the xanthophylls (yellow); red algae contain the phycobilins phycocyanin (blue) and phycoerythrin (red); and green algae contain chlorophyll b (green).

ener used in many types of microbiological media, is extracted from red algae, as is **carageenan**, a thickener used in foods and pharmaceutical agents.

GREEN ALGAE

The **green algae** are part of a lineage of eukaryotes that includes the green land plants. All of these organisms have chlorophylls a and b as their main photosynthetic pigments, produce cell walls made of cellulose, and store the products of photosynthesis as starch. Both structural evidence and comparisons of rRNA demonstrate the close relationship between the green algae and land plants, and it is commonly accepted that land plants evolved from ancestral species of green algae.

Today, the green algae consist of marine and freshwater organisms that range from single cells, like *Chlamydomonas*, to large multicellular seaweeds, like the sea lettuce *Ulva*. The smaller members of the green algae are important components of the plankton and form the basis for many food chains. Larger green algae, such as *Ulva*, are used as a food source in some parts of the world.

SLIME MOLDS

The slime molds are two interesting groups of organisms that have both fungal and amoeboid characteristics. Like fungi, they produce a stalked spore-producing structure called a **sporangium**. However, like protozoa, they can move rather quickly. Both types of slime molds are common on decaying plant matter, such as old logs and dead leaves. Although there is a superficial resemblance between the two groups of slime molds, they are not closely related.

CELLULAR SLIME MOLDS

The cellular slime molds, like *Dictyostelium discoideum*, spend most of their lives looking and acting like other amoebae. When they run out of food, however, an entire population comes together to form a giant slug that crawls as one unit and then settles down and develops into a stalked **sporangium** (Figure 12.5). The amoebae that run out of food first start this process by releasing a signaling molecule that attracts the other amoebae. When the amoebae merge into the giant slug, they move as one, but they keep their individual membranes. When the slug stops moving, they crawl over each other to form the different sections of the sporangium. Some amoebae form the base, others crawl up and become the stalk, while still others form the cap. The amoebae in the stalk secrete cellulose, which provides strength to the structure. The amoebae in the cap undergo **asexual reproduction** to form **spores**. The spores are dispersed to new environments where each can become a new amoeba.

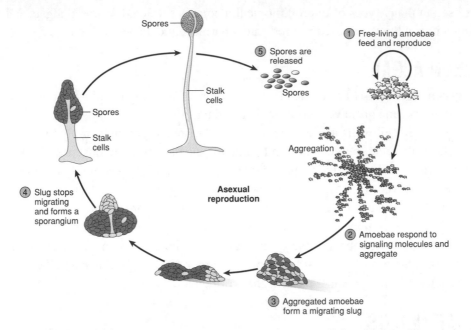

FIGURE 12.5. The life cycle of *Dictyostelium*, a cellular slime mold.

PLASMODIAL SLIME MOLDS

The **plasmodial slime molds** have an amoeboid feeding stage called a **plasmodium**. The plasmodium differs from other amoebae in that it is a mass of cytoplasm with many nuclei. The plasmodium moves by amoeboid movement, eating bacteria and organic matter. If food becomes scarce, the large plasmodium separates into many clumps of plasmodia, each of which forms a **sporangium** (Figure 12.6). Cells in the cap of the sporangium begin the process of **sexual reproduction** and form spores that contain the genetic equivalent of sperm or egg cells. The spores disperse to new environments and then release special cells called **gametes**, which have flagella and swim to find each other. When two gametes find each other, they fuse together to form a single cell. This cell develops into the feeding plasmodium, and the cycle begins again.

The plasmodium of plasmodial slime molds can get quite large; a particularly big one that made the news in Texas in 1943 was measured at 46 centimeters, which is a little bit over 18 inches! Like other amoebae, plasmodia move by **amoeboid movement**. This type of movement is possible because of interactions of the cytoskeletal proteins actin and myosin (Chapter 4), the same proteins that enable muscle contraction. Because of its large size and the presence of these proteins, the plasmodial slime mold *Physarum* has been very valuable in laboratory studies on actin and myosin. This has contributed greatly to our understanding of muscle function.

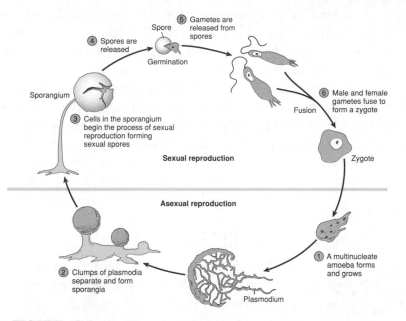

FIGURE 12.6. The life cycle of *Physarum*, an acellular slime mold.

FUNGI

Fungi are called many things, including molds, rusts, smuts, yeasts, and mushrooms. They play an important role in the environment, have economic value in the production of food, cause human disease, and produce antibiotics. In nature, fungi function as **decomposers**, breaking down dead matter and recycling nutrients. They are heterotrophs that typically grow in thin filaments called **hyphae** that have cell walls made of chitin. A mass of hyphae forms the body of the fungus, which is called a **mycelium**. The hyphae grow into their food source, secreting enzymes that break it down into smaller molecules. The small molecules can then be transported across the fungal membrane. When food is scarce, the hyphae develop into **sporangia**, producing **spores** for dispersal. Next time you see mold on your cheese, look at it closely. You will probably see some whitish growth around green colored spots. The whitish growth is made of hyphae that are spreading through your cheese, eating it before you can. The greenish spots are areas where spore production has occurred.

> **REMEMBER**
> Fungi are important decomposers, breaking down dead organisms and recycling matter so that it can be used again by living things.

Another important role of fungi in nature is as **mycorrhizae**, symbiotic partners with plants. Mycorrhizal fungi form associations with plant roots, colonizing the root exterior and spreading out into the soil. The fungi help the plants grow by gathering nutrients like nitrates and phosphates from the soil and transferring it to the plants. A single mycorrhizal fungus can spread out over a period of many years to form an enormous web of hyphae through the soil. One remarkable fungal clone measured in Washington covered over 1500 acres!

The fungi represent a eukaryotic lineage that is separate but equal to that of plants and animals. Currently, all three groups are regarded as kingdoms within the eukaryotic domain. Traditionally, three groups of terrestrial fungi, the molds, sac fungi, and club fungi, have been included in kingdom Fungi and designated, respectively, as the phyla Zygomycota, Ascomycota, and Basidiomycota. It has been more difficult to determine whether various fungal-like organisms, such as the slime molds, oomycetes, and water molds, belong in kingdom Fungi or in the Protista. Recent comparisons of rRNA sequence data indicate that the slime molds and oomycetes are not closely related to the Fungi, but that the water molds such as the chytrids are part of the fungal lineage. In this text, we will consider only the molds, sac fungi, and club fungi, as they are of greater economic and medical importance to humans.

> **REMEMBER**
> Zygomycetes, like the black mold you might see on your bread, form a zygospore when they reproduce sexually.

ZYGOMYCETES

The filamentous fungi in the Zygomycota are commonly referred to as molds. Many function as decomposers, although some are parasitic. Molds grow as masses of hyphae that elongate from the tip and periodically branch. As the hyphae grow, nuclear division occurs without the production of cell walls across the filament, so that each hypha is a tube containing many nuclei.

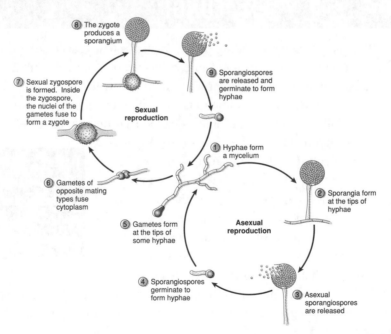

8. The zygote produces a sporangium

9. Sporangiospores are released and germinate to form hyphae

7. Sexual zygospore is formed. Inside the zygospore, the nuclei of the gametes fuse to form a zygote

Sexual reproduction

6. Gametes of opposite mating types fuse cytoplasm

1. Hyphae form a mycelium

2. Sporangia form at the tips of hyphae

5. Gametes form at the tips of some hyphae

Asexual reproduction

4. Sporangiospores germinate to form hyphae

3. Asexual sporangiospores are released

FIGURE 12.7. The life cycle of a zygomycete.

Molds can reproduce by both **asexual reproduction** and **sexual reproduction** (Figure 12.7). In asexual reproduction, hyphae grow upward and form a sporangium. Spores called **sporangiospores** develop inside the sporangium. These spores are resistant to drying and are easily dispersed. When they land in a new environment, they can germinate and begin growing as hyphae again. The spores are often brightly colored; for example, in the bread mold *Rhizopus stolonifer*, the spores are black. When molds reproduce sexually, two hyphae come together and fuse, forming a special spore called a **zygospore** (Figure 12.7). Sexual reproduction finishes inside of the zygospore, and a sporangium is produced. Sporangiospores are released and grow into hyphae. In addition to growing on bread, species of *Rhizopus* can act as opportunistic pathogens, infecting people who have diabetes or leukemia.

ASCOMYCETES

Like the zygomycetes, fungi in the Ascomycota may also be referred to as molds. They are sometimes called the sac fungi because they produce sexual spores, called **ascospores**, in a sac-like structure called an **ascus** (Figure 12.8). Ascospores germinate to produce hyphae that grow and form a mycelium. Sac fungi can also reproduce asexually when hyphae develop into sporangia called **conidiophores**. The conidiophore produce chains of **conidiospores** at their tips, making them look like little brooms (Figure 12.8).

> **REMEMBER**
> Ascomycetes, like the edible morel, produce spores in little sacs called asci (singular: ascus) when they reproduce sexually.

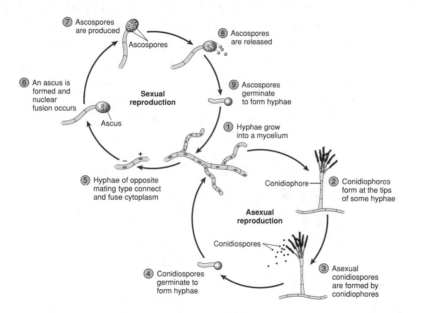

FIGURE 12.8. The life cycle of an ascomycete.

Ascomycetes have both positive and negative effects on human health. A famous conidia-producing fungus is *Penicillium chrysogenum*, the mold that produces the antibiotic penicillin. A more infamous species, *Aspergillus flavus,* grows on peanuts and grains, producing the potent cancer-causing toxin **aflatoxin**. Other species of *Aspergillus* are opportunistic pathogens, causing lung infections in people who have lung diseases such as cancer. Although **yeasts** do not typically form mycelia, they are ascomycetes and can reproduce sexually to form an ascus and ascospores. The yeast *Candida albicans* is an opportunistic pathogen of humans, causing thrush, diaper rash, and vaginitis.

Other ascomycetes are of great economic importance to humans. Yeasts such as *Saccharomyces cerevesiae* are used in the production of bread, wine, and beer. The edible morel, *Morchella esculentum*, is also an ascomycete. Some ascomycetes have had devastating effects on plants in this country. The ascomycete *Cryphonectria parasitica* was introduced to the United States from China in 1904 and destroyed the American chestnut trees that graced the forests in the northeastern United States. Similarly, American elms have been destroyed by Dutch elm disease, which is caused by *Ceratocystis ulmi*.

> **REMEMBER**
> Basidiomycetes, like common forest mushrooms, produce spores in club shaped structures called basidia (singular: basidium) when they reproduce sexually.

BASIDIOMYCETES

The fungi in the Basidiomycota include the familiar mushrooms, as well as rusts, smuts, and some mycorrhizal fungi. This group is referred to as the club fungi because it produces sexual spores in a club-shaped structure called a **basidium** (Figure 12.9).

Basidiospores are released from the basidia and germinate to form hyphae that grow into a mycelium. Asexual reproduction occurs when pieces of mycelium break off and drift to a new location. Basidiomycetes can grow unobtrusively as mycelia in the soil, becoming noticed by humans only when the mycelium forms a fruiting body like a mushroom, which is formed during the sexual reproduction of some basidiomycetes. In mushrooms, basidia are located along the gills on the underside of the mushroom cap. If a large circular mycelium enters sexual reproduction, a perfect ring of mushrooms can be formed. Some people used to think these *fairy circles* were the meeting place of fairies.

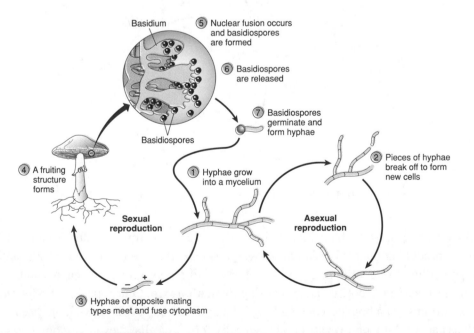

FIGURE 12.9. The life cycle of a basidiomycete.

The economic importance of basidiomycetes is obvious in the many edible species, such as *Agaricus bisporus,* which is grown commercially in many parts of the world. Another popular mushroom is the shiitake mushroom, *Lentinus edulus.* Some mushrooms produce dangerous toxins, such as phalloidin produced by *Amanita phalloides,* and hallucinogens, such as those produced by *Psilocybe.* Other basidiomycetes of economic importance include the rusts and smuts that damage cereal crops.

REVIEW EXERCISES FOR CHAPTER 12

1. Name five pathogens found within the Protista and give examples of the diseases they cause.

2. Describe amoeboid movement. Which groups of protists have amoeboid stages?

3. Name five groups of photosynthetic organisms within the Protista. What is the importance of these organisms to the ecosystems in which they live?

4. Describe the differences between cellular and acellular slime molds.

5. What are spores? Give five examples of different types of spores produced by eukaryotic microorganisms.

6. What is the difference between sexual and asexual reproduction?

7. Explain why the red, green, and brown algae are different in color.

8. Describe the differences in sexual and asexual reproduction among the zygomycetes, ascomycetes, and basidiomycetes.

9. Describe five groups of eukaryotic microorganisms that are of economic importance.

SELF-TEST

1. Which of the following protists does **not** have an amoeboid stage?

   A. *Amoeba*
   B. *Giardia*
   C. *Pelomyxa*
   D. *Trypanosoma*
   E. *Dictyostelium*

2. Which of the following groups of protists includes members that cause red tide?

    A. Foraminifera
    B. Dinoflagellates
    C. Ciliates
    D. Apicomplexans
    E. Euglenids

3. Which of the following groups of protists has a shell made of silica?

    A. Green algae
    B. Oomycetes
    C. Foraminifera
    D. Diatoms
    E. Diplomonads

4. Which of the following groups of protists is associated with oil deposits?

    A. Foraminifera
    B. Pelobiontids
    C. Diplomonads
    D. Dinoflagellates
    E. Oomycetes

5. Which of the following contains chlorophyll c?

    A. Green algae
    B. Red algae
    C. Brown algae
    D. Diatoms
    E. Both C and D are correct.

6. To which group do most edible mushrooms belong?

    A. Ascomycetes
    B. Basidiomycetes
    C. Zygomycetes
    D. All of the above are correct.
    E. None of the above is correct.

7. True or False. Yeast belong to the kingdom Fungi along with mushrooms and molds.

8. True or False. Some species of algae are eaten by humans.

9. True or False. All members of the kingdom Fungi reproduce asexually, but do not reproduce sexually.

10. True or False. The bread mold *Rhizopus* is a zygomycete.

Answers

Review Exercises

1. *Giardia* causes giardiasis, *Trypanosoma* causes African sleeping sickness, *Gonyaulax* causes paralytic shellfish poisoning, *Toxoplasma* causes toxoplasmosis, and *Entamoeba* causes amebic dysentery.

2. Amoeboid movement occurs when a cell crawls by pushing its plasma membrane outward at a point on the cell, and then pulls the rest of the cell toward that point. The groups of protists that have amoeboid stages include the amoebomastigotes, the pelobiontids, the amastigote amoebae, the diplomonads, the cellular slime molds, and the acellular slime molds.

3. Diatoms, euglenids, red algae, brown algae, and green algae are all photosynthetic. These organisms transfer energy from the sun into organic molecules and use dioxide as a carbon source. They form the basis for food webs. In other words, other organisms obtain energy by eating the photosynthesizers.

4. Although both acellular slime molds and cellular slime molds have an amoeboid stage, the amoebae are very different. In the cellular slime molds, the amoebae are single cells that feed on their own. In times of nutritional stress, these amoebae gather and form a giant slug that develops into a sporangium. In the acellular slime molds, the amoeboid feeding stage is a large multinucleate mass. If food becomes scarce, the plasmodium fragments and each fragment can form a sporangium. The spores of cellular slime molds are produced asexually, whereas the spores of acellular slime molds are produced sexually.

5. Spores are cells that are produced for reproduction and dispersal. They are often resistant to drying out. Oomycetes produce motile, flagellated cells called zoospores. Cellular slime molds produce asexual spores. Acellular slime molds produce sexual spores. Zygomycetes produce asexual sporangiospores inside a sporangium and sexual zygospores when two hyphae come together. Ascomycetes produce asexual conidiospores at the tips of hyphae and sexual ascospores inside of a sac-like structure.

6. Sexual reproduction is the production of offspring involving the exchange of genetic information between two individuals. Asexual reproduction is the production of individuals that are genetically identical to the parent.

7. All algae have chlorophyll a. The other pigments they have absorb different kinds of light and so give them different colors. The primary pigments of green algae are chlorophyll a and b, both of which appear green. In red algae, pigments called phycobilins contribute to the red or blue tones of the algae. In brown algae, chlorophyll c and xanthophylls contribute the yellow and brown tones.

8. In the zygomycetes, asexual reproduction occurs when sporangia form at the tips of hyphae and cell division leads to the formation of zygospores. When two hyphae come together, genetic information is shared, and a sexual spore called a zygospore is formed. In ascomycetes, cell division at the tips of hyphae form asexual conidiospores, which collect on the tips of hyphae. Sexual spores are formed in a sac-like structure called an ascus. In basidiomycetes, asexual reproduction occurs when bits of mycelia break off of a parent and drift away to form a new individual. Sexual reproduction occurs in a club-like structure called a basidium to form sexual spores called basidiospores.

9. Basidiomycetes include many edible mushrooms. Ascomycetes include yeast, which is used to make beer, wine, and bread. Dinoflagellates cause red tides and paralytic shellfish poisoning, which has an impact on the fishing industry. Foraminifera are associated with sediments near oil deposits and so help identify areas that might contain oil. Diatoms are part of diatomaceous earth, which is used in pool filters. *Porphyra*, or nori, is a type of red algae, which is eaten.

Self-Test

1. D	5. E	9. F
2. B	6. B	10. T
3. D	7. T	
4. A	8. T	

13

Viruses

WHAT YOU WILL LEARN

This chapter presents the fascinating world of viruses. As you study this chapter, you will:

- discover the early research into viruses;
- compare viruses and bacteria;
- examine the structure of viruses;
- learn about the multiplication of viruses inside host cells;
- explore how viruses are studied and classified today;
- find out about prions and other acellular infectious agents.

SECTIONS IN THIS CHAPTER

- History of Viruses
- Properties of Viruses
- Structure of Viruses
- Viral Multiplication Cycles
- Cultivation of Viruses
- Identification and Classification of Viruses
- Other Acellular Infectious Agents

You have a cold and just feel awful. Your head is pounding, your nose is sore and runny, and you keep having painful coughing fits. Finally, you head to the doctor's office. The doctor listens to you breathe and looks in your throat and ears. He tells you something you already know, that you have a cold, and something that you probably don't want to hear, that there's no quick fix, no cure. The doctor tells you to go home, get plenty of rest, drink lots of fluids, and wait. Colds are caused by **viruses** and, for the vast majority of viral infections, all we can do is wait for our immune systems to clear them out of our bodies.

Why don't we have any "cures" for viruses? Why aren't we supposed to take antibiotics when we have a viral infection? Viruses and bacteria both make us sick, but there is something different about them, something that affects the way we deal with them. We have already looked in detail at the structure, growth, and metabolism of bacteria (see Chapters 4, 5, and 6). To understand the differences between bacteria and viruses, we will now take a detailed look at the structure of viruses and how they reproduce themselves.

History of Viruses

We have only to look at recent history to understand the deadly power of viruses. When Spanish explorers came to Mexico in the 1500s, they brought not only weapons but also the smallpox virus, which was deadly to the native people that lived there. This same scenario was repeated when Europeans settled in what is now the United States. Entire tribes of Native Americans were virtually wiped out by smallpox to which they had no established immunity. In the 1800s, a lethal measles epidemic broke out among the native people in Hawaii. In 1918, there was an influenza pandemic that swept the planet, killing more people than died in World War I. Even before the influenza pandemic waned, polio virus was on the rise; its epidemics increased each year until 1952, when it infected 58,000 children in the U.S. alone. The development of the Salk polio vaccine and its implementation in 1954 brought polio under control in resource-rich nations, but it continued to paralyze and kill children around the globe until the Global Polio Eradication Initiative was launched in 1988. As a result of vaccination efforts by this initiative, polio has almost been eliminated from the human population, which is amazing when you consider that in 1988, polio was endemic in 125 countries and caused paralysis in more than 1000 children every day. And today we live in a world in which over 33 million people are infected with HIV, which leads to AIDS. The effects of the AIDS pandemic have been brutally severe in areas like sub-Saharan Africa, where millions of children have been orphaned and struggle to survive.

Although people have probably suffered from viral diseases for as long as there have been people, viruses weren't discovered until the late nineteenth century. The discovery of viruses occurred during a time when there were so many new discoveries in microbiology that it is now referred to as "The First Golden Age of Microbiology."

Scientists designed better microscopes and discovered better ways of growing and studying microbes. Famous scientists like Louis Pasteur and Robert Koch demonstrated connections between microbes and disease, which inspired "microbe hunters" everywhere to try to determine the causes of known diseases.

A breakthrough in the search for disease-causing microbes came in 1884 when Charles Chamberland, an associate of Pasteur, invented a porcelain filter that had holes so small that bacteria could not pass through them. Scientists used these filters to trap bacteria taken from diseased tissues (Figure 13.1). Once separated from the tissue, the bacteria could be grown and tested for their disease-causing ability. The bacteria would be introduced into a healthy organism to see whether they caused a particular disease.

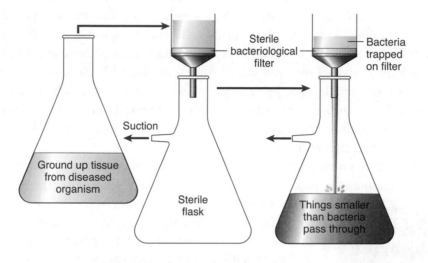

FIGURE 13.1. Using a bacteriological filter.

The holes in the filter are small enough (0.22 to 0.45 μm) that bacteria cannot pass through. Thus, bacteria can be separated from a sample by passing the sample through the filter.

One disease of interest at the time was tobacco mosaic disease. This disease caused significant economic losses to tobacco crops. In 1892, Dimitri Ivanowski was attempting to isolate the cause of the disease using a bacterial filter like the one invented by Chamberland. Ivanowski ground up tissue from diseased tobacco plants and passed it through the filter. He collected bacteria from the filter and introduced them into healthy tobacco tissue, but he could not reproduce the disease. However, when he tested the fluid that passed *through* the filter, he found that application of the fluid to plants could cause the disease. Apparently, whatever caused the disease was smaller than a bacterium. Ivanowski proposed that the disease-causing agent might be a toxin. These filterable infectious agents became known as **viruses**, which is the Latin word for poison.

> **REMEMBER**
> The first virus to be discovered was tobacco mosaic virus.

Further studies of tobacco mosaic virus revealed more about this mysterious virus. In 1898, Martinus Beijerinck found that the virus would only multiply in living plant cells. In addition, he found that it could be completely dried and yet would still remain capable of causing the disease when introduced into plant tissue. In 1935, Wendell Stanley crystallized the virus and determined that it was mostly protein. This was followed in 1936 by the work of Frederick Bawden and Norman Pirie, who determined that the viral particles were actually made of both protein and nucleic acid.

Properties of Viruses

The early studies on viruses clearly demonstrated that viruses were very different from bacteria. Bacteria can grow on their own in a wide variety of locations; viruses can only reproduce inside of cells. Bacteria are cells, and for most bacteria, if you dry them out completely, they will die. This is not true of many viruses that can exist in a dried state for long periods of time and still cause disease. Bacteria are made of four different types of macromolecules: lipids, nucleic acids, proteins, and carbohydrates. Many viruses, on the other hand, are just nucleic acid and protein. Also, as demonstrated by the early work with bacterial filters, viruses are smaller than bacteria. Clearly, viruses and bacteria are different from each other.

Just how different bacteria and viruses are from each other can be summed up in a seemingly simple statement: Bacteria are cells; viruses are not cells. The implications of this statement are huge. Bacteria can live on their own because they have ribosomes for protein synthesis and metabolic pathways for the synthesis of ATP. They have membranes, structures like flagella and pili, and hundreds of different enzymes. Viruses don't have any of this. Viruses are very tiny and very simple. The individual viral particles are called **virions**, and the simplest virions consist only of nucleic acid and a protein coat. Because they don't have any ribosomes or ATP-generating metabolism, viruses can only reproduce if they enter a living cell and steal what they need to multiply. In other words, they are **obligate intracellular parasites**. They enter a living cell, take it over, make more of themselves, and then leave, often destroying the host cell in the process. In this manner, they spread from cell to cell causing destruction and disease.

> **REMEMBER**
> Viruses are small particles of nucleic acid and protein. They lack cellular structures found in bacteria and require a host cell to reproduce themselves.

The type of cells that a particular virus can infect is its **host range**. Viruses attack every type of life on Earth. Bacteria, archaeans, plants, and nonhuman animals all get viral infections. In fact, Herpes viruses are one of the factors that are killing the world's coral reefs. For a virus to enter a cell, proteins on the surface of the virus must be able to bind to proteins on the surface of the cell. Thus, each virus is specific to the types of cells it can infect. It can only enter if the proteins on its surface fit like a key into the "lock" of a receptor on the surface of a host cell. If the protein that the virus uses for attachment binds to a receptor that is present on many cell types, it will have a broad host range. For example, the influenza virus binds to a receptor that is present on the cells of humans, birds, and pigs and can infect all three types of animals. If the attachment protein of a virus binds to a receptor that is present on only a few cell types, the virus will have a narrow host range. Smallpox had a narrow host range because it could only infect humans. Once a virus is inside its host, it can only infect certain types of cells. This **tissue tropism** is also controlled by the match between viral proteins and receptors on the surface of the cells. For example, cold viruses show tissue tropism for cells of the upper respiratory tract, while hepatitis viruses show tissue tropism for the liver.

Structure of Viruses

All viruses are made of at least two things, nucleic acid surrounded by a protein coat. The nucleic acid may be DNA or RNA, and it may be double-stranded like the double helix or single-stranded like mRNA. Thus, there are double-stranded DNA viruses, single-stranded DNA viruses, double-stranded RNA viruses, and single-stranded RNA viruses. There are no known viruses that have both DNA and RNA. Also, viruses typically have much smaller nucleic acid molecules with fewer genes than cells. With so few genes, viruses must manage with very few proteins. To build the **capsid**, the protein coat that surrounds and protects the nucleic acid, viruses use many repeats of proteins called **capsomeres**. Some viruses have only one type of capsomere in their capsid; others have a few types.

> **REMEMBER**
> Viral genomes can be RNA or DNA and are protected by a protein coat called a capsid.

Some viruses have an additional layer surrounding the capsid called an **envelope** that is made of phospholipid, protein, and carbohydrate. The envelope is taken from the host cell as the virions exit. During viral infection, viral proteins typically insert themselves into the host plasma membrane. When the virions exit the cell, they are wrapped in this modified plasma membrane that contains both viral and host proteins. Some viral proteins may project from the surface of the envelope, forming **spikes**. Envelopes and spikes are important in the attachment and entry of viruses into host cells.

Viruses are categorized by the shape of their virions, which depends on the presence of an envelope and the shape of their nucleic acid (Figure 13.2). Because viruses are so small, their shapes can only be seen with an electron microscope. Viruses that have a very regular, geometrical capsid are called **polyhedral** viruses (a polyhedral virus with twenty sides is called an icosahedral virus). Viruses that have a thin, linear shape are called **helical** viruses. **Enveloped viruses** do not have a regular shape, although the capsid within the envelope may have different shapes. Some viruses have many specialized structures and are referred to as **complex** viruses.

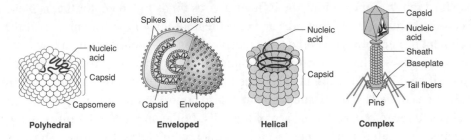

FIGURE 13.2. Viral shapes (morphology).

Viral Multiplication Cycles

Much of what we know about how viruses multiply in cells, we learned by studying viruses that attack bacterial cells. A virus that attacks a bacterium is called a **bacteriophage** or **phage**. From the study of bacteriophage, we learned that there are two basic cycles for viral infection, one that results in immediate multiplication of the virus and destruction of the host cell and another that may include a period of time in which the virus is inactive or dormant. The essential features of viral multiplication are the same for bacteriophage and viruses that infect eukaryotic cells. Thus, we will first consider the multiplication cycles of bacteriophage in detail to learn these basic principles. Then, a few significant variations between these and eukaryotic cell viruses will be presented. (Archael viruses have only recently been discovered and are just beginning to be studied.)

LYTIC CYCLE

When a bacteriophage enters a bacterium, immediately reproduces itself, and then bursts out, destroying the host cell, it is called the **lytic cycle** (Figure 13.3). To enter the cell, the virus must first **attach** and use its surface proteins to bind to the host cell receptors. Once it is attached, the virus **penetrates** the host cell by injecting its DNA into the cell. The viral DNA will essentially reprogram the bacterial cell, forcing it to become a factory for the manufacture of more virions. The bacterial DNA is destroyed, and all of the host cell resources are taken over, beginning the process of viral **biosynthesis**. Viral nucleic acids are produced using nucleotides from the bacterial cell, and the bacterial ribosomes are used to build viral proteins. All of this is accomplished using ATP that was produced by the bacterial cell. After the manufacture of viral parts is complete, **maturation** occurs, and the virions assemble themselves, packaging nucleic acid into capsids. The host cell is lysed, and the virions are **released** to infect another cell.

> **REMEMBER**
> The lytic cyle of bacteriophage leads rapidly to the production of new virions and the destruction of the host cell.

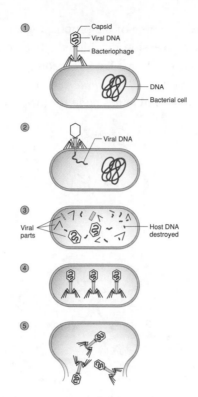

FIGURE 13.3. The lytic cycle of a T-even bacteriophage.

1. Attachment: viral proteins attach to receptors on the surface of the cell. 2. Penetration: viral DNA enters the host cell. 3. Biosynthesis: the phage DNA directs the synthesis of new viral parts using resources from the host cell. 4. Maturation: the virions put themselves together (self-assemble). 5. Release: the host cell is destroyed and the virions exit the cell.

LYSOGENIC CYCLE

For some bacteriophage, the lytic cycle may be interrupted by dormant periods during which the virus is not actively making new virions. This type of multiplication cycle is called the **lysogenic cycle** (Figure 13.4). It begins the same way as the lytic cycle, with attachment and penetration of the bacterio- phage. After the viral DNA has penetrated the cell, however, **recombination** occurs, and the viral DNA is inserted into the DNA of the host cell, forming a dormant phage called a **prophage**. Most of the prophage genes are repressed, and new virions are not made. Although new virions are not being made, the viral DNA is being copied along with the host DNA every time the bacterial cell divides, creating a population of virally infected cells. Environmental signals may trig- ger the phage to exit from the host DNA and resume the lytic cycle with biosynthesis, maturation, and release. During the lysogenic cycle, when virions are not being pro-

> **REMEMBER**
> Lysogenic bacteriophage can insert their genetic material into the genome of the host cell and may remain dormant within the host cell for long periods of time.

duced, the bacterial cell may use some of the prophage genes to make new proteins. This can lead to a **lysogenic conversion** in which the bacterial cell develops new traits, such as the ability to make a toxin. Some bacteria require the presence of a prophage in order to cause disease, for example, *Vibrio cholerae* only causes cholera if it is infected by a lysogenized phage because viral genes code for the cholera toxin.

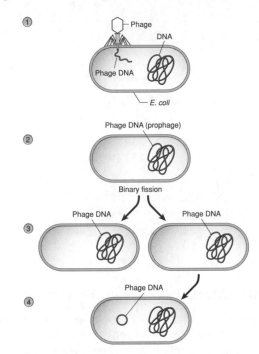

FIGURE 13.4. The lysogenic cycle of bacteriophage λ in *E. coli*.

1. Attachment and penetration: viral proteins attach to host receptors, and the viral DNA enters the cell. 2. Recombination: phage DNA may form a circle and then become integrated into the host DNA. 3. As the host cell divides, viral DNA is replicated along with host DNA. 4. Environmental signals may cause the phage to exit host DNA and return to the lytic mode. Biosynthesis, maturation, and release would then occur as in the lytic cycle. (See steps 3–5 in Figure 13.3)

ANIMAL VIRUSES

Viruses that infect animal cells show the same two patterns of infection as those of bacteriophage. However, the terminology that is used for animal cell viruses is a little bit different from that used for bacteriophage. Viruses that infect and immediately produce new virions typically result in rapid, intense diseases called **acute** infections. Influenza and the common cold result from acute viral infections. Viruses that are capable of inserting their nucleic acid into that of eukaryotic cells typically result in **latent** diseases that have periods of inactivity, such as herpes. One of the reasons AIDS is so hard to cure is because HIV can remain latent in cells. While the virus is latent, it is unaffected by antiviral drugs and remains in the body.

In addition to demonstrating the same two basic patterns of infection as bacteriophage, animal cell viruses also demonstrate the same basic steps of viral multiplication. Essentially, all viruses must do attachment, penetration, biosynthesis, maturation, and release. However, many viruses that infect animal cells are enveloped viruses, and this affects the way some of these steps are done. After an enveloped virus has attached to a host cell, penetration occurs by **fusion** (Figure 13.5.1) or **endocytosis** (Figure 13.5.2). In endocytosis, the host cell surrounds the virus with plasma membrane and brings it into the cell in a vesicle. In fusion, the viral envelope becomes part of the plasma membrane of the host cell. Both endocytosis and fusion result in the viral capsid entering the host cell along with the viral genetic material. In order for viral multiplication to begin, the capsid must be removed. This is called **uncoating**, and it may be accomplished by viral or host enzymes, depending on the virus. Finally, enveloped viruses must pick up a new envelope as they exit the cell. During biosynthesis, viral proteins are inserted into the plasma membrane of the cell. After the virions are assembled during maturation, **budding** (Figure 13.6) occurs. During budding, the virions push out through the modified plasma membrane and are wrapped in a new envelope as they exit the cell.

> **REMEMBER**
> Animal viruses enter host cells by endocytosis or fusion.

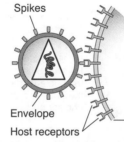

A. The viral envelope merges with the host plasma membrane.

B. The virion enters the cell by fusion.

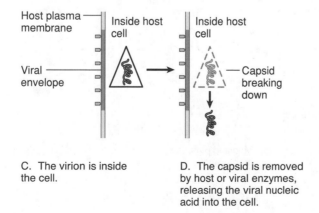

C. The virion is inside the cell.

D. The capsid is removed by host or viral enzymes, releasing the viral nucleic acid into the cell.

FIGURE 13.5.1. Penetration and uncoating of eukaryotic cell viruses.

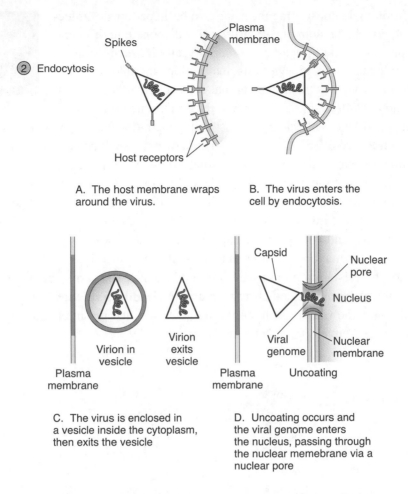

(2) Endocytosis

A. The host membrane wraps around the virus.

B. The virus enters the cell by endocytosis.

C. The virus is enclosed in a vesicle inside the cytoplasm, then exits the vesicle

D. Uncoating occurs and the viral genome enters the nucleus, passing through the nuclear memebrane via a nuclear pore

FIGURE 13.5.2. Penetration and uncoating of eukaryotic cell viruses.

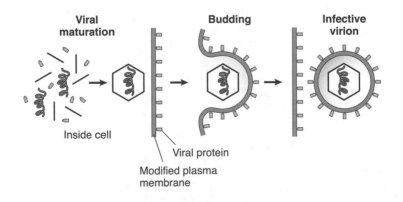

FIGURE 13.6. Budding.

After maturation, the assembled virion pushes into the modified plasma membrane and is wrapped in an envelope.

Another difference between bacteriophage and animal cell viruses results from differences in the viral genetic material. Bacteriophage are DNA viruses. Replication of their DNA typically occurs by the DNA polymerases of the bacterial cell that they infect in the same way that bacterial DNA would be replicated (see Chapter 7). However, there is a great deal more diversity in the genetic material of animal viruses, resulting in more variety in the way that the viral genetic material is copied (Table 13.1). Animal viruses may have double-stranded DNA, single-stranded DNA, double-stranded RNA, or single-stranded RNA. Replication of the genetic material may occur in the nucleus or the cytoplasm of the host cell, and it may involve host enzymes or viral enzymes, or both.

> **REMEMBER**
> Many animal viruses are highly variable due to the high mutation rates of viral polymerases.

TABLE 13.1. Biosynthesis of Animal Viruses

Type of Nucleic Acid	Method and Location for Copying Nucleic Acid	Enzymes Involved	Example Disease and Viral Family
Double-stranded DNA	DNA to DNA in nucleus DNA to DNA in virion DNA to RNA in nucleus, then RNA to DNA in cytoplasm	Host DNA polymerase Viral DNA Polymerase Host RNA Polymerase, then viral reverse transcriptase (RNA-dependent DNA polymerase)	Herpes, Herpesviridae smallpox, Poxviridae hepatitis B, Hepadnaviridae
Single-stranded DNA	DNA to DNA in nucleus	Host DNA polymerase	Common cold, Adenoviridae
Single-stranded RNA	RNA to RNA in cytoplasm RNA to DNA in cytoplasm	Viral RNA-dependent RNA polymerase Reverse transcriptase (RNA-dependent DNA polymerase)	Polio, Picornaviridae rabies, Rhabdoviridae AIDS, Retroviridae
Double-stranded RNA	RNA to RNA in cytoplasm	Viral RNA-dependent RNA polymerase	Respiratory and digestive tract infections, Reoviridae

In general, viral enzymes that copy nucleic acids do not have the same proofreading ability as the DNA polymerases made by cells. Thus, replication of viral genetic material often results in many mutations and thus changes in the viral strains. This makes it harder for the body to establish immunity to the virus or for antiviral drugs to remain effective. For example, high mutation rate and strain variability is the reason we have new flu epidemics every year. It is also one of the reasons that HIV is so difficult to treat.

Cultivation of Viruses

Because viruses cannot multiply without a host cell, they cannot be grown in sterile media the way bacteria can. Host cells must be provided for the virus to infect. Bacteriophage may be cultured with bacterial cells in liquid culture or on plates. To culture on plates, bacteria and bacteriophage are combined in melted agar and then poured into plates. The bacterial cells divide and evenly cover the surface of the plate forming a **lawn**. Wherever bacterial cells have been destroyed by the bacteriophage, clearings called **plaques** will appear in the lawn.

Because it is harder to culture animal cells than it is for most bacteria, it is also more difficult to grow animal viruses in the lab. Some animal cells will grow in culture dishes much as bacteria do and will form a single layer on the bottom of a dish that contains a suitable growth medium. Viruses that can infect these cells can be grown along with them. Animal cells that are infected by the virus will display **cytopathic effects**, such as visible deterioration. Some viruses whose host cells cannot be grown in culture can be grown in

> **REMEMBER**
> Viruses that are cultivated in the lab must be grown inside host cells.

living animals such as rabbits, mice, and guinea pigs. Other viruses can be grown in fertilized chicken eggs. Because some viruses that are raised for vaccines are cultured in fertilized eggs, people may be asked if they have egg allergies before receiving the vaccines because some egg proteins might be present in the vaccine. All of these culture methods together have enabled us to grow, study, and develop vaccines for quite a few viruses. However, because some viral host ranges are highly specific (each virus infects only certain cell types) and because we cannot grow all cells in culture, we cannot yet grow all viruses. This makes the study of certain viruses very problematic.

Identification and Classification of Viruses

Because viruses are difficult to grow and study, our understanding of the relationships between viruses is still not complete. Currently, viral classification is based upon three factors: the type of nucleic acid a virus has, how the nucleic acid is replicated, and the morphology of the virus. Viral species are defined by their genetic information and by their host range. They are generally referred to by common names that reflect the disease they cause. For example, the virus that causes AIDS is referred to as the human immunodeficiency virus. Numbers represent different strains of a species. For example, a strain of HIV is designated as HIV-1. The genus of a virus always ends in *–virus* and may be different than the common name for the virus. For example, the virus that causes AIDS is in the genus *Lentivirus*. Viral genera are grouped into families that end in *–viri-*

dae. For example, HIV is in the family Retroviridae. Our classification system for viruses will continue to develop as we learn more about viruses and their relationships.

Because viral genetic material largely defines its identity, viral identification relies upon molecular techniques. Viral nucleic acids are often isolated and sequenced (see Chapter 8) to determine the order of nucleotides. After the genetic code of a virus has been determined, it can be compared to the genetic code of other known viruses to determine its identity. Antibodies to known viruses can also be used to identify an unknown virus, just as they are used to identify bacteria (see Chapters 9 and 21). Antibodies are proteins that bind very specifically to molecules called antigens, such as viral proteins. Because there must be an exact match for binding to occur, antibodies known to react with a certain virus can be used to detect whether the virus is present in a sample.

Other Acellular Infectious Agents

Viruses are not the only disease-causing entities that are not made of cells. Naked RNA molecules called **viroids** cause some diseases of crop plants, while infectious proteins called **prions** cause some animal diseases. An infamous example of a disease that may be caused by prions is bovine spongiform encephalopathy (BSE), also known as Mad Cow disease. BSE seems to be related to a disease called scrapie in sheep and to variant Creutzfeldt–Jacob disease (vCJD) in humans. All three diseases result in the development of large open spaces in the brain and thus severe neurological degeneration.

One hypothesis for the development of Mad Cow disease in cattle and vCJD in humans is that it is caused by eating food contaminated with neurological tissue from infected animals. To encourage more rapid growth of cattle, their feed is sometimes supplemented with animal tissue. The Mad Cow epidemic in Britain during the 1980s seemed to be related to changes made in the late 1970s in the system for rendering livestock carcasses into animal feed. These changes may have allowed more neurological tissue to enter the animal feed. This idea is supported by the fact that a ban on meat-and-bone meal in the late 1980s was followed by a decrease in BSE cases in the early 1990s.

> **REMEMBER**
> Prions are infectious proteins that are capable of causing disease.

About seven years after the outbreak of BSE, humans in Britain began to display a variant form of Creutzveld–Jacob disease. Typically, vCJD occurs in older people and recurs in certain families. The variant form of the disease occurred in younger people and did not seem to have a genetic component. Because the epidemic of vCJD in Britain had a very similar time course to the epidemic of BSE, a link between the two was proposed. The hypothesis is that humans contracted vCJD by eating beef contaminated with neurological tissue. This type of contamination occurs when carcasses are compressed to yield a paste of *mechanically recovered meat*. This paste may be used in meat products like sausages, lunch meats, and hot dogs.

The exact mechanism by which prions cause disease is not known. Prions are believed to be abnormally folded forms of proteins that normally exist on the surfaces of cells. According to the **prion hypothesis**, when the prions enter the body, they are thought to convert the normal proteins into abnormal ones. The abnormal proteins form clusters called **plaques**, resulting in damage to nervous tissue.

While scientists continue to research prions and try to understand their links to vCJD and other diseases, many countries are struggling with decisions about how to prevent BSE. Some countries have instituted bans on feed that is supplemented with animal tissue and have developed regular testing protocols for cattle. Changes in the way meat is rendered for packing might also help to prevent the disease. When meat is torn off carcasses by machines, it is difficult to prevent contamination by neurological tissue. These changes result in short-term costs to ranchers and meat processors, and so may be resisted by some. The long-term costs of allowing BSE to spread through herds, however, must also be considered. For example, many countries will not import cattle from countries known to have BSE in their herds. In a situation like this, where the scientific data is not yet conclusive, there appears to be a threat to public health, yet it is very difficult to resolve the conflicts between the economic costs and the risk to public health. The measures adopted in Britain seemed to halt the epidemic of vCJD in the British population and can serve as a model for other countries.

REVIEW EXERCISES FOR CHAPTER 13

1. State the basic properties of viruses.

2. Create the following table and fill in the information.

Characteristic	Bacteria	Viruses
Can they reproduce on their own?		
Are they cells?		
Do they divide by binary fission?		
Do they have both DNA and RNA?		
Do they have ribosomes?		
Can they make their own ATP?		
Are they sensitive to antibiotics?		

3. Define virion, lytic cycle, lysogenic cycle, latent infection, and acute infection.

4. State the basic parts of viruses and describe the function of each.

5. Define host range and explain why it is always specific for a particular virus.

6. Name and describe the steps of the lytic cycle of a virus.

7. Name and describe the steps of a lysogenic cycle of a virus.

8. Describe how the multiplication cycle of enveloped animal viruses differs from that of bacteriophage.

9. Compare biosynthesis in DNA and RNA viruses.

10. Explain why viruses generally have higher mutation rates than bacteria.

11. State the different methods for viral cell culture.

12. State the major methods for identifying viruses.

13. Distinguish between viruses, viroids, and prions.

SELF-TEST

1. The early work on viruses revealed all of the following characteristics of viruses **except**:

   A. Viruses are smaller than bacteria.
   B. Viruses contain nucleic acid and protein.
   C. Viruses can reproduce in complete media.
   D. Viruses can cause disease.
   E. Viruses can be dried but still remain capable of causing disease.

2. During which phase does replication of viral genetic material occur?

   A. Penetration
   B. Maturation
   C. Attachment
   D. Biosynthesis
   E. Release

3. The outer coat of a virus is called a

   A. virion.
   B. capsule.
   C. capsomere.
   D. capsid.
   E. spike.

4. True or False. A virus that integrates its genetic material into that of its host is undergoing the lysogenic cycle of multiplication.

5. All of the following are true of proteins on the surfaces of viruses **except**:

   A. They attach to host cell receptors.
   B. There are usually many different kinds.
   C. They determine the host range of the virus.
   D. They make up the capsid.
   E. They may project as spikes.

6. True or False. Animal viruses typically have a lower mutation rate than do animal cells.

7. True or False. Viruses may be cultured in fertilized chicken eggs.

8. The viral enzyme reverse transcriptase

   A. makes DNA from RNA.
   B. makes RNA from DNA.
   C. makes protein from mRNA.
   D. makes mRNA from protein.
   E. has a low mutation rate.

9. Enveloped animal viruses may enter animal cells by

   A. fusion.
   B. endocytosis.
   C. budding.
   D. Both A and B are correct.
   E. A, B, and C are correct.

10. Mad Cow disease is caused by a

    A. virus.
    B. bacterium.
    C. infectious RNA molecule.
    D. infectious protein.
    E. infectious DNA molecule.

11. *Varicellovirus* (*Varicella-Zoster* virus) is an enveloped DNA virus. Initially, *Varicellovirus* causes chicken pox, an acute illness resulting in itchy, round, fluid-filled blisters. In some individuals, the virus may reemerge later in life as shingles. Shingles is characterized by tingling, itchiness, or stabbing pain on the skin, followed by a rash that appears as a band or patch of raised dots on the side of the trunk or face. The rash develops into small, fluid-filled blisters, which begin to dry out and crust over within several days. When the rash is at its peak, symptoms can range from mild itching to extreme and intense pain. Name and describe the multiplication cycle of *Varicellovirus*.

12. Some drugs help lessen the symptoms of flu if they are taken very early after infection. One of these drugs stops the virus from uncoating. What is uncoating and why does this drug lessen the symptoms of flu?

Answers

Review Exercises

1. Viruses are acellular (not cells), obligate intracellular parasites (need a host) that are made of nucleic acid and protein and can be passed from cell to cell.

2.

Characteristic	Bacteria	Viruses
Can they reproduce on their own?	Yes	No
Are they cells?	Yes	No
Do they divide by binary fission?	Yes	No
Do they have both DNA and RNA?	Yes	No
Do they have ribosomes?	Yes	No
Can they make their own ATP?	Yes	No
Are they sensitive to antibiotics?	Yes	No

3. A virion is a viral particle. The lytic cycle leads to immediate destruction of the host cell. In the lysogenic cycle, the viral DNA incorporates into the host DNA. A latent viral infection has periods of inactivity. A virus capable of a latent infection can incorporate its DNA into the DNA of a eukaryotic host. An acute viral infection develops rapidly and is of short duration.

4. The capsid is made of protein and protects the nucleic acid. The nucleic acid contains the information for the replication and transmission of the virus. Envelopes are phospholipid bilayers that are involved in entry and exit from the host cell. Protein spikes may attach to receptors on host cells.

5. The host range is the number of cell types a virus can infect. It is specific to each virus because it depends on what types of receptors a virus can bind with its surface proteins.

6. First, the virus, using surface proteins, attaches to host cell receptors. Second, the viral genetic material penetrates, or enters, the host cell. In biosynthesis, the virus takes over the host cell resources and makes more viral parts. During maturation, the virions put themselves together. Finally, the virus destroys the host cell and is released.

7. First, the virus attaches to host cell receptors using its surface proteins. Second, the viral genetic material penetrates, or enters, the host cell. The viral DNA can then integrate into the host chromosome. As the host cell divides, the virus is copied along with the host genetic material. Environmental signals can trigger the virus to exit the host DNA and resume replication. Biosynthesis, maturation, and release would then occur as for the lytic cycle.

8. Enveloped animal viruses differ from bacteriophage in a number of ways. First, because of the presence of the envelope, enveloped animal viruses enter the host cell by endocytosis or fusion. In endocytosis, host membrane wraps around the virus and brings it into the cell in a vesicle. In fusion, the viral envelope merges with the host membrane. After the virus is inside the cell, it must remove its capsid, which it does by uncoating. Because animal viruses may be DNA or RNA viruses, biosynthesis may involve a variety of viral enzymes. RNA viruses may copy their RNA directly with RNA-dependent RNA polymerases or may make a DNA copy of their RNA using the viral enzyme reverse transcriptase. DNA

viruses may use a viral DNA polymerase or a host DNA polymerase. Another difference is that when enveloped viruses exit the cell, they exit by budding, in which the assembled virion is wrapped in modified plasma membrane as it leaves the cell. The host cell may not lyse.

9. RNA viruses may be single-stranded or double-stranded. Some copy their RNA directly with RNA-dependent RNA polymerases. Others make a DNA copy of their RNA using the viral enzyme reverse transcriptase. The DNA may direct biosynthesis or sometimes may be incorporated into the host DNA. DNA viruses may also be single-stranded or double-stranded and may use a viral DNA polymerase or a host DNA polymerase to make copies of their DNA.

10. Bacteria copy their DNA with DNA polymerase III, which has the ability to detect mismatched base pairs and correct them. In addition, DNA polymerase I can also repair mismatched DNA. Viral enzymes do not have the same proofreading ability as do the bacterial enzymes and so the error rate, or mutation rate, is higher.

11. Viruses must be cultured in host cells. This can be done in a number of ways. Bacteriophage can be grown in cultures of bacteria. Eukaryotic cell viruses can be grown in fertilized chicken eggs, in cultured cells, or in animals.

12. Viruses are identified by observing their effects on host cells, by sequencing their nucleic acids, or by using specific antibodies to test for viral antigens.

13. All three are acellular (not cells) infectious agents. Viruses have genetic material surrounded by a protein coat. Viroids are infectious RNA, and prions are infectious proteins.

Self-Test

1. C	3. D	5. B	7. T	9. D
2. D	4. T	6. F	8. A	10. D

11. Multiplication cycle:
    1. Attachment—viral proteins attach to receptors on the host cell
    2. Penetration—the virus enters the host cell by either endocytosis or fusion
    3. Uncoating—viral capsid is separated from viral DNA
        *Recombination—viral DNA integrates into host DNA
        *Multiplication—viral DNA is copied along with host DNA as host cells divide
    4. Biosynthesis—virus uses host resources to make copies of its genome and proteins
    5. Maturation—virions assemble
    6. Release—virions exit host cell by budding (virions push out of plasma membrane, taking an envelope with them)
    Note: Asterisks indicate stages during the latent period that occur in people who get shingles. If the virus causes chicken pox only, these stages do not occur.

12. Uncoating is the removal of the viral capsid after it enters the host cell. If a drug blocked uncoating, it would prevent the viral nucleic acid from entering the cell cytoplasm. The virus would not be able to multiply and affect the host cells, lessening the signs and symptoms of the disease.

Environmental Microbiology

14

WHAT YOU WILL LEARN

This chapter presents the role of microorganisms in the environment. As you study this chapter, you will:

- discover the role of microorganisms in ecosystems;
- examine the importance of microbial diversity in the cycling of matter;
- look at the role of gradients in the structuring of microbial communities;
- explore microbial habitats;
- learn how microbes are used in sewage treatment and the cleanup of pollution.

SECTIONS IN THIS CHAPTER

- Microbial Ecosystems
- Biogeochemical Cycles
- Microenvironments and Gradients
- Microbial Habitats
- Microorganisms and Water Quality
- Sewage Treatment
- Bioremediation

Microbes are everywhere. They are inside our intestines, in the soil, in the sea, in the saltiest lake, and in boiling hot springs. Microbes live and grow, using nutrients and releasing wastes that change the environment around them. They were the first life on Earth, at a time when Earth was hot and had no oxygen in the atmosphere. Microbes transformed the Earth, inventing oxygenic photosynthesis and releasing oxygen gas to the atmosphere. They are still transforming the Earth, producing oxygen, fixing carbon and nitrogen, and recycling nutrients. Without microbes, life on Earth would not exist.

Microbial Ecosystems

METABOLIC DIVERSITY

Microbes inhabit so many different types of environments because of their great metabolic diversity (Chapter 5). Photosynthetic microbes use light energy and carbon dioxide (CO_2) to make sugar ($C_6H_{12}O_6$), which is used for carbon and energy by themselves and the organisms that eat them. Chemolithoautotrophic microbes also make sugar from carbon dioxide, but they obtain energy from inorganic molecules like reduced iron and sulfur compounds. The microbes that eat organic molecules have a dazzling array of enzymes that allow them to break down a huge variety of carbon sources, including numerous man-made chemicals that are not normally found in nature.

As all of this wonderful, complex metabolism chugs merrily along, electrons come into microbial cells, and electrons go out. Microbes can obtain electrons from organic molecules like sugars, fatty acids, alcohols, aromatic compounds, and amino acids, as well as from inorganic molecules like hydrogen sulfide (H_2S), ammonia (NH_3), hydrogen gas (H_2), and carbon monoxide (CO). After electrons are used in energy transfer reactions, they are given to electron acceptors, forming waste products that may be released from the cell. Microbial electron acceptors include a variety of compounds such as oxygen (O_2), nitrate (NO_3^-), sulfate (SO_4^{2-}), and pyruvate (CH_3COCOO^-). When these acceptors receive electrons, they are reduced to a variety of waste products such as water (H_2O), nitrogen gas (N_2), hydrogen sulfide (H_2S), organic acids, and alcohols. Just as you breathe in oxygen, eat food, and release carbon dioxide and water, so the microbes take in molecules from their environment and transform them to obtain energy and building blocks for growth.

MICROBIAL INTERACTIONS IN ECOSYSTEMS

Microbes are important members of the **ecosystems** in which they live. Ecosystems, like soils, lakes, forests, and rivers, are **communities** of organisms interacting with each other and their environment. A community is made up of a variety of **populations**, each of which is a group of related organisms. For example, you are a member of the human population in your hometown. The human population of your hometown is part of a larger community that also includes the populations of cats, birds, trees,

E. coli, and other types of organisms that surround you. You and all of these other organisms interact with each other, use resources from the environment, and produce waste products, thus forming an ecosystem.

All living things in ecosystems must have a source of energy. Energy from the sun or the Earth flows into ecosystems and is captured by **primary producers**. Primary producers use this energy to convert carbon dioxide (CO_2) into sugar ($C_6H_{12}O_6$), building organic matter. **Consumers** use organic matter from the producers as their source of energy and carbon. For example, in many terrestrial ecosystems like the one you live in, plants act as primary producers and humans are consumers of plants. **Decomposers** are like consumers except that they use dead organic matter as their source of energy and carbon. Microbes act as primary producers, consumers, and decomposers in ecosystems.

> **REMEMBER**
> Microbes can act as producers, consumers, or decomposers in ecosystems.

Microbes that act as primary producers include the photosynthetic bacteria and the photosynthetic algae. Photosynthetic bacteria can obtain electrons for photosynthesis from either water (H_2O) or sulfur compounds such as hydrogen sulfide (H_2S). Those that obtain electrons from water perform **oxygenic photosynthesis** (Chapter 5), producing oxygen as waste as in the following reaction:

$$\text{Sunlight} + 6CO_2 + 6H_2O \rightarrow C_6H_{12}O_6 + 6O_2$$

Those that obtain electrons from sulfur compounds perform **anoxygenic photosynthesis** (Chapter 5) and do not release oxygen as waste, as in the following reaction:

$$\text{Sunlight} + 6CO_2 + 6H_2S \rightarrow C_6H_{12}O_6 + 6S$$

Photosynthetic algae perform oxygenic photosynthesis.

Chemolithoautotrophic bacteria and archaea (Chapter 5) are also primary producers. They oxidize reduced inorganic compounds such as hydrogen sulfide (H_2S), elemental sulfur (S^0), ammonia (NH_3), nitrite (NO_2^-), and ferrous iron (Fe^{2+}). The energy and electrons from this oxidation are used to make ATP by an electron transport chain. Energy from ATP and electrons from inorganic or organic molecules are used to convert carbon dioxide (CO_2) into sugar ($C_6H_{12}O_6$). Waste products from the oxidation of inorganic compounds may be released as molecules such as elemental sulfur (S^0), sulfate (SO_4^{2-}), nitrite (NO_2^-), nitrate (NO_3^-), and ferric iron (Fe^{3+}).

Microbial consumers include the chemoheterotrophs (Chapter 5) and photoheterotrophs. Chemoheterotrophs are just like us: They obtain their carbon and energy from organic matter. In other words, they eat other microbes, decaying organic matter, or organic pollutants in their environment. Photoheterotrophs obtain their carbon from organic molecules, but they get their energy from the sun (check out the extreme halophiles in Chapter 11). Many heterotrophic microbes use cellular respiration (Chapter 5) to break down organic molecules. Some, like us, use oxygen as the final electron acceptor in this process, performing **aerobic respiration** as in the following reaction:

$$C_6H_{12}O_6 + 6O_2 \rightarrow 6H_2O + 6CO_2$$

Other microbes can substitute other inorganic molecules for oxygen, using molecules like nitrate (NO_3^-) or sulfate (SO_4^{2-}) as electron acceptors in **anaerobic respiration**. In addition, many heterotrophic microbes perform **fermentation** (Chapter 5), oxidizing glucose and producing a variety of products such as acids and alcohols. Fermentation does not require oxygen, and some of the fermentative heterotrophs are obligate anaerobes. Fermentation is discussed in more detail in Chapter 5, and examples of fermentation products are given in Table 5.4 and Chapter 15.

Biogeochemical Cycles

In all of the types of microbial metabolism, microbes take in molecules, use them in chemical reactions, and release other molecules as waste. However, a molecule that one microbe releases as *waste*, might serve as an excellent energy *source* for a different microbe. In other words, *one microbe's trash is another one's treasure*. This is true for all organisms that interact in ecosystems, not just microbes. For example, you eat food and breathe out carbon dioxide as waste. A nearby plant takes in the carbon dioxide and makes food. These types of cycles, which move elements from one form to another in living things and the environment, are called **biogeochemical cycles**. And although all organisms are part of these cycles, the metabolic diversity of microbes makes them especially important to these processes.

A compost pile is a familiar example of biogeochemical cycling in action. Organic matter in the form of food scraps and yard waste is added to the compost pile. Microbes from the yard clippings and soil go to work, breaking down the organic matter. The pile is turned regularly to keep the organic matter and microbes evenly mixed and to provide oxygen. Over time, the food scraps and yard waste seem to disappear, and all that remains is a much smaller amount of rich compost. However, organic matter does not really disappear. The molecules that made up the food and yard waste were taken in by microbes and used in their metabolism. Much of the waste from this process was released back to the atmosphere as carbon dioxide (CO_2) or nitrogen gas (N_2).

CARBON CYCLE

The simplest biogeochemical cycle to understand is the **carbon cycle** (Figure 14.1). Carbon dioxide (CO_2) in the atmosphere and oceans is taken in by autotrophic organisms and converted into organic molecules. This process is called **carbon fixation** and is part of both photosynthesis and chemolithoautotrophy. As consumers and decomposers perform cellular respiration, they release carbon dioxide back to the atmosphere. Some carbon is temporarily removed from the cycle and stored in the form of coal, methane gas, oil, and calcium carbonate. When fossil fuels are burned, the carbon is released as carbon dioxide back to the atmosphere. Calcium carbonate deposits, for example those that make up the White Cliffs of Dover, England, made by protozoa (Chapter 12) are relatively stable.

> **REMEMBER**
> During the carbon cycle, autotrophs convert CO_2 to $C_6H_{12}O_6$, then heterotrophs convert $C_6H_{12}O_6$ back to CO_2.

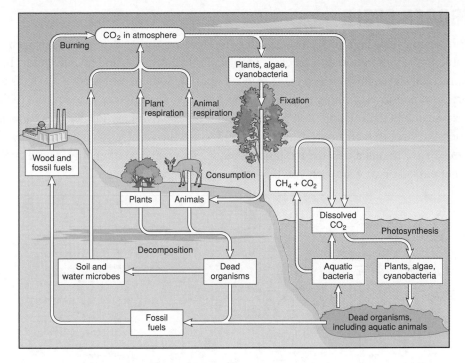

FIGURE 14.1. The carbon cycle.

Respiration breaks down organic molecules, releasing carbon dioxide as waste. Carbon dioxide can also be released to the environment by the burning of fossil fuels. Autotrophic organisms, like bacteria, plants, and algae, fix carbon dioxide and convert it back to organic molecules.

The production and consumption of methane (CH_4) is also part of the carbon cycle. Some microbial consumers and decomposers are **methanogens** (Chapter 11) that produce methane as a by-product of their metabolism. These microbes, which live in anaerobic environments such as the guts of animals, sewage, rice paddies, and landfills, are estimated to contribute 400 million metric tons of methane to the atmosphere each year. Other microbes, called **methanotrophs**, can utilize methane as a source of energy. Aerobic methanotrophic bacteria grow in natural environments like the surface sediments of lakes, oceans, and wetlands, and are also found in sediments and soils that contain sewage. **Anaerobic oxidation of methane** (**AOM**) also occurs as a result of a partnership between anaerobic methane-oxidizing archaens and anaerobic sulfide-oxidizing bacteria. The two prokaryotes live together in clusters of cells in anaerobic methane-rich environments, such as on the ocean floor above methane seeps. The archaens oxidize methane, producing methyl sulfide as waste, which the sulfide-oxidizing bacteria happily use as their energy source. Together, these microbes oxidize up to 90 percent of the methane that would otherwise be released into the ocean. Thus, they play an extremely important role in the global carbon cycle. And, as methane is a greenhouse gas, they also help prevent global warming.

NITROGEN CYCLE

The **nitrogen cycle** (Figure 14.2) is critical to living things because it makes nitrogen available for incorporation into necessary molecules like proteins and nucleic acids. In a very practical sense, the nitrogen cycle is also important to humans because it affects agriculture. Usable forms of nitrogen, like ammonia (NH_3), nitrate (NO_3^-), and nitrite (NO_2^-), are very scarce in the soil and limit plant growth. This is why many farmers have to add so much fertilizer to their fields. Chemical fertilizers, which typically contain nitrates, are very expensive to make because the process requires a great deal of energy. These energy costs get passed on to you when you buy your food and gasoline.

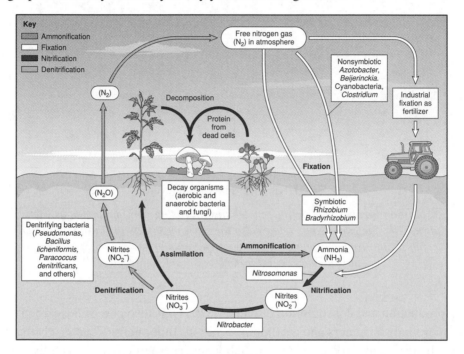

FIGURE 14.2. The nitrogen cycle.

The largest reservoir of nitrogen is the nitrogen gas in the atmosphere. Nitrogen-fixing bacteria convert nitrogen gas into ammonia. Ammonia undergoes nitrification, producing nitrate. Nitrates are removed from the soil by denitrification, resulting in the return of nitrogen gas to the atmosphere. Nitrogen that is assimilated by plants and animals is released back to the environment by decomposition and ammonification.

Nitrogen gas makes up approximately 80 percent of the Earth's atmosphere. However, for most cells, this form of nitrogen is completely unusable. Thus, life on Earth depends heavily on the activity of **nitrogen-fixing bacteria** (Chapter 10) and archaea (Chapter 11). These prokaryotes are able to capture nitrogen gas (N_2) from the atmosphere and convert it to usable forms of nitrogen like ammonia (NH_3). Many of the N_2-fixing bacteria are free-living in the soil and water, while others form symbiotic associations with plant roots. The symbiotic nitrogen-fixing bacteria colonize the roots of leguminous plants like peas, beans, alfalfa, and clover, reproducing in the root cells and causing the formation of **root nodules**. It is interesting to note that long before anyone knew these bacteria existed or knew what these bacteria were doing, farmers realized that they could improve the quality of their soil by rotating their crops and periodically including a leguminous crop on their fields. Today, many farmers will plant a cover crop of alfalfa and then plow it under to put as much usable nitrogen into the soil as possible.

> **REMEMBER**
> Nitrogen-fixing bacteria convert atmospheric N_2 into forms of nitrogen that other types of cells can use.

As a result of nitrogen fixation, ammonia (NH_3) becomes available to plants and microbes in the environment. Ammonia is a reduced form of nitrogen, which makes it an excellent source of energy and electrons for chemolithotrophic bacteria. These bacteria oxidize the ammonia (NH_3), taking electrons, and then release the nitrogen as nitrite (NO_2^-). Although the nitrite (NO_2^-) represents a waste product to these bacteria, other chemolithoautotrophic bacteria can use the nitrite (NO_2^-), further oxidizing it for electrons and energy. This second group of chemolithoautotrophs release nitrate (NO_3^-) as their waste product. Because these processes lead to the production of nitrate (NO_3^-) in the soil, they are referred to as **nitrification**. The two groups of chemolithoautotrophs that perform these reactions are called **nitrifying bacteria**. Examples of nitrifying bacteria are given in Figure 14.2 and in Chapter 10.

Just as the process of adding nitrate (NO_3^-) to the soil is called nitrification, the microbial processes that *remove* nitrate (NO_3^-) from the soil are called **denitrification**. Denitrification occurs as a result of anaerobic respiration by **denitrifying bacteria** (Figure 14.2, and in Chapter 10). These bacteria can use nitrate (NO_3^-) instead of oxygen (O_2) as the final electron acceptor in their electron transport chains. When nitrate (NO_3^-) accepts electrons, it is reduced to nitrite (NO_2^-). Other denitrifying bacteria can use the nitrite (NO_2^-) as their final electron acceptor, reducing it to nitrous oxide (N_2O) or nitrogen gas (N_2). Nitrate can also be removed from the soil by plants and incorporated into their organic molecules; this process is called **assimilation**.

Finally, the nitrogen that has been incorporated into the molecules of living organisms is returned to the environment when these organisms die. Microbial decomposers break down the proteins of the dead, removing amino groups and releasing them into the environment as ammonia (NH_3). The process of removing amino groups from amino acids is called **deamination**. When ammonia is released to the soil by this process, it is called **ammonification**.

SULFUR CYCLE

Many of the processes that affect the nitrogen cycle also affect the **sulfur cycle** (Figure 14.3). Sulfate (SO_4^{2-}) can be taken up by plants and microbes and **assimilated** into their organic molecules. When organisms die and are decomposed, their proteins are **desulfurylated**, and the sulfur is released back to the environment as hydrogen sulfide (H_2S). Sulfate is also removed from the environment by **sulfate-reducing bacteria**, which use sulfate (SO_4^{2-}) as an electron acceptor during anaerobic respiration. Examples of bacteria that produce hydrogen sulfide (H_2S) as waste are given in Figure 14.3 and Chapter 10.

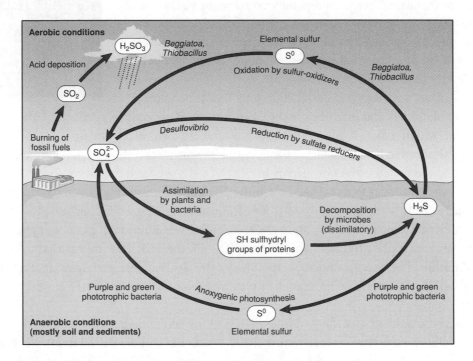

FIGURE 14.3. The sulfur cycle.

Hydrogen sulfide is oxidized to sulfur and sulfate by chemolithoautotrophs and anoxygenic photosynthetic bacteria. Sulfate is reduced to hydrogen sulfide by anaerobic respiration. Sulfur that was assimilated by plants and animals is released back to the environment by decomposition and dulsulfurylation.

Hydrogen sulfide (H_2S) is a reduced form of sulfur and thus makes a good electron donor. Chemolithoautotrophic bacteria can oxidize hydrogen sulfide (H_2S), using it as a source of energy and electrons. The oxidation of hydrogen sulfide (H_2S) produces elemental sulfur (S^0), which can also be oxidized by chemolithoautotrophic bacteria. The oxidation of sulfur produces sulfate (SO_4^{2-}) as waste. The chemolithoautotrophs that perform these reactions are called **sulfur-oxidizing bacteria**. Examples of sulfur-oxidizing bacteria are given in Figure 14.3 and Chapter 10. Hydrogen sulfide (H_2S) and elemental sulfur (S^0) also serve as electron donors for bacteria that perform anoxygenic photosynthesis. Like the chemolithoautotrophs, these bacteria produce elemental sulfur (S^0) and sulfate (SO_4^{2-}) as waste.

Microenvironments And Gradients

Biogeochemical cycling is carried out by the many microbial communities that live on planet Earth. As we have seen, these communities exist in all types of environments. Because of our size, humans might look at a forest soil or a freshwater pond and think of it as one type of environment. But to a microbe, the picture is very different. In the words of microbiologist T. S. Brock, "Microbes are small; their environments are also small." Thus, a patch of forest soil might contain many different **microenvironments** (Figure 14.4), each of which is suited to a different group of microbes. Each microbe in a community may have very specific requirements for oxygen, light, temperature, and nutrients.

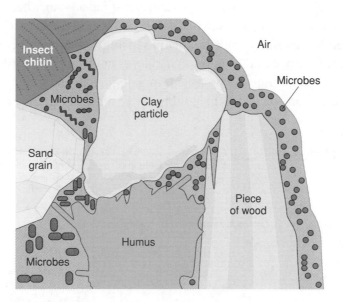

FIGURE 14.4. Microbes in the soil.

A soil aggregate showing the location of microorganisms. Most microbes in this ecosystem are attached to the surfaces of soil particles.

When we consider an environment such as forest soil or a freshwater pond from the point of view of a microbe, we see that conditions are very different depending on where you are. **Gradients** of light, oxygen, and various chemicals exist in microbial ecosystems. In a pond, for example, light is greatest at the surface of the water and decreases as you go deeper into the water. Oxygen is also greatest at the surface and decreases as you go deeper. As you follow these gradients from the surface of the water to the bottom of the pond, different species of microbes might be found growing all along the gradient, each growing in the specific microenvironment that is optimal for it.

BIOFILMS

Surfaces are also important in the establishment of microbial communities. For example, ions and water may adhere to the surfaces of soil particles, creating a nutrient-rich microenvironment. Microbes establish themselves in these surface microenvironments, forming communities called **biofilms**. Biofilms form almost everywhere there is a surface and some water. If you have ever felt the surface of a slick dock piling or tried to cross a stream by stepping on a slippery rock, you have felt a biofilm. You grow your own biofilm on the surface of your teeth every night.

> **REMEMBER**
> Biofilms are microbial communities living in a slimy matrix.

Biofilms form when certain types of bacteria settle on surfaces and begin to produce sticky polysaccharides (Figure 14.5). These bacteria also release signaling molecules that induce other bacteria of the same type to start producing polysaccharides. When bacteria sense each other and coordinate their behavior, it is called **quorum sensing**. Other microbes colonize the surface or become trapped in the polysaccharide, establishing the biofilm community. Microbes that live in biofilms are protected from drying out and from antimicrobial compounds. Some bacteria that form biofilms are of medical significance (Chapter 20).

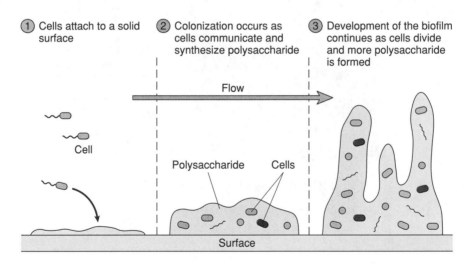

FIGURE 14.5. The formation of a biofilm.

Microbial Habitats

SOIL

The number of microorganisms present in the soil depends most strongly on the amount of nutrients and the availability of water, although temperature and physical structure of the soil also play a role. It is estimated that some soils have as many as 1,000,000,000 (or 1×10^9) microbial cells per gram of soil (dry weight). Microbes are important in the establishment of soils, releasing acids that help break down rocks into smaller particles. They contribute to the nutrients in soils by decomposing organic matter and fixing nitrogen. Fungi in the soil form symbiotic associations with plant roots called **mycorrhizae** (Chapter 12) that create a greater range for absorption of nutrients and water for plants, which is very important to the health of forests.

> **REMEMBER**
> Soil contains minerals, organic matter, and lots of microbes.

Microbial activity is highest near the surface of soils, where organic matter is most plentiful. Most microbial growth is found on the surfaces of soil particles (Figure 14.4) that are located in the **rhizosphere**, the soil microenvironment that surrounds plant roots. In soils that have abundant organic material, such that carbon is readily available, the amount of inorganic nutrients such as nitrogen and phosphorous determines the amount of microbial growth. In environments like this, nitrogen and phosphorous are referred to as **limiting nutrients**.

Besides nutrient availability, the other major factor that affects microbial growth in the soil is the availability of water. In the soil, water may attach to surfaces of soil particles (Figure 14.4) or may exist as thin sheets of free water between soil particles. In well-drained soils, air is able to penetrate freely, and there is plenty of oxygen to support rapid microbial growth. In waterlogged soils, the amount of oxygen is very low and can be quickly used up. The lack of oxygen kills some decomposers and allows others to survive, thus changing the types of microbial activity in the soil. For example, nitrification may not occur in waterlogged soils, resulting in the loss of useful nitrogenous compounds from the soil.

FRESHWATER

Just as in the soil, the number of microorganisms found in aquatic ecosystems depends on the availability of nutrients. Water with abundant nutrients, like marine estuaries or sewage-contaminated rivers, has large microbial populations. In low-nutrient environments, microbes tend to grow on surfaces where nutrients are likely to be more available. The microbial populations in aquatic environments are also strongly affected by the gradients of light and oxygen. Light is most available at the surface of the water and decreases as you go deeper. Likewise, oxygen is not very soluble in water, and its concentration decreases as you go deeper into a body of water.

Freshwater environments like lakes are good examples of the effect of gradients on microbial communities. A freshwater lake contains many different microenvironments, each of which supports a unique microbial community (Figure 14.6). Near the surface of the lake, in the **limnetic zone**, light and oxygen are plentiful. Photosynthetic microbes like algae and cyanobacteria grow nearest to the surface, acting as primary producers for the rest of the lake community, which includes microbes, fish, and other aquatic life. In the limnetic zone, there are also consumers that require oxygen and are good at attaching to surfaces, including members of the genera *Pseudomonas* and *Caulobacter* (Chapter 10).

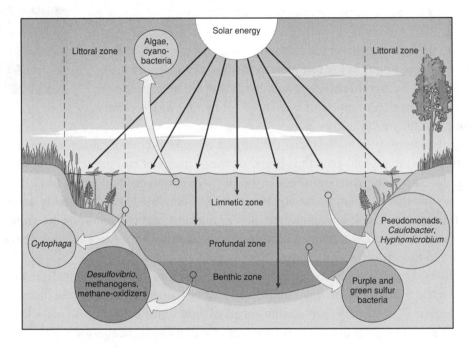

FIGURE 14.6. The zones of a freshwater lake.

*Light and oxygen are most abundant at the surface
of the lake and decrease toward the bottom.*

Just below the limnetic zone is the **profundal zone**, which contains light but very little oxygen. This zone is preferred by the photosynthetic purple and green sulfur bacteria, which are generally anaerobes (Chapter 10). These bacteria perform anoxygenic photosynthesis, using hydrogen sulfide (H_2S) as a source of electrons. The hydrogen sulfide used by the purple and green sulfur bacteria diffuses up from the **benthic zone**, the zone at the bottom of the lake that contains the sediments. Anaerobic sulfur-reducing bacteria (Chapter 10) like *Desulfovibrio* live there, using sulfate (SO_4^{2-}) as the final electron acceptor in anaerobic respiration. The reduction of sulfate (SO_4^{2-}) produces the hydrogen sulfide (H_2S), which diffuses up into the profundal zone, creating yet another gradient that influences microbial growth.

Plants grow along the shores of the lake in the **littoral zone**. As the plants live and die, they contribute organic matter to the lake. Among their roots live cellulose-degrading bacteria such as those in the genus *Cytophaga*. As organic detritus makes its way into the benthic zone, anaerobic consumers such as those in the genus *Clostridium* break it down. These bacteria ferment the organic matter, producing organic acids that serve as food for other bacteria like *Desulfovibrio*. Methane prokaryotes also flourish in the benthic zone. Methanogens break down the organic detritus, releasing methane as a waste product. Methane-oxidizing prokaryotes can then use the methane as an energy source.

MARINE ENVIRONMENTS

With the exception of coastal waters and hydrothermal vents, marine ecosystems like the open ocean are very low-nutrient environments. Marine communities depend heavily upon the action of the **phytoplankton**, small photosynthetic organisms that drift along at the ocean's surface and serve as the primary producers for marine ecosystems. The phytoplankton include cyanobacteria (Chapter 10), and photosynthetic protists such as diatoms (Chapter 12), and green algae (Chapter 12). These organisms serve as food for the **zooplankton**, which consists of protozoa and small animals. Zooplankton serve as the food for larger species such as fish, and so on, up the food chain. Because low nutrients limit the growth of the phytoplankton, the open ocean community is much less dense than communities in coastal waters or terrestrial environments.

Although open oceans are not densely populated, they cover approximately 70 percent of the planet and have an enormous impact on terrestrial ecosystems in terms of the weather and availability of food. For example, the importance of marine phytoplankton to marine communities and the entire planet is demonstrated by their photosynthetic output. As measured by the amount of oxygen waste they produce, estimates of the contribution of marine phytoplankton to the world's photosynthesis range from 50 to 80 percent! This indicates that the oceans are responsible for at least half of the primary production on Earth. The high levels of primary productivity occur in the **photic zone** of the ocean, where light is available.

> **REMEMBER**
> Microbes perform over half the world's photosynthesis.

As you move away from the surface of the ocean water, the available light decreases, and phytoplankton can no longer survive. Organisms that live in the dark, **aphotic zone** must either travel upward to find food or rely on the organic detritus, called *marine snow*, that drifts down from the surface. Because there is less organic matter, fewer organisms can live in this zone. The deep ocean is also very cold and is under enormous pressure. Organisms that live there have evolved strategies to deal with these conditions. Bacteria and archaea that live in the deep ocean are cold-loving, or **psychrophilic**, and pressure-loving, or **barophilic**. **Bioluminescent** bacteria, which perform a chemical reaction that emits light, are symbiotically associated with many deep-sea animals. One deep-sea fish, the anglerfish, hangs a lure containing these bacteria in front of its mouth. When curious little fish come to see what the glowing dot is, the anglerfish has them for lunch!

HYDROTHERMAL VENTS

On the ocean floor, where there is no sunlight and the pressure is intense, there are vast spaces where it seems like nothing except some slow-growing barophilic bacteria live. However, this vast space is punctuated by small, thriving communities of bacteria and animals that cluster around hydrothermal vents. These vent ecosystems rely upon primary production by chemolithoautotrophic bacteria that oxidize the reduced inorganic materials that come spewing out of the vents. Animals such as giant tube worms, clams, and mussels that live around the vents depend upon the carbon that is fixed by these chemolithoautotrophic bacteria. In fact, these animals actually contain symbiotic chemolithoautotrophs inside of them. The worms absorb hydrogen sulfide (H_2S) and carbon dioxide (CO_2) from the environment, which is used by the bacteria to produce organic carbon. The bacteria release some of this carbon to the worm, which it uses as food.

> **REMEMBER**
> Chemolithoautotrophs use reduced gases from the Earth and carbon dioxide to make food, forming the basis of hydrothermal vent communities.

Hydrothermal vent communities come and go. If something happens to close the vent, the entire community dies. Remnants of these dead communities can be seen clustered around old chimneys on the ocean floor. One of the great mysteries of the deep sea is how these communities establish themselves when new vents open. Some people believe that hyperthermophilic bacteria, which tolerate extremely high temperatures, can actually travel under the Earth's surface within the hot minerals. However, given the vast spaces of inhospitable ocean between some vents, how do the animals of these communities colonize a new vent site?

COASTAL WATERS

Because coastal waters receive run-off from the land, they are much more nutrient-rich than the open ocean and support abundant communities of marine life. The availability of light and nutrients supports the growth of phytoplankton and the familiar red, green, and brown algae (Chapter 12). Primary production by bacteria and algae supports a vast community of zooplankton and animals such as shellfish, fish, birds, and marine mammals. **Estuaries**, where rivers empty into the marine water, are particularly important to the health of marine life because these areas serve as nurseries for many species of fish, including important sources of food for the human population.

Increases in the human population continue to push development into proximity with coastal waters, which is putting these ecosystems at great risk. Oceanfront real estate is usually very valuable, and there is great short-term economic incentive to develop at these locations. However, because coastal waters are so important to the health of the oceans and the sustainability of the world's fish populations, there is a very real long-term economic cost to developing near-shore environments. The decline in the world's fisheries as a result of overfishing and the destruction of coastal habitat is evidence of this cost. Unfortunately, it is very difficult to find a balance between the rights of individual property owners and the rights of the general population. Currently, the declining health of the world's oceans suggests that the long-term costs of human development are not being adequately considered.

MICROORGANISMS AND WATER QUALITY

Human impacts on water quality occur everywhere around the world. Anywhere there are people, for as long as there have been people, there has been the problem of human excrement. If human excrement contaminates drinking water, a number of diseases can result (Table 14.1, see also Chapter 26). The pathogens that cause these diseases reproduce inside human hosts, exit the body in the feces, and then are taken in orally by a new host. This is called **fecal–oral transmission**. In the United States alone, about 900,000 people become sick each year from waterborne diseases. Worldwide, diarrheal diseases are among the leading killers of children younger than 5 years of age and are responsible for over 2 million deaths each year.

> **REMEMBER**
> Fecal contamination of water supplies can spread microbes that cause diarrheal diseases.

TABLE 14.1. Selected Waterborne Diseases

Pathogen	Disease
Bacterial pathogens	
E. coli O157:H7	Bloody diarrhea, hemolytic uremic syndrome
Vibrio cholera	Cholera
Salmonella	Salmonellosis
Shigella	Shigellosis
Toxoplasma	Toxoplasmosis
Legionella	Legionellosis
Protozoan pathogens	
Cryptosporidium	Cryptosporidiosis
Cyclospora	Cyclosporiasis
Giardia	Giardiasis
Viral pathogens	
Norwalk virus	Norwalk virus infection
Hepatitis A	Hepatitis

In addition to the problems of human excrement, water quality is also impacted by chemical runoff from agriculture, industry, urban populations, and mining operations. Because of the extraordinary metabolic diversity of microbes, they have become our partners in our efforts to clean up our wastes. **Sewage treatment** uses microbes to degrade the organic matter in human excrement, and **bioremediation** uses microbes to clean up pollution.

POLLUTION

For humans to survive and be healthy, they must have clean water. Clean water, or **potable water**, is water that is free of all objectionable material. In other words, it looks good, smells good, tastes good, and doesn't contain anything that could make you sick like pathogens, toxins, or radioactive material. **Pollution** is anything in the water that makes it nonpotable. **Freshwater** is water that does not contain salt. Sewage refers to human wastes and garbage that gets dumped into sewers. Wherever there is a human population, humans produce wastes that have the potential to affect their water quality.

In addition to the possible presence of pathogens in sewage, sewage and some pollutants are rich in nutrients. Some of these nutrients, like nitrogen and phosphorous, are often in short supply in natural ecosystems. The growth of microbes such as cyanobacteria and algae are limited by the low levels of nitrogen and phosphorous, which act as **limiting nutrients**. When raw sewage, agricultural runoff, or certain industrial pollutants are released into water, nitrogen and phosphorous become abundant. This leads to the **eutrophication** of lakes and streams.

> **REMEMBER**
> Pollution can cause eutrophication of lakes and streams, ultimately leading to a decrease in oxygen, fish kills, and dead zones.

Eutrophication begins with **algal blooms**, huge bursts in the growth of photosynthetic microbes. As the algae die, they are decomposed by bacteria that use aerobic respiration. The bacterial decomposition uses up the oxygen in the water, creating anoxic conditions. The lack of oxygen results in the death of oxygen-requiring organisms, including fish. This causes huge **fish kills**. The organic matter is degraded and settles to the bottom of the water, filling in the lake or stream.

Nutrients and pollutants can be removed from water as it passes through **wetlands**. Wetlands are areas that are between permanently flooded environments and dry land, such as marshes, swamps, forested wetlands, and bogs. As water flows through wetlands, the microbes and plants utilize the nitrogen and phosphorous for growth, as well as break down many pollutants. Runoff from human populations that passes through wetlands is much cleaner when it empties into natural waters.

In addition to the value of wetlands for water quality, these areas have many other impacts on ecosystems and human economics. Like coastal waters, wetlands serve as the habitat for many aquatic birds and economically important fish species. Wetlands can absorb extra water during periods of heavy rainfall and help decrease destructive flooding, which can destroy people's homes. Despite their importance, draining and filling for new developments are rapidly destroying wetlands. It is estimated that approximately 50 percent of the wetlands that existed in the United States in colonial times has already been lost, with the greatest losses occurring in southern states like Louisiana and Florida. In the late 1970s and early 1980s, the estimated rate of wetland loss was 300,000 acres per year. Federal legislation known as the Clean Water Act and state wetland protection programs have helped to decrease wetland losses to an estimated 70,000 to 90,000 acres per year.

> **REMEMBER**
> Wetlands act as natural filtration systems for water because microbes and plants in the wetland eliminate pollutants from runoff before it returns to natural waterways.

Sewage Treatment

HISTORY OF SEWAGE TREATMENT

The earliest evidence of humans trying to deal with their waste are the remains of sewer systems that were built around 3000 B.C. in Scotland, India, and Pakistan. These communities used channels and open drains that carried human waste away from habitation. In the first century A.D., the ancient Romans built large communal toilets over flowing water that carried the wastes away. However, as human populations became denser, the problem of sewage increased. In many European cities in the 1700s and 1800s, people defecated and urinated in chamber pots that were then dumped out onto the street every morning. In London in the 1800s, huge cesspits of waste were located under the buildings. These cesspits stank horribly and sometimes contaminated the groundwater, leading to outbreaks of diseases such as cholera and typhoid fever.

Technological innovations like the invention of inexpensive iron pipes combined with scientific advances in the understanding of disease led to development of the first sewage treatment systems in the early 1900s. The earliest systems passed human waste through sand filter beds and were widely used in Europe around 1894. Until very recently in the United States, most cities either had little or no sewage treatment facilities. Many cities simply dumped their raw sewage into a nearby waterway. As populations grew, the abilities of these natural ecosystems to absorb and process the wastes began to fail, and problems such as eutrophication resulted.

Then in the 1970s, thanks to pressure from the environmental movement of the 1960s, a major piece of legislation was passed with the aim of protecting the right of all Americans to clean water. Known as the Clean Water Act, this legislation established regulations for discharging pollutants into the waters of the United States and funded the construction of sewage treatment plants. After the Clean Water Act took effect in 1972, the percentage of the nation's waters deemed clean enough for fishing and swimming nearly doubled. However, the conflict between the needs of development, industry, and conservation continue to raise challenges to the Clean Water Act.

BIOCHEMICAL OXYGEN DEMAND

The amount of organic matter in sewage and other water is measured by the **biochemical oxygen demand (BOD)**. This test measures how much aerobic respiration by bacteria it takes to break down the organic matter in a sample. The greater the amount of organic matter, the more work the bacteria will have to do to break it down. Because aerobic respiration requires oxygen, the more work the bacteria must do, the more oxygen they will use up. So, to measure the BOD of a sample, the sample is well aerated, mixed with the appropriate bacteria, and placed in the dark for 5 days. The decrease in the amount of oxygen during those 5 days is measured. The greater the amount of organic matter, the more the aerobic respiration, the greater the BOD.

PRIMARY SEWAGE TREATMENT

Sewage treatment (Figure 14.7) typically begins with physical treatment of the sewage in a process called **primary sewage treatment**. During this process, screens are used to remove large floating material, small particles like sand settle out, and skimmers are used to remove floating oil and grease. The sewage travels to sedimentation tanks where more solid matter settles out, forming **sludge**. After primary treatment, about 25 to 35 percent of the BOD has been removed.

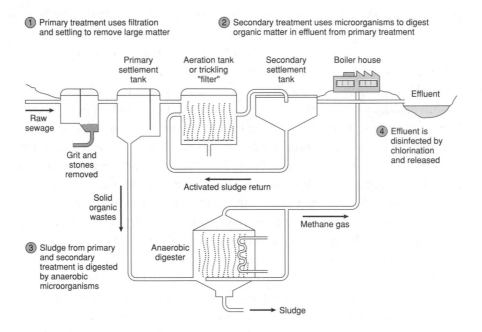

FIGURE 14.7. Sewage treatment.

Aerobic and anaerobic digestion by microorganisms removes organic matter from sewage.

SECONDARY SEWAGE TREATMENT

Secondary sewage treatment uses microbes to further break down the organic matter in the sewage. Most of the organic matter that remains in the sewage is dissolved in the **primary effluent**, the liquid portion that remains from the primary treatment. During secondary treatment, the primary effluent is well aerated to favor the quick degradation of the organic matter by aerobic bacteria. The two major designs for secondary sewage treatment plants are the **trickling filter system** and the **activated sludge system**.

In trickling filter systems, the primary effluent is sprayed over rocks, wooden blocks, or plastic honeycomb. There are plenty of airspaces in these arrangements so that oxygen is readily available to the biofilms of organisms that coat the surfaces of these materials. As the effluent trickles over them, the microbes oxidize the organic matter, removing 80 to 85 percent of the BOD. The effluent is then disinfected and released.

In activated sludge systems, the primary effluent is aerated and placed in a tank. Sludge from the primary treatment, called **activated sludge**, is added to the effluent in order to inoculate it with sewage-degrading microbes. The organisms in the activated sludge multiply and begin breaking down the organic matter. Some of the bacteria, such as members of the genus *Zoogloea*, form masses of bacteria called **flocs**. Organic matter gets trapped in these flocs and settles to the bottom of the tank as sludge. This removes 75 to 95 percent of the BOD from the effluent, which is then disinfected and released.

> **REMEMBER**
> Secondary sewage treatment uses microbes to break down organic matter in the sewage.

Solid matter, or sludge, that has settled from the primary and secondary treatment is further broken down in an **anaerobic sludge digester**. Anaerobic sludge digestion uses microbes such as methane-producing bacteria to break down the organic matter in the sludge. This converts the sludge to soluble substances and gases such as methane and carbon dioxide. The remaining solids are dried and may be used in landfills or as fertilizers if they are free of human pathogens and heavy metals.

DISINFECTION AND RELEASE

In most areas of the United States, secondary sewage treatment is used, and the effluent is disinfected with chlorine and released into natural waters. This effluent still contains organic material and may contain some pathogens, particularly viruses and spore-forming protozoa that might survive passage through secondary treatment. If the sewage was contaminated with heavy metals, these usually end up in the sludge. As long as the amount of effluent that is discharged does not exceed the ability of the ecosystem to break down the BOD that remains in the water, then eutrophication will not occur. However, as human populations grow, the impact on natural systems becomes more severe, and water quality can be compromised.

TERTIARY SEWAGE TREATMENT

In situations where secondary treatment is no longer adequate, additional strategies need to be employed to maintain water quality. In some parts of the United States where there is strong economic incentive to preserve water quality, such as at resort lakes, additional **tertiary treatment** is applied to the water. Tertiary treatment uses filtration and the addition of chemicals to remove the remaining BOD, phosphorous, and nitrogen from the water. After tertiary treatment, the water is potable. Although tertiary treatment is expensive, some countries routinely use it to treat their sewage. As an alternative to tertiary treatment, changes may be made to secondary treatment systems so that effluent is released into the soil. As the effluent is passed through the soil, the microorganisms and plants further purify the water.

Bioremediation

When you consider the metabolic diversity of microbes and the amazing efficiency of biogeochemical cycling, you might start to think that everything that goes into the ground gets broken down. Unfortunately, that is not the case. Industrialized societies produce a wide variety of compounds that are resistant to microbial degradation and dangerous to the health of humans and other organisms. These pollutants include pesticides, petroleum products, plastics, and heavy metals. Some of these pollutants are man-made compounds called **xenobiotics**, which do not normally occur in nature. For example, the insecticide dichlorodiphenyltrichloroethane (DDT) can persist in the environment for four years, or even longer. It is passed from one organism to the next in the food chain, eventually reaching lethal concentrations in the top predators.

The good news is that many of these pollutants *can* be broken down by microbes and converted into harmless products. This has opened up a whole area of research into **bioremediation**, the use of microbes to clean up pollution. Some of the most well-known attempts at bioremediation have been to clean up oil spills, like that of the *Exxon Valdez* in 1989. Oil-eating bacteria like *Pseudomonas* occur naturally in the environment, but they normally exist at such low numbers that they do not break down the oil very rapidly. After the *Exxon Valdez* spill, scientists sprayed nitrogen and phosphorous fertilizers onto a test beach. This increased the numbers of oil-eating bacteria and the speed at which oil was removed from the beaches. Although bioremediation hasn't yet been used in response to the BP oil spill in the Gulf of Mexico in April 2010, scientists discovered the presence of naturally occurring oil-eating bacteria growing in the oil plumes.

Although bioremediation of the *Exxon Valdez* oil spill illustrates the possible benefits of bioremediation, it also points out some of the difficulties. Even when pollution-eating bacteria exist, they may break down the pollutants at very slow rates, particularly in low-oxygen environments. Many pollutants, including petroleum, are hydrophobic and will coat the surfaces of soil particles, making it difficult for microbes to adhere to these surfaces. If fertilizers are used to increase the microbial population in a polluted area, it makes bioremediation very expensive.

Another approach to increasing the numbers of the pollution-eating bacteria is to grow them in large numbers in the lab and add them to the polluted environment. This process is called **bioaugmentation** or seeding. One potential problem with bioaugmentation is making sure the microbes access the pollution. For example, if the pollution is deep in the soil or in the groundwater, how do you get the microbes down there? One possible answer is to dig up the soil, mix in the microbes, and return this mixture to the same location. Obviously, this is going to greatly add to the cost of the bioremediation.

Even with all of these challenges yet to be solved, the possibility of bioremediation is a very exciting one. As the Earth becomes increasingly crowded with people, our need to clean up our messes only grows more intense. Researchers in bioremediation are developing strains of faster growing pollution-eating bacteria, using genetic engineering to create new strains that can break down xenobiotics, and inventing biodegradable plastic. All of these approaches give hope for the future of bioremediation.

REVIEW EXERCISES FOR CHAPTER 14

1. Describe the relationship between producers, consumers, and decomposers in an ecosystem.

2. How does microbial metabolism contribute to the cycling of carbon in the environment?

3. How does microbial metabolism contribute to the cycling of nitrogen in the environment?

4. How does microbial metabolism contribute to the cycling of sulfur in the environment?

5. Explain the statement, "Microbes are small; their environments are also small."

6. How would bacteria that live in a flowing stream benefit from being part of a biofilm on a rock, rather than growing in the open water?

7. What factors influence microbes living in the soil?

8. What factors influence microbes living in freshwater?

9. What factors influence microbes living in marine environments?

10. Describe the interactions among organisms in a hydrothermal vent community.

11. Describe and give an example of fecal–oral transmission.

12. Describe the effect of dumping untreated sewage into a pond.

13. Describe primary and secondary treatment of sewage (wastewater).

SELF-TEST

1. Photosynthetic cyanobacteria would be most abundant in which part of a freshwater lake?

   A. Profundal zone
   B. Benthic zone
   C. Limnetic zone
   D. All of the above are correct.
   E. None of the above is correct.

2. True or False. The reduction of nitrate (NO_3^-) to nitrite (NO_2^-) occurs in the absence of oxygen.

3. The reduction of sulfate (SO_4^{2-}) to hydrogen sulfide (H_2S) represents which of the following metabolic processes?

   A. Anoyxgenic photosynthesis
   B. Anaerobic respiration
   C. Chemolithoautotrophy
   D. Protein catabolism
   E. None of the above is correct.

4. True or False. Biofilms can protect organisms from antimicrobial compounds like antibiotics.

5. True or False. Mycorrhizae are fungi that are parasitic on plants.

6. Biological degradation occurs during which type of sewage treatment?

   A. Primary treatment
   B. Secondary treatment
   C. Tertiary treatment
   D. Both A and B are correct.
   E. A, B, and C are correct.

7. The release of untreated sewage into water can cause all of the following **except**:

   A. Disease
   B. Increased biological oxygen demand
   C. Eutrophication
   D. Decrease in algal growth
   E. All of the above are correct.

8. Bioremediation is the use of microorganisms to

   A. produce useful compounds.
   B. degrade pollutants in the environment.
   C. degrade organic matter in sewage.
   D. cycle nitrogenous compounds.
   E. determine the amount of organic matter in sewage.

9. Hydrothermal vent communities rely upon

   A. organic detritus from the photic zone.
   B. photosynthesis by deep-sea algae.
   C. primary production by chemolithoautotrophs.
   D. aerobic respiration by vent worms.
   E. All of the above are correct.

10. Bioaugmentation refers to

   A. adding microbes to the environment to speed up bioremediation.
   B. adding nutrients to the environment to encourage the growth of microbes for bioremediation.
   C. adding microbes to sewage to speed up the breakdown of organic matter.
   D. adding oxygen to sewage to speed up the breakdown of organic matter.
   E. None of the above is correct.

Answers

Review Exercises

1. Producers are at the bottom of every food chain. They are autotrophs that make organic molecules from carbon dioxide. Consumers eat the producers and other consumers. As producers and consumers die, they are broken down by decomposers.

2. Photosynthetic microbes convert carbon dioxide into sugars. Chemoheterotrophic microbes catabolize sugars and release carbon dioxide back to the environment.

3. Nitrogen-fixing bacteria capture nitrogen gas from the atmosphere and convert it into ammonia. Ammonia is oxidized by chemolithoautotrophic bacteria into nitrite and nitrate. Nitrite can also be oxidized by chemolithoautotrophs into nitrate. Nitrate is reduced by bacteria performing anaerobic respiration to nitrite and nitrogen gas. Nitrite can also be reduced by anaerobically respiring bacteria. Decomposers catabolizing dead matter that contains proteins release ammonia into the environment.

4. Anoxygenic photosynthetic bacteria use hydrogen sulfide as an electron donor, producing sulfur or sulfate as waste. Chemolithoautotrophic bacteria can also oxidize hydrogen sulfide and produce sulfur or sulfate as waste. Bacteria that perform anaerobic respiration can utilize sulfate as a final electron acceptor, reducing it to sulfur and hydrogen sulfide. Decomposers that are catabolizing dead matter that contains proteins may release hydrogen sulfide into the environment.

5. What seems like a single environment to us may in fact contain many subtle gradients of nutrients, light, water, and oxygen. Because microbes are so small, they are affected by these minute differences. The immediate environment of a microbe is a very small piece of a larger system.

6. A bacterium in a biofilm attached to a rock is stable and protected from the environment by polysaccharides produced by the biofilm. Nutrients may become trapped in the polysaccharide. In the open water, nutrients would be less plentiful, and the environment of the bacterium would not be as protected.

7. Microbes living in the soil are affected by the availability of nutrients, light, oxygen, and water. They are also affected by temperature and physical structure of the soil. Microbial activity is greatest near the surface of soils, where nutrients are most plentiful.

8. Microbes living in freshwater are affected by the availability of nutrients and the gradients of light and oxygen. Oxygenic photosynthetic bacteria live close to the surface where oxygen and light are plentiful. Aerobic consumers are also found near the surface. Deeper in the lake, as oxygen becomes scarce, anaerobic bacteria like anoxygenic photosynthetic bacteria and anaerobic chemoheterotrophs can be found.

9. The availability of light strongly influences microbes living in marine environments. Photosynthetic algae and bacteria that live near the surface of the ocean form the basis of the ocean's food webs. As you go deeper and deeper into the ocean, light becomes less available and then totally absent, and pressure increases. The microbes that live in the deeper ocean may be decomposers that rely upon organic detritus from above. Many of them are bioluminescent.

10. Hydrothermal vent communities depend upon the primary production of chemolithoautotrophic bacteria that oxidize reduced inorganic compounds that seep out of the Earth. Many vent animals harbor these bacteria as symbionts and receive organic carbon from them.

11. Fecal–oral transmission occurs when an enteric pathogen exits the intestines of one person and is transferred into the food or water ingested by another person. *Salmonella, E. coli, Shigella*, and many other diarrheal pathogens are transmitted this way.

12. When untreated sewage is released into freshwater, it greatly increases the amount of nutrients, especially nitrogen and phosphorous. This promotes the rapid growth of algae. As the algae bloom and die, microbial decomposers grow rapidly and break down the algae. The decomposers utilize oxygen during aerobic respiration, depleting the oxygen from the water. This kills oxygen-requiring organisms such as fish. In addition, untreated sewage may contain pathogens.

13. Primary treatment of wastewater involves physical filtration of large debris combined with settling of organic matter. Secondary treatment takes the effluent from primary treatment and further degrades the organic matter using microorganisms. The effluent is well aerated to encourage the rapid growth of aerobic microorganisms that catabolize the organic matter. Settled organic matter (sludge) from both the primary and secondary treatments is further broken down by anaerobic microbes in an anaerobic sludge digester.

Self-Test

1. C	3. B	5. F	7. D	9. C
2. T	4. T	6. B	8. B	10. A

Applied Microbiology

WHAT YOU WILL LEARN

This chapter presents some of the useful ways that humans use microbes in industry. As you study this chapter, you will:

- discover how scientists unraveled the secrets of fermentation;
- learn how the food industry prevents foodborne diseases;
- examine the many foods produced with help from microbes;
- explore how microbes are used to produce medicines and industrial products.

SECTIONS IN THIS CHAPTER

- History of Fermentation
- Food Microbiology
- Microorganisms and Food Production
- Industrial Microbiology

The earliest uses of microbes by humans were probably in the preparation of fermented beverages. Some people think these beverages have been produced by humans since farming was started about 10,000 years ago. Of course, when ancient civilizations like those of the Sumerians, Egyptians, Mesopotamians, and Greeks were producing beer and wine thousands of years ago, they didn't know that microbes were involved. Indeed, many cultures thought that beer was a gift from the gods, and that intoxication was a divine experience. Evidence for the importance of fermented beverages to early societies is found in their laws, which regulated the price of beer and its minimum alcohol content. In addition, brewers sometimes enjoyed a position of privilege in society.

History of Fermentation

An understanding of the microbial basis of fermentation did not begin until after the discovery of microbes by van Leeuwenhoek in 1673. In 1779, the French Academy of Sciences offered a prize of 1 kilogram of gold to anyone who solved the mystery of fermentation. This inspired the physicist Charles Cagniard-Latour to try his hand at biology and investigate the properties of yeast cells. He observed the distinctive uneven division of yeast, called budding, and noted differences between yeast isolated from wine and beer. (The offer of the gold prize was withdrawn in 1793 owing to the political unrest in France at that time.) In 1837, Theodor Schwann showed that yeast was required for fermentation to occur and that, if airborne contaminants were excluded from fermentable material, the fermentable material remained unchanged. In 1856, Louis Pasteur, who had been called in by French wine merchants to solve their problems with the souring of wine, showed that different types of microbes make different fermentation products.

As biologists began to unravel the secrets of yeast, chemists at the same time were also studying fermentation and beginning to discover the function and importance of enzymes. These chemists did not see the link between their work and the work of the biologists. Indeed, the Swedish chemist J. J. Berzelius, who published an important work on enzyme catalysis in 1836, harshly criticized the work of the biologists. The chemists favored the explanation put forth by J. Liebig in 1839 that decaying organic matter transmitted vibrations to fermentable sugar, causing it to decompose into carbon dioxide and ethanol. In 1839, the organic chemist F. Wohler went so far as to publish an insulting article in which he made fun of the work of the biologists by pretending to describe microscopic observations of little animals that were shaped like distilling flasks!

REMEMBER
Theodor Schwann showed that the presence of living cells (yeast) was required for fermentation to occur.

What the overly confident chemists failed to realize, of course, was that both they and the biologists were correct. The first person to realize this was Moritz Traube, a wine merchant who also made contributions to chemistry and physiology. Traube studied chemistry with Liebig but did not fully accept Liebig's views on fermentation. Rather, in 1858, he proposed that if Schwann's observations on yeast were valid, then further biological research should show that yeast contain the enzymes necessary for fermentation. Because of the lack of cooperation between biologists and chemists, this idea wasn't followed up on until many years later. Then in 1897, Eduard Buchner published the results of his work with cell-free extracts of yeast cells. Buchner put brewer's yeast into a hydraulic press and used pressure to squeeze the contents out of the cells. When the extracts of the yeast cells were mixed with sugar, fermentation began. The observations of the chemists and biologists on fermentation were finally united, and the science of biochemistry was born!

Today, humans use microbes to make a wide variety of products, including foods, vitamins, enzymes, antibiotics, flavorings, and alcohols. In addition to feeding us and helping us fight disease, microbial products stonewash our jeans, tenderize our meat, and get greasy dirt out of our clothing. Microbial proteins from thermophilic bacteria, which are already used in molecular biology, are being examined for their possible uses in general industry. The light-absorbing bacterial protein, bacteriorhodopsin, which is also found in members of the Archaea, is being used to make a new type of computer chip that processes information at the speed of light. Complex inks made using bacteriorhodopsin as the pigment are also being made as a way to combat counterfeiting.

Food Microbiology

FOODBORNE DISEASE

When it comes to food, the relationship between microbes and humans definitely has its ups and downs. On the one hand, we use microbes to produce many of our favorite foods. On the other hand, we continually have to fight to preserve our food from microbes. Yet even in resource-rich countries, where modern conveniences like refrigeration, canning, and chemical preservatives are generally successful at preserving food quality, epidemics of foodborne disease routinely occur. In the United States alone, the Centers for Disease Control and Prevention (CDC) estimates that foodborne disease affects 76 million people per year, hospitalizing 300,000 and killing 5000 people. Table 15.1 lists many of the common foodborne diseases (see also Chapter 26).

TABLE 15.1. Some Common Foodborne Diseases

Pathogen	Common Food Source	Disease and Symptoms
E. coli O157:H7	Undercooked hamburger, alfalfa sprouts, salami, unpasteurized milk and juice	Bloody diarrhea, hemolytic uremic syndrome
Campylobacter	Raw poultry, undercooked poultry, raw milk	Campylobacteriosis: fever, diarrhea, nausea, vomiting
Salmonella	Raw poultry, eggs, beef	Salmonellosis: diarrhea, fever, cramps, headache
Norwalk-like virus	Shellfish, salads	Vomiting, diarrhea
Clostridium botulinum	Home-canned foods with low acid content	Botulism due to botulinum toxin: double vision, slurred speech, dry mouth, weak muscles
Listeria monocytogenes	Raw hot dogs, deli meat	Listeriosis: fever, aches, vomiting, diarrhea

FOOD PRESERVATION

Any change in the appearance, taste, or smell of a food that makes a person not want to eat it is termed **food spoilage**. Spoiled food will not necessarily make a person sick; in fact, what is spoiled to a person from one culture may be appetizing to a person from another. For example, slightly rotten cabbage is the basis for the Korean food kim chee, and expensive restaurant steaks often come from sides of beef that have been aged until they are covered by a thick layer of mold. However, if food is perceived to be spoiled and if people don't want to buy it, this causes economic losses to the farmers and distributors of the food.

The shelf life of food is largely determined by its water content. **Perishable foods**, like milk, meat, and fresh fruits, contain high amounts of water and spoil quickly. **Semiperishable foods**, like potatoes and nuts, have a lower water content and spoil more slowly. **Nonperishable foods**, like sugar, flour, grains, and dry beans, have a very low water content and can be stored at room temperature for long periods of time.

> **REMEMBER**
> The more water a food contains, the more quickly it will spoil.

Perishable foods are routinely preserved by refrigeration and freezing. Cold temperatures lower the metabolic rate of many microbes and stop them from growing, although they do not usually kill them. Refrigeration at temperatures of about 4°C stops most pathogens from growing but allows the growth of the psychrotrophic bacteria and fungi that will eventually spoil your food if you do not eat it fast enough. Typical household freezers have temperatures of about –20°C, which is enough to stop microbial growth.

Pickling can preserve many perishable foods, which involves the addition of acetic acid (vinegar) and sugar or salt. The acetic acid lowers the pH of the food, preventing most microbes from growing (Chapter 6). Sugar and salt create a hypertonic solution, decreasing water availability and further inhibiting microbial growth (Chapter 6). A wide variety of vegetables are pickled, the most familiar of which is pickled cucumbers. Meats and fish can also be pickled. Sometimes, microbial growth occurs during the pickling, partially fermenting the food and producing acids. This is the origin of fermented foods such as yogurt, sauerkraut, kim chee, and sour cream.

Even without vinegar, the addition of sugar and salt or drying can preserve food by decreasing the amount of water that is available for microbial growth (Chapter 6). The salting of meat and fish is an ancient form of food preservation. The addition of sugar to fruit, combined with boiling to remove water, creates a jam or jelly that is much more resistant to food spoilage than is the fresh fruit. Drying is one of the oldest means of preserving foods known to man and is still a common method for food preservation in warmer climates. Foods like fruit or thin strips of meat are placed in a warm dry environment, either in nature or in a food dehydrator.

Chemicals can also be added to food to resist food spoilage. These include chemical preservatives, such as sodium propionate, nitrites, and sulfites, as well as antibiotics. Sodium propionate, which is commonly included in breads, has been used for many years and is not known to have any negative effects on human health. Nitrites are included in meats such as deli meats and sausages to prevent discoloration and spoilage by *C. botulinum*. However, in recent years nitrites have been discovered to be precursors to cancer-causing chemicals (carcinogens), prompting their removal from some foods. Sulfites are commonly used to prevent discoloration and spoilage of dried fruits and vegetables and wine; however, some people are highly allergic to this chemical. Vitamin C (ascorbic acid) can be used as a substitute for sulfite, but it is slightly more expensive than sulfite and does not last as long.

The preparation of pickles, jams, and jellies is often accompanied by **canning**. During canning, foods are placed in a jar or a can and heated to temperatures that will kill all potential pathogens. The correct time and temperature for canning depends on the size of the jar and the type of the food. Because high acid foods resist the growth of *Clostridium botulinum*, they can be canned safely by heating to boiling temperatures. Low acid foods, however, must be heated to 121°C, which is sufficient to kill the endospores of *C. botulinum*. This temperature is achieved by the use of a pressure canner.

> **REMEMBER**
> Commercial sterilization uses steam and pressure to kill microbes, including bacterial endospores.

INDUSTRIAL FOOD CANNING

Industrial food canning follows the same principles as home canning, but on a much larger scale. Foods are softened with steam and then loaded into cans (Figure 15.1). The food undergoes **commercial sterilization**: The cans are sealed and placed in large chambers called **retorts** that use steam and pressure to achieve the temperatures necessary to kill *C. botulinum* spores. In addition to the endospores, commercial sterilization destroys all other significant spoilage or pathogenic bacteria. Although some thermophilic bacteria may survive commercial sterilization, they are typically not capable of growing below 45°C, and so do not usually cause food spoilage or any threat to human health.

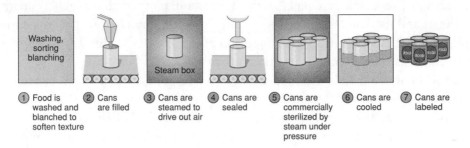

FIGURE 15.1. Industrial food canning.

   Thermophilic anaerobic spoilage occasionally occurs when canned foods are left in high-temperature environments for extended periods of time. These conditions allow the growth of thermophilic bacteria that survive the commercial sterilization. These bacteria produce acid and gas as by-products of their metabolism, causing the cans to swell and the food to have a sour odor. A few thermophilic bacteria are capable of causing spoilage without gas production; this is called **flat sour spoilage**.

ASEPTIC PACKAGING

As an alternative to canning, increasing numbers of foods are being packaged in **aseptic packaging**. Huge rolls of packaging materials of laminated paper or plastic are fed into machines. As the packaging materials travel through the machine, they are sterilized by chemicals, UV radiation (Chapter 22), or high-energy electron beams. The machine aseptically folds the sterile packaging into containers, fills them with liquid foods that have been conventionally sterilized with heat, and seals the package.

RADIATION

The safety of some foods can be improved by the use of **ionizing radiation** (Chapter 22). Gamma rays generated from radioactive elements are delivered in a dose designed to kill pathogens that are typically associated with the food to be treated (Table 15.2). The food itself is not made radioactive by the treatment, and there is no evidence to suggest that radiation makes the food harmful to humans. Because radiation does cause some changes in the chemistry of the food, it can affect flavor. Thus, except for some military applications, radiation is not typically used to sterilize food completely. Irradiated food in the United States is marked with a radura symbol, although in some special cases the food may be labeled with *cold pasteurization* rather than *irradiation*.

TABLE 15.2. Food Irradiation

Food	Purpose of Irradiation
Flour	Insect control
Fresh pork	Control of *Trichinella spiralis*
Fresh fruits and vegetables	Inhibit sprouting, prevent ripening
Dried herbs and spices	Microbial control
Refrigerated or frozen meat	Microbial control
Fresh or frozen poultry	Microbial control
Packaged frozen meat for NASA	Sterilization
Animal feed and pet food	Microbial control

MICROORGANISMS AND FOOD PRODUCTION

Microbes contribute the unique flavors of many of our favorite foods (Table 15.3). Particular strains of yeast are used to make different varieties of beer and wine. Different combinations of milk, bacteria, and fungi are used to create over 2000 different kinds of cheese in the world. Sourdough bread owes its unique flavor to a combination of products from yeast and lactobacilli. Today, the desired microbes are often added from a pure culture, such as the package of yeast you might buy at the grocery store. However, even today we sometimes do things as our ancestors did, saving a bit of bread dough as a starter for the next batch or placing a cheese to be ripened in a particular cave known to produce just the right flavor.

TABLE 15.3. Fermented Foods

Food	Ingredients	Microorganisms
Cheese	Milk curd	*Streptococcus, Leuconostoc*
Kefir	Milk	*Streptococcus lactis, Lactobacillus bulgaricus, Candida*
Yogurt	Milk, milk solids	*Streptococcus thermophiles, Lactobacillus bulgaricus*
Dry sausages	Pork, beef	*Pediococcus cerevisiae*
Cocoa beans	Cacao fruit	*Candida krusei*, Chocolate *Geotrichum*
Coffee beans	Coffee cherries	*Erwinia dissolvens, Saccharomyces*
Kim chee	Cabbage	Lactic acid bacteria
Miso	Soybeans	*Aspergillus oryzae, Zygosaccharomyces rouxii*

TABLE 15.3. (continued)

Food	Ingredients	Microorganisms
Soy sauce	Soybeans	*Aspergillus oryzae, Aspergillus soyae, Zygosaccharomyces rouxii, Lactobacillus delbrueckii*
Olives	Green olives	*Leuconostoc mesenteroides, Lactobacillus plantarum*
Sauerkraut	Cabbage	*Leuconostoc mesenteroides, Lactobacillus plantarum*
Poi	Taro root	Lactic acid bacteria
Bread	Wheat flour	*Saccharomyces cerevisiae*

CHEESE

Cheese making can start with any type of milk, but the milk of cows and goats is very commonly used. A starter culture of a lactic acid bacterium known to produce the desired flavors is added along with an enzyme called **rennin**, which hydrolyzes or splits the milk protein **casein**. The lactic acid bacteria ferment the milk sugar **lactose**, producing lactic acid. The combination of the acid environment and the action of the rennin causes the casein to coagulate into a **curd**. At this point, the developing cheese is very similar to cottage cheese: A thin liquid called whey surrounds lumps of protein curd. The whey is removed from the curd, sometimes by the application of pressure, and used in a variety of foods. Depending on how much moisture is removed, the curd can give rise to a **hard**, **soft**, or **semisoft cheese**.

> **REMEMBER**
> Cheese production begins with the fermentation of milk which causes the milk protein casein to coagulate into a curd.

To finish the cheese, the curd is ripened for various periods of time in carefully controlled environmental conditions. Just prior to ripening, it is inoculated with special bacteria or molds specific to the type of cheese to be made. For example, the mold *Penicillium* is used to make both blue cheeses and camembert. A bacterium in the genus *Propionibacterium* is used to make Swiss cheese. In soft cheeses like camembert, enzymes from the mold diffuse into the cheese, partially breaking it down. Microbes, particularly the lactic acid bacteria (Chapter 10), are also used to produce a variety of other fermented dairy products, including buttermilk, cultured sour cream, yogurt, and kefir.

BREAD

The basic ingredients of **bread** are yeast, water, and flour from wheat or other grains. The grains contain proteins called **glutens**, which form long stringy masses when bread dough is kneaded, giving it elasticity. Kneading also introduces oxygen into the dough. The yeast breaks down the sugar in the grain, releasing carbon dioxide (CO_2)

and some ethanol. As the carbon dioxide gas is released, it is trapped in the elastic dough, causing the dough to rise. When the bread is baked, the glutens denature, and the bread hardens. Ethanol in the dough typically evaporates during baking.

BEER

Grains, usually barley, can be fermented by yeast in anaerobic conditions to produce **beer** and **ale**. Because the yeast cannot ferment the starch in the grain, it must first be converted into the simple sugars, maltose and glucose. This process, called **malting**, involves letting the grain sprout in a warm dark place. As the seedlings begin to grow, they release enzymes that convert the starch into simple sugars. The sprouted grains, or **malt**, are dried and crushed. Malt may also be roasted to create the rich flavors that distinguish different kinds of beer.

The next step in brewing is called **mashing**. The malt is infused with hot water and allowed to steep, almost as if you were making a big cup of tea. The water draws the starches and enzymes out of the grain, so that the starches get broken down into simple sugars. The **wort**, or liquid portion of the **mash**, is separated from the spent grain. **Hops**, flowers that are used to flavor beer, are added to the wort, and the mixture is boiled. Many homebrewers cut the malting and mashing process short by purchasing **malt extract**, which is a syrupy solution of concentrated wort.

After the wort is boiled and cooled, it is transferred to a large container for **fermentation**. Yeast is added, or **pitched**, and the mixture is aerated to provide optimal growing conditions for the yeast. After the yeast begins to multiply, it rapidly uses up the oxygen and begins to ferment the malt sugars by **ethanol fermentation** (Chapter 5), producing carbon dioxide and ethanol as by-products. Fermentation can go on for 5 to 10 days. When fermentation is complete, the beer is transferred to bottles, kegs, or cans. A small amount of finishing sugar may be added at this point to restart fermentation and create carbonation in the bottles. Large commercial brewers typically skip this step and just pump in carbon dioxide. Commercial beers that are bottled or canned are usually pasteurized.

> **REMEMBER**
> Beer and wine are produced using yeast, which does ethanol fermentation.

WINE

Ethanol fermentation by yeast is also used to produce **wine**. Wine production begins by releasing juice from fruit, usually grapes. Yeast is added and begins fermenting the simple sugars in the juice. After the fermentation is complete, the liquid wine is separated from the solids. The wine is transferred to vats so that small particles can settle out of the solution. It is then filtered and transferred to kegs for aging. After the wine has aged, it is bottled.

SPIRITS

Distilled spirits are made from a variety of fermented foods. For example, vodka is made from fermented potatoes, tequila is made from fermented agave, and rum is made from fermented molasses. Once fermentation is complete, the alcohol is distilled.

Industrial Microbiology

Industrial microbiology utilizes microorganisms to produce useful products on a large scale. Since its origins in the production of beer and wine, industrial microbiology has expanded to include the production of pharmaceuticals, food additives, enzymes, biofuels, and other chemicals. Originally, industrial microbiology was used to produce large quantities of compounds that were naturally made by microbes. For example, microbes that naturally make the fermentation product acetone, which is needed to make gunpowder, helped the Allies win World War I. Today, recombinant DNA technology (Chapter 8) has given rise to the field of *microbial biotechnology*, which engineers microbes to make compounds that they would not normally make. Examples of compounds made by industrial microbiology are given in Table 15.4.

TABLE 15.4. Selected Products Made by Industrial Microbiology

Product	Uses
Amino acids	
Glutamic acid	Monosodium glutamate (MSG)
Phenylalanine	Aspartame (Nutrasweet)
Aspartic acid	Aspartame (Nutrasweet)
Lysine	Food additive
Citric acid	Flavor enhancer, emulsifier in dairy products, additive to adjust food pH
Enzymes	
Amylase	Corn syrup, paper, fructose for food sweetener
Glucose isomerase	Low-gluten flours for baked goods, meat tenderizers
Proteases	Detergents, cheese production (rennin)
Vitamins	
B-12	Nutritional supplements
Riboflavin	Nutritional supplements
C	Nutritional supplements
Pharmaceuticals	
Antibiotics	Fight infections
Vaccines	Disease prevention
Steroids	Anti-inflammatory drugs, oral contraceptives
Vinegar	Food production, pickling

The microbes used for industrial microbiology are typically yeasts, molds, and some bacteria, particularly the actinomycetes *Streptomyces*. The most important concern in choosing microbes for the production of commercial products is **yield**. After a microbe has been selected or engineered to make a product, further work is done to choose the strain that makes the most product per volume of culture medium. The selected microbes may be exposed to mutagens to cause changes in their DNA and then grown in conditions that favor strains that have the desired qualities. For example, the original strain of *Penicillium* mold used for penicillin production made about 5 milligrams of penicillin per liter of culture medium. After mutation and selection of strains, the strains used today produce 60,000 milligrams of penicillin per liter. After a high-yield strain has been developed, it becomes very valuable to the company that has proprietary rights to it.

There are several other qualities that are very desirable in an organism used in industrial microbiology. Preferred microorganisms grow rapidly in inexpensive media, such as whey, an inexpensive waste product from the dairy industry. The production of spores or other reproductive cells make it easier to store the microbe and to inoculate new batches of media. Also, in addition to the ability to make a lot of the product, the microbe should be able to make it quickly. Finally, because these microbes are grown in large volumes and may be released in waste to the environment, they should not be human pathogens or environmental hazards.

> **REMEMBER**
> Bioreactors are big tanks used to do fermentation on a large scale.

Growth of the microbe and production of the product occur in enormous containers called **bioreactors** (Figure 15.2), some of which hold 500,000 liters! These bioreactors are typically made of stainless steel, which is nonreactive and can be sterilized by heat. After the media and microbes are added, the reactor is sealed to protect the culture from contamination. However, there are many ports and controls so that environmental conditions such as oxygen availability, pH, and temperature can be continuously monitored and adjusted. These environmental conditions can greatly affect the yield of the product, and adjustments are often made during the fermentation process. For example, bioreactors can be aerated and stirred (Figure 15.2) when the goal is to increase the number of microbes; then aeration is slowed to enhance the production of the fermentation product.

After the reaction is complete, the product is harvested and purified, and the remaining culture material is disposed of as waste. This waste often contains large quantities of organic nutrients that can pollute the environment if released untreated. Treatment of waste to remove the organic matter is expensive and decreases the profit from sale of the product. Environmental regulations like the Clean Water Act require certain levels of treatment before these wastes are released. However, lobbyists from industry often persuade legislators to introduce legislation to modify these rules.

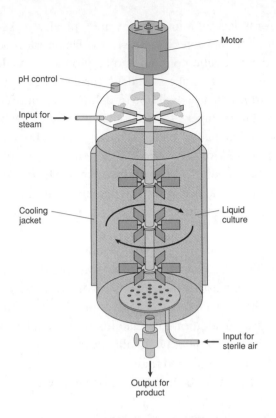

Motor

pH control

Input for
steam

Cooling
jacket

Liquid
culture

Input for
sterile air

Output for
product

FIGURE 15.2. Bioreactors.

The metabolic products made by microbes and other organisms fall into two general categories: **primary metabolites** and **secondary metabolites**. Primary metabolites are required for growth of the organism and are produced during the log phase of growth (Figure 15.3, see also Chapter 6). Primary metabolites include things like amino acids and fermentation products. Secondary metabolites, which are often complex chemical structures, are not required for growth of the organism. They are not produced until growth reaches the stationary phase (Figure 15.3). Secondary metabolites are sometimes made from primary metabolites and include things like defensive compounds such as antibiotics that inhibit the growth of other organisms. Because primary metabolites are used for growth, it is easier to get a microbe to overproduce a secondary metabolite.

Industrial microbiology also produces microbes themselves as products. Yeast is produced and sold to the baking and brewing industries. Nitrogen-fixing bacteria are grown and sold for agricultural use, where they are mixed with the seeds of leguminous plants to ensure good establishment of the symbiosis (Chapter 14). *Bacillus thuringiensis* is also grown and sold for agricultural use as a natural pesticide. This bacterium, which is commonly called *B.T.*, produces a protein toxin that kills certain insect pests.

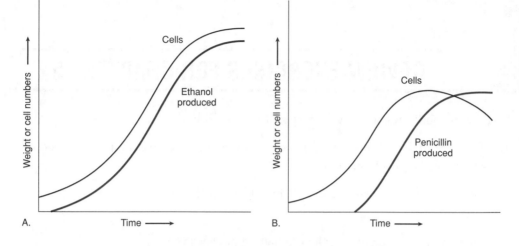

FIGURE 15.3. Primary and secondary metabolites.

A. Alcohol is a primary metabolite. Its production closely follows cell growth.
B. Penicillin is a secondary metabolite. It is not produced until the cells reach
mid-log phase, and the main production occurs when cells reach stationary phase.
(See Chapter 6 for growth phases.)

Industry is currently exploring the uses of other microbial products. The enzymes from bacteria that survive extreme environments (Chapter 6), called **extremozymes**, are of particular interest because of their tolerance to extremes of temperature, salinity, and pH. For example, industrial processes can sometimes become quite hot, which destroys most proteins and could block product formation. The use of extremozymes could greatly reduce the need for temperature control. As petroleum products become scarce, we may need to turn to microbes to help us find new sources of energy or products that are typically derived from petroleum. Development of methods to produce fuels using microbes are currently underway. Different types of microbes can use biomass from agricultural waste, sewage, or food waste to make energy-rich molecules such as methane (natural gas), ethanol, and even diesel. The production of **biofuels** by microbes is already occurring on a small scale, but scientists and industrial engineers are working to develop large-scale approaches. Some day, our machines and vehicles may be powered by methane produced by methanogens, ethanol produced by fermentative microbes, or biodiesel made by genetically engineered *E. coli*. Other industrial products like acetone, which we currently derive from petroleum, may also be obtained in the future by microbial fermentation.

REVIEW EXERCISES FOR CHAPTER 15

1. How is food categorized with respect to perishability?

2. Name the major methods of food preservation and give an example of a food that is preserved by each method.

3. What is commercial sterilization? How does it differ from other types of sterilization?

4. What are the major steps in the production of cheese?

5. What are the major steps in the production of beer?

6. What are the major steps in the production of bread?

7. What are the major steps in the production of wine?

8. What are the advantages of bioreactors used in the large-scale production of microbial products?

9. Distinguish between primary and secondary metabolites.

SELF-TEST

1. True or False. The greater the water content of a food, the faster it will spoil.

2. True or False. Foods can be preserved with salt because salt lowers the pH of the food.

3. True or False. Commercial sterilization completely sterilizes food.

4. True or False. Ultraviolet radiation can be used to prevent microbial growth in food.

5. Which of the following foods does **not** involve microbial fermentation in its production?

    A. Bread
    B. Wine
    C. Cheese
    D. Chocolate
    E. Pasta

6. All of the following are products of industrial microbiology **except**:

   A. Enzymes
   B. Amino acids
   C. Vitamins
   D. Antibiotics
   E. All of the above are correct.

7. Which of the following is the **most** important criterion for choosing a microbe that will be used in an industrial fermentation to produce a useful product?

   A. Growth in inexpensive media
   B. Yield of product
   C. Production of spores
   D. Not a human pathogen
   E. Fast production of product

8. True or False. Secondary metabolites are produced during the stationary phase of growth.

9. Which of the following organisms does not cause a foodborne disease?

   A. *E. coli* O157:H7
   B. *Salmonella*
   C. *Saccharomyces cerevisiae*
   D. *Clostridium botulinum*
   E. *Listeria monocytogenes*

10. True or False. The definition of spoiled food is that it makes someone sick.

Answers

Review Exercises

1. Perishable foods contain high amounts of water and spoil quickly. Semiperishable foods have a lower water content and spoil more slowly. Nonperishable foods have a very low water content and can be stored at room temperature for long periods of time.

2. Refrigeration and freezing slows microbial metabolism and is used to preserve a wide variety of foods, including fresh meat, cheese, and milk. Pickling slows microbial growth by lowering the pH and is used to preserve vegetables and meats. Drying makes water unavailable for microbial growth and is used to preserve meats and fruits. Salting and sugaring both create a hypertonic environment, making water unavailable for microbial growth. Salting is used to preserve meats, and sugaring is used to preserve fruits in jams and jellies. Chemical preservatives inhibit microbial growth and are included in many foods, including breads, deli

meats, dried fruits, and wine. Canning uses high temperature or temperature and pressure to kill microbes. It is used to preserve fruits, vegetables, and meats.

3. Commercial sterilization uses pressure and steam to generate temperatures high enough to kill the spores of *Clostridium botulinum*. It does not completely sterilize food as thermophilic bacteria may survive.

4. Lactic acid bacteria and the enzyme rennin are added to milk. The rennin breaks down the milk protein, and the bacteria ferment the milk sugar and produce acids. This causes the protein to coagulate into a curd. The curd is separated from the fluid whey and ripened.

5. Grain is sprouted and crushed into malt. The malt is infused with hot water, producing a dark liquid called wart that contains sugars from the grain. The wort is separated from the spent grain. Hops are added to the wort to add flavor. Yeast is added, and fermentation begins. When fermentation is complete, the beer is transferred to bottles or kegs.

6. Yeast, flour, and water are combined to make a dough. The dough is kneaded to bring the wheat proteins together and give the dough elasticity. As fermentation occurs, the carbon dioxide released by the yeast causes the dough to rise. When fermentation is complete, the dough is baked.

7. Fruit is crushed to release juice. Yeast is added, and fermentation begins. When fermentation is complete, the liquid is separated from the solids and transferred to kegs. The wine is aged in kegs and bottled.

8. Bioreactors can hold great volumes for large-scale production of microbial products. In addition, bioreactors have many sensors so that environmental conditions can be continually monitored and adjusted to maintain optimal conditions during the fermentation process. The bioreactors also have built-in stirring mechanisms to keep cultures well aerated and mixed.

9. Primary metabolites are required for growth of the organism and are produced during the log phase of growth. Secondary metabolites are not required for growth of the organism and are not produced until the organism reaches the stationary phase of growth.

Self-Test

1. T	5. E	9. C
2. F	6. E	10. F
3. F	7. B	
4. F	8. T	

The Study of Disease

WHAT YOU WILL LEARN

This chapter introduces the subject of infectious disease. As you study this chapter, you will:

- investigate the history of the human understanding of disease;
- learn how scientists can use Koch's postulates to prove an infectious agent causes a disease;
- discover how your normal microbiota helps protect you from disease;
- review the terminology that's used to describe disease;
- examine the stages of disease in the body;
- explore how diseases are transmitted.

SECTIONS IN THIS CHAPTER

- History of the Study of Disease
- Pathology
- Normal Microbiota
- Classifying Infectious Disease
- Development of Disease
- Spread of Infections
- Nosocomial Infections

Imagine for a moment that you live in the small seaside town of Messina, Italy. The year is 1347. Twelve ships of Italian sailors arrive in port one day, coming home from the East. Some of the sailors are very sick; many are dead. The friends and families of the sailors take their sick loved ones home to care for them, and the dead are buried. Very quickly, the sickness starts to spread to the families of the sailors and then to the other people of your town. You hear of people dying a terrible death. At first they feel chilly and stiff and then tingling like the points of arrows under their skin. Many develop hard swollen boils under their armpits or in their groins that cause agonizing pain. Their skin shows reddish blotches that later turn black. Their breath is foul, and their whole bodies have an intolerable stench. Their urine is thick and sometimes black or red. Most people who get the sickness die within three days. It is even said that some who go to bed healthy are dead before dawn.

Fear grips your town. People say prayers and ring church bells, or purchase magical charms and eat certain foods. Some try locking themselves within their homes. Physicians come to treat the sick. They wear wax-treated clothing that covers them from head to toe and strange masks over their faces. In the beak-like noses of their masks, they keep posies of flowers to protect themselves from the stench of the sick. They release blood from the sick and wash them in vinegar or urine. Some prescribe herbal medicines, but it is no use. The physicians themselves become ill, some dying before their patients. The sickness spares no one. The young, the old, the healthy, and the infirm all seem as likely to die. You think of fleeing the town, but reports are coming back that other towns are turning away strangers, that the sickness is spreading into the countryside.

> **REMEMBER**
> To be able to control infectious disease, we need to know what causes the disease, how it's transmitted, and what treatments are effective.

The year was 1347, and it was the beginning of the black death, or the bubonic plague, in Europe. Over the next 5 years, recurring outbreaks of the plague killed one third of the population of Europe and had enormous impacts on society. No one knew what caused the plague, where it came from, or how to stop it. They didn't understand what it did to the body, and they didn't have any means to treat it. To control disease, we need the answers to many questions like these. What causes the disease? Where does it come from? How does it spread to people? What does it do to the body? What treatments are effective against it? Today, we are fortunate to know much more about many diseases. However, there are diseases about which we know very little, or others that we cannot cure. And there is always the possibility that we will encounter something new. There is always one more nagging question: Could it happen again?

History of the Study of Disease

The people who lived in Europe in the Middle Ages thought disease was sent as a punishment from God or that it was caused by an imbalance of the humors of the body (fire, earth, water, air). However, common experience also showed them, as it does us, that one person could catch an illness from another person. Eventually, people began to question whether there might be something that passed from person to person that caused disease. In 1550, the Italian physician Fracastaro proposed that invisible *contagions* could pass from person to another, causing disease. In 1673, when van Leeuwenhoek showed the world his animalcules, humans had their first glimpse at what those contagions might be. It wasn't until almost 200 years later, however, that someone proved beyond a doubt that microbes could cause disease. This idea is called the **germ theory of disease** and a physician named Robert Koch proved it in 1872.

Before Robert Koch provided his proof, several other people suggested that microbes caused certain diseases. In 1835, Agostino Bassi proposed that a disease in silkworms was caused by a fungal infection. In 1845, during the Irish potato blight, the Reverend M. J. Berkeley observed that the fungus on the potato plants was similar in appearance to a fungus that he had observed on other plants with similar symptoms and proposed that the fungus might be the cause of the blight. In 1865, Louis Pasteur was asked to help study and combat another silkworm disease. From careful observation of diseased and healthy silkworms, he determined that a protozoan parasite was always associated with the diseased silkworms and not with the healthy worms. All of these suggestions that microbes caused disease were based on observations and correlations between the appearance of a pathogen and a disease. Robert Koch took these observations one step further and demonstrated an elegant experiment to determine whether a particular microbe causes a particular disease.

Robert Koch was a German country doctor who had an interest in microbes. To satisfy his curiosity, he began studying cattle that were affected by the disease anthrax. He examined samples of the diseased cattle's blood and found that a large rod-shaped bacterium was always associated with the blood of diseased cattle and not with the blood of healthy cattle. He developed a system for culturing these bacteria in the sterile interior of an ox eyeball. He achieved a pure culture of the suspect bacteria. Then he took some of the bacteria and introduced them into healthy cattle. When the healthy cattle developed anthrax, Koch took fresh samples of their blood and confirmed that the same bacterium was present by growing it in pure culture and examining it again. The two samples of bacteria matched, and Koch had his proof that this bacterium was the causative agent of anthrax. The bacterium is *Bacillus anthracis,* and the series of steps that Koch demonstrated to prove that a particular organism causes a particular disease is called **Koch's postulates** (Figure 16.1).

> **REMEMBER**
> To follow Koch's postulates, associate the microbe with the disease, isolate the microbe into pure culture, inoculate the microbe into a healthy organism, and reisolate the microbe.

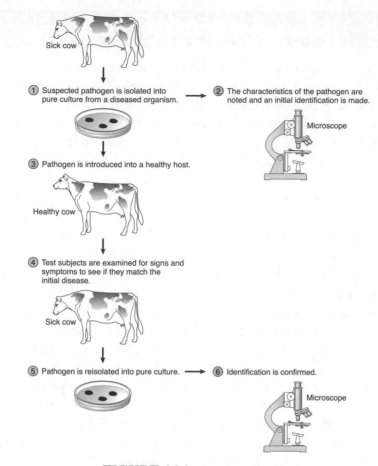

Sick cow

① Suspected pathogen is isolated into pure culture from a diseased organism. → ② The characteristics of the pathogen are noted and an initial identification is made.

Microscope

③ Pathogen is introduced into a healthy host.

Healthy cow

④ Test subjects are examined for signs and symptoms to see if they match the initial disease.

Sick cow

⑤ Pathogen is reisolated into pure culture. → ⑥ Identification is confirmed.

Microscope

FIGURE 16.1. Koch's postulates.

To follow Koch's postulates, first the microbe must be associated with the infected organism at all stages of disease. Second, the microbe must be isolated into pure culture and characterized. Third, when a healthy organism is inoculated with the suspect microbe, the original signs and symptoms must be reproduced. Finally, the suspect microbe must be reisolated from the newly infected organism and the identification confirmed.

Pathology

Koch's work represented the culmination of a major shift in the way we view disease and helped spark a frenzy of research into different diseases and treatments. The study of disease is called **pathology**, and a microorganism that can cause disease is called a **pathogen**. The science of pathology seeks to understand the cause, or **etiology**, of a disease as well as the development, or **pathogenesis**, of the disease. During the first golden age of microbiology (1850–1920), the causative agents for many diseases were discovered, including those that are listed in Table 16.1. Also during this time, many advances were made in our understanding of immunity and vaccination.

TABLE 16.1. Some Causative Agents Discovered
During the First Golden Age of Microbiology

Disease	Causative Agent
Gonorrhea	*Neisseria gonorrhoeae*
Tuberculosis	*Mycobacterium tuberculosis*
Cholera	*Vibrio cholerae*
Diphtheria	*Corynebacterium diphtheriae*
Tetanus	*Clostridium tetani*
Malaria	*Trypanosoma cruzi*

When we are searching for the causes of disease, it is not always possible to apply Koch's postulates fully. When studying human disease, most people today would probably agree that it is unethical to infect healthy humans with suspected pathogens. Also, it has proved very difficult to grow some human pathogens in pure culture. The situation can also get complicated when more than one pathogen causes the same signs and symptoms, as is the case with pneumonia. Additionally, some pathogens cause more than one disease, depending on where they are introduced into the body. When Koch's postulates cannot be followed absolutely, we sometimes have to rely on statistical observations of the patterns of disease. This is part of the science of epidemiology, which will be considered in Chapter 17.

Normal Microbiota

When a disease-causing microorganism tries to colonize your body, one of your primary defenses is that your body is already colonized. For every body cell you have, there are about ten microbes living on your surface or on the surfaces of your respiratory, gastrointestinal, and genitourinary tracts. You have approximately 10^{13} or 10,000,000,000,000 body cells, so that's 10^{14} or 100,000,000,000,000 resident microbes! These microbes are called your **normal microbiota** (normal flora), and they began colonizing you at birth, rapidly establishing unique microbial communities in the different environments of your body.

Your normal microbiota protect you from colonization by **transient microbiota**, bacteria that might live temporarily on your body. In order for transients to establish themselves, they must compete with your normal residents for space and the available nutrients. The metabolic activities of your normal microbiota may change the pH of body environments, preventing colonization by certain organisms. Some members of your normal microbiota might even fight against certain bacteria by producing **bacteriocins**, chemicals that prevent bacterial growth. If you have ever had diarrhea or a vaginal yeast infection after taking antibiotics, you have witnessed firsthand the effects of disrupting your normal microbiota. In

> **REMEMBER**
> Your normal microbiota help protect you from infection by outcompeting potential pathogens.

addition to killing the pathogen that was making you sick in the first place, the antibiotics killed off members of your normal microbiota. This may upset the function of the intestines or allow yeast to multiply in the vagina.

The relationship between you and your normal microbiota is a **symbiosis**, which literally means *living together*. Because you and your microbes live together for a significant portion of your life, you are considered symbiotic with each other. Different kinds of symbioses can exist between partners. In a situation where both partners provide a benefit to each other, the symbiosis is called a **mutualism**. For example, *E. coli* in your large intestine synthesizes vitamin K, which you need and can absorb into the blood. In return, you provide *E. coli* with a perfect environment with lots of food. If only one partner derives a benefit, but the other is not harmed, the symbiosis is a **commensalism**. If one partner derives a benefit at the expense of the other, the relationship is called a **parasitism**. Any time a microbe has made you sick, multiplying itself at the expense of your cells, it was a parasitism.

To apply these terms properly, we need to study closely the relationship between the organisms. Also, all of these relationships exist in a delicate balance. If conditions change, relationships can change. For example, if *E. coli* from your large intestine is introduced into your urinary tract, you could develop a urinary tract infection. In one environment of the body, *E. coli* is a mutualist; in another, it is a parasite. Changes like this can also happen if the body is compromised in some way, perhaps by injury or illness. Organisms that are harmless in one circumstance but that can cause disease under certain conditions are called **opportunistic pathogens**. As demonstrated by the *E. coli* example, even members of your normal microbiota can be opportunistic pathogens.

Classifying Infectious Disease

Any colonization of the body by a pathogenic microbe is an **infection**, but not all infections result in disease. **Disease** occurs when there is an abnormal change in the physiology of the body. An infection that does not result in any noticeable illness is called a **subclinical infection**. An infection that directly leads to disease is called a **primary infection**. If the body's defenses are weakened by a primary infection, then disease may result from a **secondary infection** by an opportunistic pathogen. This is what happens as a result of HIV infection. HIV weakens the human immune system and makes it susceptible to secondary infections. Death from AIDS typically results from one of these secondary infections.

When disease occurs, the patient feels **symptoms**. Symptoms are subjective and may not be observable by a health care worker. The physiological changes that can be observed are the **signs** of the disease. The combination of signs and symptoms that are associated with a particular disease is called the **syndrome**. If a person has an infectious disease that can be passed to another person, the disease is **communicable**. Many infectious diseases, including sexually transmitted diseases (STDs), diarrheal diseases, and respiratory infections are communicable. If the disease can be spread *easily*, it is **contagious**. Measles, mumps, and chicken pox are all contagious diseases. Diseases that cannot be spread are **noncommunicable**. Gangrene, which is caused when *Clostridium perfringens* is introduced into a deep wound, is an example of a noncommunicable infectious disease.

Infectious disease can also be classified according to the severity and duration of the signs and symptoms. **Acute** diseases like the common cold typically develop rapidly but last for only a short period of time. **Chronic** diseases like hepatitis B develop more slowly but persist over a long interval of time. The duration of **subacute** diseases is between that of acute and chronic diseases; they may not develop rapidly, but they don't necessarily last very long. **Latent** diseases like herpes display periods of inactivity followed by a recurrence of signs and symptoms.

The degree to which a pathogen spreads through the body can also be characterized. **Local infections** such as those that cause abscesses are confined to a very limited area of the body. **Systemic infections** result from the spread of the pathogen throughout the body by the blood or the lymph. For example, in some people *Neisseria meningitidis* can initially infect the upper respiratory tract but then spread to the blood and even cross the blood–brain barrier to cause meningitis. A pathogen may initially cause a local infection but then spread through the blood or lymph to specific regions of the body, resulting in a **focal infection**. For example, miliary tuberculosis begins as an infection by *Mycobacterium tuberculosis* in the lungs but then spread by the blood to other areas of the body. As pathogens spread through the body, their presence in the blood is noted by terms that end in –*emia*, which means in the blood. **Bacteremia** indicates that bacteria are present in the blood, **toxemia** indicates that toxins are present in the blood, and **viremia** indicates that viruses are present in the blood. If bacteria are actually multiplying in the blood, it is termed **septicemia**.

Development of Disease

As the pathogen multiplies in the host, the signs and symptoms of disease begin. The development of the disease tends to follow a common pattern (Figure 16.2). First, there is an **incubation period** during which the pathogen is multiplying, but the signs and symptoms have not begun. When the host begins to feel mild symptoms, the **prodromal period** has begun. The pathogen continues to multiply throughout the prodromal period, and signs and symptoms worsen. The **period of illness** begins when signs and symptoms become severe. The pathogen continues to multiply. If the defenses of the body are not able to defeat the pathogen, or medical help is not available, then **death** may result. If the pathogen is successfully fought off, the numbers of the pathogen will begin to decline, and the **period of decline** will begin. Signs and symptoms may still be severe for part of this period, but eventually they will begin to lessen. Gradually, signs and symptoms become milder as the pathogen continues to decline and the **period of convalescence** begins. The period of convalescence is not over until all of the

> **REMEMBER**
> Although a person may be most infectious at certain times during the development of a disease, it's possible for the infection to be spread any time pathogens are in the body.

pathogens are gone from the body. During the entire development of disease, as long as pathogens are present in the body, it is possible for the disease to be spread to another person. In fact, for many diseases, a person is most contagious during the early stages of the disease.

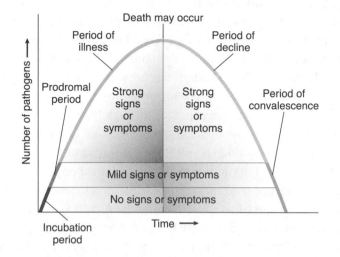

FIGURE 16.2. The stages of disease.

Spread of Infections

In order for a person to catch an infectious disease, the pathogen must come from somewhere and infect the person's body. The places where pathogens live and multiply are called **reservoirs of infection**, and the way in which they arrive at a new host is the **mode of transmission**.

RESERVOIRS OF INFECTION

It should come as no surprise that humans can be reservoirs for pathogens. After all, we all know we can catch something from someone who is sick. In fact, for some diseases, like polio and smallpox, humans are the only reservoir. These are both viral diseases, so a host is required, and human cells are the only possible hosts for these viruses. People who do not appear to be sick can also be reservoirs. An opportunistic pathogen can colonize some people without causing disease in their bodies and can then be spread to another person. A person who harbors a pathogen without showing signs of the disease is called a **carrier**. For example, some people carry *Staphylococcus aureus* in their noses. Poor hygiene can lead to the transfer of this opportunistic pathogen to other people where it can cause a variety of diseases, including simple abscesses, scalded skin syndrome, toxic shock syndrome, and food poisoning.

Nonhuman animals can also serve as reservoirs for disease. If a disease can spread from an animal to a human, it is called a **zoonosis**. Some examples of zoonotic diseases include rabies, West Nile virus, and the bubonic plague. Rabies virus can multiply in a variety of mammals, from dogs to bats. West Nile virus multiplies in birds and *Yersinia pestis*, the bacterium that causes the plague, multiplies in rodents.

Some pathogens can multiply outside of living things, in the soil or water. The soil is a reservoir for *Clostridium botulinum* and *Clostridum tetanii*. Many diarrheal pathogens, like *Vibrio cholerae* and *Salmonella typhi*, can multiply in the water.

MODES OF TRANSMISSION

> **REMEMBER**
> Carriers are people that are infected with a pathogen and can transmit it to others, but they don't have any signs or symptoms of disease themselves.

There are many ways that pathogens can travel from reservoirs to human hosts. Humans may contact the reservoir, either directly or indirectly. **Direct contact transmission** occurs when the host actually touches the reservoir. Touching, kissing, or sexual intercourse are all examples of direct contact and can lead to the spread of STDs, colds, and streptococcal infections. **Indirect contact transmission** occurs when the pathogen is transferred from the reservoir to a nonliving object. The host is colonized when it touches the object, which is called a **fomite**. Toothbrushes, drinking glasses, hypodermic needles, and bedrails are all examples of possible fomites. Finally, **droplet transmission** occurs when pathogens are transmitted in mucus as it travels short distances owing to sneezing and coughing.

Humans may also acquire diseases without actually contacting the reservoir. In **vehicle transmission**, a vehicle such as water, food, or air may bring the pathogen from the reservoir to the person. Many diarrheal diseases are transmitted by *waterborne transmission* and *foodborne transmission*. This can happen if people who are either ill or carriers of a diarrheal disease prepare food without adequately washing their hands. Because the diarrheal pathogen is being transmitted from the feces of one person, and then consumed by a second person, this is also termed the **fecal–oral route of transmission**. *Airborne transmission* is common for respiratory illnesses, such as flu, tuberculosis, and the common cold.

Another way that pathogens can be brought from the reservoir to people is by the action of animal **vectors**. The most common vectors are insects and arthropods such as ticks and fleas. In **mechanical vector transmission**, the pathogen is carried on the surface of the vector, just hitchhiking a ride. The classic example of this is when pathogens are carried from feces to food by flies. If the pathogen actually enters the body of the vector and is transmitted to the new host by biting, defecation, or vomiting of the vector, this is termed **biological vector transmission**. Diseases that are spread by the bite of mosquitoes or ticks fall under this category. For example, West Nile virus is spread to humans by the bite of mosquitoes that have also bitten infected birds. Fleas that first bite infected rats and then bite humans spread the bubonic plague.

Understanding the reservoir and mode of transmission for a disease is essential to the control of the disease. After we know where a disease comes from and how it spreads, we can act to break the chain of transmission. For example, health care workers routinely wear masks and gloves to prevent the spread of opportunistic pathogens from their bodies to their patients and to protect themselves from the pathogens their patients may harbor.

Nosocomial Infections

Even with our understanding of disease transmission and the use of universal precautions in health care settings, we do not yet have complete control over the spread of disease in hospitals. Five to fifteen percent of hospital patients will acquire a **nosocomial infection (hospital-acquired infections)**, an infection that they acquire while in the hospital. There are three main reasons that so many nosocomial infections occur. First, hospitals are obviously filled with sick people, so there are many reservoirs of disease. Second, patients in the hospital may be **compromised**. Their immune systems may be weakened by a primary infection, or surgery or burns may have broken their skin. Finally, as health care workers travel from patient to patient, there are lots of opportunities for transmission of disease.

Hospitals take many precautions to lower the numbers of nosocomial infections. Patients who are severely compromised or who harbor especially contagious or dangerous pathogens may be isolated. Health care workers are expected to wash their hands frequently and wear gloves and masks when appropriate. Equipment is routinely disinfected and sterilized. Sterile instruments such as catheters must be handled with aseptic technique, so they are not contaminated before entering a patient's body. Routine monitoring of the hospital environment to determine the levels of contamination is often performed. When followed, these procedures effectively break the chain of transmission between reservoirs and hospital patients and can reduce the number of nosocomial infections.

> **REMEMBER**
> Nosocomial, or hospital-acquired, infections occur because of the combination of compromised hosts, the presence of many microbes, and the potential for transmission from patient to patient by health care workers.

REVIEW EXERCISES FOR CHAPTER 16

1. Name and describe each of the steps you would take, using Koch's postulates, to prove an organism causes a particular disease.

2. Describe situations in which Koch's postulates cannot be applied exactly and suggest how they might be modified.

3. Define the following terms: pathology, pathogenesis, pathogen, opportunistic pathogen, infection, disease, and normal microbiota.

4. Describe the circumstances under which microbes might become opportunistic pathogens.

5. Describe the benefits to health of the normal microbiota.

6. Define the following terms: symptoms, sign, syndrome, communicable disease, noncommunicable disease, contagious disease, acute disease, chronic disease, subacute disease, latent disease, local infection, focal infection, systemic infection, bacteremia, toxemia, viremia, septicemia, primary infection, secondary infection, and subclinical infection.

7. Name and describe the stages in the development of a disease.

8. Name and describe the reservoirs that exist for microbes.

9. Name and describe the methods by which microbes are transmitted from reservoirs to hosts.

10. Define nosocomial infection and describe the factors that contribute to their spread.

SELF-TEST

1. If you have bacteria multiplying in your blood, you have

    A. bacteremia.
    B. septicemia.
    C. viremia.
    D. toxemia.
    E. bloodemia.

2. Which of the following pairs is mismatched?

    A. Disease—colonization of the body by a microbe
    B. Pathogen—an organism that is capable of causing disease
    C. Sign—an observable change in physiology
    D. Acute—rapid and short disease
    E. Local—an infection confined to a very limited area of the body

3. If you were to place the procedures below in order to follow Koch's postulates, which one would occur **third** in the list?

    A. Observe characteristics of pathogen for initial identification.
    B. Isolate suspected pathogen from host.
    C. Grow pathogen in pure culture.
    D. Introduce pathogen into healthy host.
    E. Observe test subject for signs and symptoms.

4. All of the following are beneficial consequences of your normal microbiota **except**:

    A. They may produce vitamins for your cells.
    B. They may change the pH of their environment preventing other microbes from growing.
    C. They may produce bacteriocins to prevent other microbes from growing.
    D. They may compete for nutrients with your cells.
    E. All of the above are correct.

5. True or False. Symbiotic organisms are always mutually beneficial to each other.

6. True or False. The leading preventative measure for the prevention of nosocomial disease is handwashing.

7. If you get the flu because someone sneezed near you, this is an example of

   A. direct contact.
   B. indirect contact.
   C. droplet transmission.
   D. airborne transmission.
   E. waterborne transmission.

8. What do all zoonotic diseases have in common?

   A. They are passed by biological vectors.
   B. They are passed by mechanical vectors.
   C. They are passed by direct contact.
   D. They are passed from animals to humans.
   E. They are passed from reservoirs to animals.

9. True or False: The difference between mechanical and biological vectors is that mechanical vectors are machines that may carry disease whereas biological vectors are organisms that may carry disease.

10. The stage of disease during which signs and symptoms are severe but the number of microorganisms is declining is the

    A. prodromal stage.
    B. incubation stage.
    C. period of illness.
    D. convalescence.
    E. period of decline.

11. Quentin is a 5-year-old boy who attends a Montessori kindergarten. On January 13, his best pal Kristie went home from school early because of the appearance of itchy red blotches on her stomach. The next day a note went out to the parents saying that their children may have been exposed to chicken pox. Sure enough, on January 23, Quentin had a slight fever and was extremely crabby in the morning, so his parents kept him home. Two days later, January 25, Quentin woke up with some flat red spots on his chest. The spots soon become raised with itchy, round, fluid-filled blisters and finally crusted over. Over the next few days, Quentin continued to develop spots and blisters. Finally, on January 29, his mom was glad to notice that no new spots appeared, and Quentin seemed marginally less crabby. He continued to improve over the next couple of weeks, and, by February 13, all existing spots had cleared up. Identify the stages of disease for this incident of chicken pox and give the dates for each.

12. Lyme disease is caused by the spirochete *Borrelia burgdorferi*. The spirochete multiplies inside the bodies of both deer and field mice. Nick caught Lyme disease when he was out hiking after he was bit by a tick that was feeding on field mice. What is the reservoir and mode of transmission in this example?

13. *Poliovirus* causes polio in humans. A person contracts polio by drinking contaminated water. What is the reservoir and mode of transmission in this example?

Answers

Review Exercises

1. First, isolate the suspected pathogen from a diseased organism. Grow the pathogen in pure culture and identify. Then, introduce the pathogen into a healthy host and examine for same signs and symptoms as original disease. Re-isolate the pathogen, grow it in a pure culture, and confirm identification.

2. It would not be ethical to inoculate a human host with a suspected pathogen. Instead, pathogenesis of some human diseases can be studied in animal models. Also, some pathogens cannot be grown in pure culture. For example, viruses must be cultured inside of host cells.

3. Pathology is the study of disease. Pathogenesis is the development of the disease in the body. A pathogen is an organism capable of causing disease. Opportunistic pathogens cause disease in weakened hosts. An infection is the colonization of the body by a microbe. Disease is an abnormal change in the state of health. Normal microbiota are the microorganisms that normally live in and on the body without causing disease.

4. Some organisms can become pathogenic when the host is weakened. Hosts may be weakened by other infections, burns, surgery, or poor general health owing to poor nutrition or age.

5. The normal microbiota takes up space and nutrients on the body, preventing transient microbiota from colonizing. Some members of the normal microbiota can produce compounds that inhibit growth of bacteria (bacteriocins) or may change the pH so that other organisms cannot grow. They may also benefit the body by synthesizing vitamins.

6. Symptoms are the subjective things a patient feels. Signs are objective things that can be measured. Syndrome is the collection of signs and symptoms associated with a particular disease. Communicable diseases can be spread from one person to another. Noncommunicable diseases cannot be spread. Contagious diseases are spread easily. Acute diseases develop rapidly and are of short duration. Chronic diseases develop more slowly and last a long time. Subacute diseases are of intermediate duration between acute and chronic. Latent diseases have periods of inac-

tivity. Local infections are limited to one area of the body. Focal infections begin in one area but then spread through the blood or lymph. Systemic infections spread throughout the body through the blood or lymph. Bacteremia indicates the presence of bacteria in the blood. Toxemia indicates the presence of toxins in the blood. Viremia indicates the presence of viruses in the blood. Septicemia indicates bacteria multiplying in the blood. A primary infection is when the initial infection causes the disease. A secondary infection is when a disease is caused by an opportunistic pathogen after the host is weakened by a prior infection. A subclinical infection has no detectable signs.

7. During incubation, microbes multiply in the body but do not cause signs or symptoms. Mild symptoms begin in the prodromal phase as the microbes continue to multiply. During the period of illness, signs and symptoms are severe, and the microbes continue to multiply. When microbes begin to decrease in number, this marks the start of the period of decline, although signs and symptoms can still be severe. Finally, signs and symptoms lessen, and the number of microbes decreases to zero during convalescence.

8. Humans, both sick and carriers, are reservoirs for disease. Animals, soil, and water can also be reservoirs.

9. Disease can be transmitted by direct contact with a person, by indirect contact with a contaminated object (fomite), or by sneezing and coughing (droplet transmission). It is also possible to contract a disease through vehicles such as soil, water, and air. Vectors such as ticks and fleas can also carry pathogens to new hosts.

10. Nosocomial infections are acquired in the hospital. The three main factors that contribute to their spread are the existence of many reservoirs in a hospital, the chain of transmission between hospital personnel and patients, and the presence of weakened hosts.

Self-Test

1. B	3. C	5. F	7. C	9. F
2. A	4. D	6. T	8. D	10. E

11. January 13 (or before) to January 23—Incubation
January 23 to January 25—Prodromal
January 25 to January 29—Disease
January 29 to February 13—Decline and convalescence

12. The reservoir was the field mice, and the mode of transmission was biological vector transmission.

13. The reservoir is humans (this is a viral disease; it can only multiply in humans). The mode of transmission is vehicle transmission (waterborne).

17

Epidemiology and Emerging Infectious Disease

WHAT YOU WILL LEARN

This chapter explores epidemiology, the science of tracking disease, and looks at its relationship to some recent events. As you study this chapter, you will:

- discover the factors that enable us to control infectious disease;
- examine the role of infectious disease around the world;
- explore the roots and fundamental principles of epidemiology;
- investigate how genetic change in pathogens leads to outbreaks of infectious disease;
- consider how environmental change can lead to outbreaks of infectious disease;
- learn how herd immunity can protect you from disease.

SECTIONS IN THIS CHAPTER

- History of Epidemiology
- The Science of Epidemiology
- Origins of Emerging Infectious Disease
- The Virulence of Epidemics

At the turn of the last century in the United States of America, the leading three causes of death were pneumonia, tuberculosis, and diarrheal disease. At the turn of this century, the leading three causes of death were heart disease, cancer, and stroke. Clearly, deaths from infectious disease in the United States have declined dramatically in the last century. This is largely the result of good public sanitation and water treatment systems. Vaccination programs and access to health care have also had a great effect on reducing death from infectious disease. As a result, people in the United States are living longer than they did a hundred years ago and are more likely to develop diseases that are associated with aging.

The decline in death from infectious disease in the United States might suggest that infectious disease is not as big a concern as it once was. Unfortunately, if we look at the global situation, we see that this is not the case. In 1998, according to the World Health Organization (WHO), infectious disease accounted for the 25 percent of deaths in the human population worldwide. In low-income African and Asian countries, infectious disease caused 45 percent of deaths. For children, the statistics are even worse:

REMEMBER
Infectious disease is still the leading killer of children in the world, and many of these deaths are caused by preventable diseases.

Infectious disease is responsible for 65 percent of the deaths in children younger than age 5 worldwide. The leading causes of death caused by infectious disease are respiratory infections, including pneumonia and influenza, AIDS, diarrheal diseases, tuberculosis, malaria, measles, and hepatitis B. Respiratory illness and diarrheal diseases are the two leading killers of children.

Why is infectious disease still so prevalent globally? The answers to this question are very complex and involve economics, culture, and education. Diarrheal diseases are still rampant in many areas because access to clean water is not available. Sanitation systems such as those that exist in the United States are expensive and do not exist in many areas of the world. War can destroy existing systems and force people into unsafe conditions in refugee camps. War also disrupts organized health care and vaccination programs. Education and cultural issues may also be barriers to vaccination or other control methods. For example, sub-Saharan Africa has the highest level of HIV infection in the world, in part because of a lack of understanding of HIV and how it is transmitted. In addition, in some areas, there are cultural barriers to acknowledging the disease and instituting safe-sex practices. In areas where these barriers are not as strong, it still may not be possible to gain access to the expensive drug regimen needed to treat HIV.

Even in developed nations like the United States, infectious disease is still a concern. Global travel makes it possible for people to carry dangerous pathogens to all countries. Development around the world is driving people into new areas, where they may contact new pathogens that can be spread around the globe. In urban areas, people live in high-density conditions ideal for the rapid spread of these infectious organisms. In recent years in the United States, there have been numerous epidemics of pathogenic *E. coli* from contaminated food sources such as hamburger, sprouts, and juice. Because so much food is processed centrally in the United States, contaminated products spread widely and can affect large numbers of people before anyone realizes that the food is contaminated. In countries where antibiotics are used to treat infections, antibiotic-resistant bacteria have become a problem. After an antibiotic-resistant strain arises in one country, it is only a matter of time before it is spread to other countries.

The solutions to these problems are complex and controversial. They involve not just science and medicine but also economics and politics. Organizations like the World Health Organization are trying to work on these issues. For example, the WHO is trying to eradicate polio by vaccinating people all over the world. Making vaccines available to children around the world is also a goal of UNICEF and the Bill and Melinda Gates Foundation. In addition, cooperation between governments and epidemiological agencies has been evident in the recent pandemics of influenza and severe acute respiratory syndrome (SARS). Because pathogens do not respect political boundaries, global efforts like these are crucial to the health of the humans on planet Earth.

History of Epidemiology

The Soho neighborhood of London in the 1800s was a crowded, dirty place. Human waste drained to cesspits located beneath buildings and was sometimes located right next to wells for drinking water. This situation was ideal for the transmission of intestinal pathogens, such as *Vibrio cholerae,* which causes cholera. Between 1831 and 1854, four cholera outbreaks in London resulted in the deaths of tens of thousands of people. The last outbreak was the worst, killing 127 people in the first 3 days alone.

A doctor who lived in the area, John Snow, noted that most of these deaths occurred in people who lived around Broad Street. He interviewed families in the neighborhood and made a map of the location of all of the victims. He found that nearly all of the victims lived in the area right around the same well on Broad Street. Dr. Snow examined the water from the well and found that it contained white particles. He believed that this was the source of the cholera and convinced the authorities to remove the handle from the Broad Street pump so that no one could drink out of the well. As soon as the handle was removed, the incidence of cholera began to drop dramatically, and the epidemic ended. Through his careful observations and analysis, Dr. Snow arrived at a conclusion about the cause of a disease and determined how to stop it. This was the birth of the science of epidemiology.

The Science of Epidemiology

Epidemiology is the science of monitoring disease and understanding its causes (**etiology**) and transmission. Epidemiologists monitor the **incidence** or numbers of new cases that arise in populations during a particular time period. They are also concerned with the **prevalence** of diseases, that is, how many people have the disease any one given time. By monitoring the incidence of disease, epidemiologists can determine

> **REMEMBER**
> The incidence of a disease is the number of new cases over a given time period. When incidence is on the rise, the risk of disease for the population is increasing.

when **epidemics** are occurring (Figure 17.1). During an epidemic, the number of new cases of a particular disease is greater than what is normally expected in the population. When epidemics spread around the globe and affect more than one continent or population, they are termed **pandemics**. In some areas of the world, certain diseases are constantly present at a relatively stable level. These diseases are said to be **endemic** to that region (Figure 17.1). Diseases that occur occasionally within a population are **sporadic**.

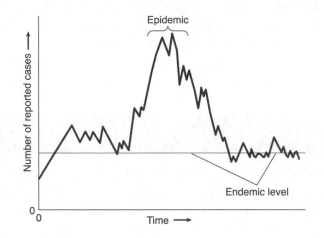

FIGURE 17.1. Incidence of disease.

When the number of new cases of a disease rises above the expected level, it is an epidemic. If a disease exists at a relatively constant level in a population, that is the endemic level of the disease.

To monitor diseases in this way, epidemiologists need sources of data. For certain **notifiable diseases**, health care workers are required by law to report any cases of the disease to public health agencies. Notifiable diseases may be of concern to epidemiologists because they are highly contagious, they are very virulent, or there are public health measures in place to control that disease. Examples of notifiable diseases include AIDS, measles, gonorrhea, tetanus, typhoid, chlamydia, hepatitis, and hemorrhagic *E. coli*. Epidemiologists can also obtain information from insurance questionnaires, clinical studies, and interviews with patients. They may also consult databases on known pathogens when trying to identify the cause of an epidemic.

TYPES OF EPIDEMIOLOGICAL ANALYSIS

After epidemiologists collect data about the incidence and prevalence of a disease, they use statistics to look for clues as to the cause and transmission of the disease. For example, they may look for patterns like association of the disease with occupation, age, sex, or personal habits of the people who have the disease. If there is a local epidemic of a foodborne or waterborne pathogen, they will try to determine the common water or food source for the affected people. For some diseases, they might look to see if the disease is associated with a particular season or weather pattern. Epidemiologists use all of this data to create a picture that describes the occurrence of the disease, including who was affected, where and when it occurred, and any factors that were associated with contracting the disease. The type of data collection and comparison that looks back at an epidemic that has already occurred is called **descriptive epidemiology**.

In addition to descriptive epidemiology, epidemiologists do other types of analysis. **Analytical epidemiology** involves making comparisons between groups of people in order to try to determine the cause of a disease. For example, people with a disease may be compared to people who do not have the disease. In **experimental epidemiology**, an experimental treatment is administered to one group of people while another group acts as a control and does not receive the treatment. This is commonly part of the process of developing new drugs.

After epidemiologists have determined the cause (etiology) and mode of transmission for a disease, they make recommendations on how to break the chain of transmission and prevent the disease. For example, thanks to the work of epidemiologists, many people today understand the risks of exposure to STDs by sexual intercourse and practice "safe sex" by utilizing condoms. The WHO monitors worldwide epidemiological data and works closely with national public health agencies to try and control disease.

CENTERS FOR DISEASE CONTROL

One very important national public health agency is the Centers for Disease Control and Prevention in the United States. The CDC collects public health data from the states, monitors the national trends in disease, and makes recommendations for the control of disease. When there are large-scale epidemics and pandemics, the CDC works closely with laboratories in the Unites States and elsewhere to identify the causative agents. In addition, it provides a great deal of public health information to health care professionals and the general public through a series of informative publications. One of these is the *Morbidity and Mortality Weekly Report (MMWR)*. In the *MMWR*, the CDC provides data on the **morbidity**, the number of new cases of notifiable diseases, and **mortality**, the numbers of deaths caused by notifiable diseases. Another valuable publication is the *Emerging Infectious Disease (EID)*. This journal presents information on **emerging infectious diseases**, diseases that are on the rise (incidence is increasing) either because they are new to an area or because a pathogen has somehow changed. In addition to these rather technical publications, the CDC also provides many concise fact sheets on pathogens of interest to health care professionals and the general public. All of these excellent resources on disease are available at the CDC web site, *www.cdc.gov*.

Origins of Emerging Infectious Disease

New diseases constantly seem to appear and spark concern among the public. In recent years, we have seen the spread of West Nile virus to the United States, heard about the dangers of antibiotic-resistant bacteria, and watched as respiratory infections like SARS and new strains of influenza caused epidemics in Asia and elsewhere. Where do these

REMEMBER
Emerging infectious diseases are diseases whose incidence is increasing in the human population. Diseases commonly emerge due to genetic or environmental change.

emerging infectious diseases come from? Why do they spread so rapidly? And why are they sometimes so deadly?

There are two main causes of emerging infectious disease. Many *new* diseases are really modifications of an old one as a result of genetic change in a pathogen. Basically, an old pathogen learns new tricks and becomes more dangerous or harder to control. The rise of antibiotic-resistant bacteria or epidemics of new flu strains are both examples of this. Some emerging diseases really are new, at least new to a particular population. Changes in the environment of a pathogen may cause it to encounter a new population of hosts. This can happen as a result of increases in the population of a disease reservoir, or when reservoirs migrate to new areas. The emergence of West Nile virus in the United States is an example of this; the virus was brought to the United States by birds that are reservoirs for the virus. Another example is the periodic epidemics of Hantavirus that are caused by changes in the rodent populations that act as reservoirs.

INFLUENZA: AN EXAMPLE OF GENETIC CHANGE IN A PATHOGEN

The influenza virus is a significant human pathogen and an excellent example of how genetic change of a pathogen leads to epidemics and the emergence of new strains. The influenza virus is an enveloped RNA virus (Figure 17.2). Two spike proteins, **hemagglutinin** and **neuraminidase**, project off the envelope and are involved in entry and exit of the virus from the host cell. The RNA of the virus consists of eight separate pieces of RNA. When the virus undergoes biosynthesis inside of a host cell, a viral enzyme copies the viral RNA. This enzyme, an RNA-dependent RNA polymerase, has a high mutation rate. Because of this, the genetic material of influenza virus is highly variable, and new strains are constantly being generated. Of the three influenza types that infect humans—type A, B, and C—type A is the most significant pathogen. Within type A influenza, subtypes are determined based on the major differences in the hemagglutinin (H) and neuraminidase (N) proteins (Table 17.1). Within subtypes, further differences in the flu proteins leads to the creation of different strains (Table 17.1).

In humans, the influenza virus causes an acute upper respiratory infection, which is usually accompanied by high fever, aches, and general malaise. Influenza is usually defeated by the immune response of the body. When the influenza virus enters the body, its proteins act as **antigens** and stimulate the production of defensive proteins called **antibodies**. The defensive antibodies bind to the antigens on circulating flu virions (Figure 17.3), preventing them from attacking any more cells and marking them for destruction by the body. This stops the virus from multiplying and rids the body of the infection. At the start of flu season, many people do not have immunity to the flu. These people get sick and transmit the disease to others. Eventually, however, people in the population develop resistance to further attack by the same strain because they can make the correct antibodies. When enough people in the population become resistant, the flu season ends.

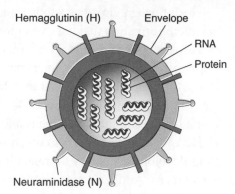

FIGURE 17.2. Structure of the influenza virus.

Influenza is caused by an enveloped RNA virus. The viral genome consists of eight separate segments of RNA.

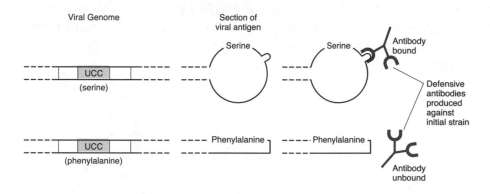

FIGURE 17.3. Antigenic drift.

Errors in replication of the viral genome by the viral enzyme lead to small changes as the virus multiplies. These small changes in the genome lead to changes in antigens (antigenic drift). As a result, previously established immunity fails, and disease recurs.

If the influenza virus were genetically stable, this would be the end of the story. Everyone who had the flu would become immune and never get the flu again. However, the influenza virus is not genetically stable. Every time a virion attacks a cell and multiplies, its RNA-dependent RNA polymerase induces point mutations into the viral genetic code. These small changes in the RNA lead to changes in the viral proteins, or **antigenic drift** (small changes in the antigens). In particular, they lead to small changes in the hemagglutinin proteins, which are the most important for triggering antibody production. Thus, when a new flu season begins, new influenza strains with slightly different hemagglutinin proteins are present in the population. Many of the antibodies that the body learned to produce in response to the old flu strain will not bind to these new antigens (Figure 17.3), and people can get sick all over again. In other words, antigenic drift of the pathogen leads to recurring epidemics.

The influenza virus has another way of changing its genetic information that has even more dangerous implications for people. If different influenza viruses infect the same cell at the same time, they can exchange pieces of RNA, forming recombinant strains that may be very different from the originals (Figure 17.4). This can result in major changes in the flu antigens and is called **antigenic shift**. Antigenic shift can also result when an influenza virus develops the ability to transmit directly from a non-human animal, like a bird, into humans. Antigenic shift in flu can lead to human flu strains that have entirely new hemagglutinin proteins that are very different from those in existing human flu strains. There is no established immunity in the population to these new hemagglutinins, and the flu spreads very easily from person to person, resulting in a pandemic. The recent history of flu outbreaks indicates that these pandemics occur regularly (Table 17.1). One of the most infamous is the 1918 flu pandemic, which killed 20 million to 40 million people, more people than died in World War I. Scientists have recovered genes from the 1918 influenza virus from the bodies of victims who were buried in the permafrost in Alaska. They are studying these genes to try to determine why the 1918 flu was so deadly.

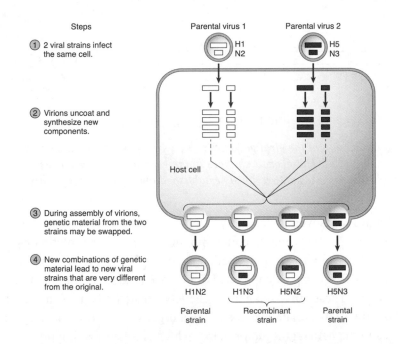

FIGURE 17.4. Recombination between influenza virions.

Gene exchange between viral strains can lead to strains that have entirely new combinations of antigens (antigenic shift). There may be no preexisting immunity in the population to these strains, and a pandemic could result.

TABLE 17.1. A Recent History of Influenza Outbreaks

Some strains of Influenza A

Date	Strain	Subtype	Extent of Spread
1918	A (this outbreak occurred before modern surveillance methods were established)	H1N1	Pandemic of "Spanish" flu, 20 million to 40 million dead
1957	A/Singapore/57	H2N2	Pandemic of Asian flu, 1 million dead
1962	A/Japan/62	H2N2	Epidemic
1964	A/Taiwan/64	H2N2	Epidemic
1968	A/Aichi/68	H3N2	Pandemic of Hong Kong flu, 700,000 dead
1976	A/New Jersey/76	H1N1	Epidemic
1997/1998	A/Hong Kong/97	H5N1	Fast public health response combined with inability of virus to transmit effectively person to person blocked a possible pandemic
2009	Swine flu A/Hamburg/7/2009	Recombinant H1N1	Pandemic

Recombination between very different flu strains is possible because there is more than one reservoir for flu. In addition to humans, the influenza virus can multiply in birds and pigs. The bird reservoir in particular is host to a very diverse population of influenza viruses with hemagglutinin proteins that have never been seen in human strains. Avian influenza viruses (bird flus) do not typically infect humans very easily. However, both avian influenza viruses and human influenza viruses can infect pigs. When avian and human influenza viruses infect pigs at the same time, recombination between the viruses is possible. This may create influenza strains that have the ability to infect humans easily and that may contain new hemagglutinin proteins. When this happens, the results can be far-reaching and severe.

> **REMEMBER**
> Antigenic shift in a pathogen can cause pandemics.

Awareness of the deadly possibilities of an influenza pandemic has resulted in careful monitoring of the influenza virus by both the CDC and the WHO. It is also the reason that a great deal of attention has been paid to certain avian flu strains that emerged in 1997 in Hong Kong. This flu strain infected and killed several people. Analysis of the strain revealed that it contained a type of hemagglutinin, type 5, that had not been seen in human influenza before. The neuraminidase present in this strain was type 1, making the subtype of this virus H5N1. Based on a descriptive epidemiological analysis that traced the recent activities of people who contracted the H5N1 flu, it was believed that the H5N1 flu strain had been contracted directly from chickens that had the disease. Because of the possibility of a pandemic, the WHO and CDC recommended that all the chickens in Hong Kong be destroyed. The Hong Kong authorities agreed, and there were no further cases of the H5N1 flu strain that year. During the winter of 2003–2004, however, more cases of infection and death caused by an H5N1 virus were reported in several Asian countries. The CDC and WHO continue to monitor the situation very closely.

An even more recent public health scare due to influenza occurred with the outbreak of an H1N1 "swine" flu in 2009. Although the swine flu was subtype H1N1, a subtype already common in the human population, this particular flu strain was a "quadruple recombinant" of influenza genes from four different influenza strains, including one pig strain and one bird strain. Thus, this strain contained antigens that were unfamiliar to humans, enabling it to bypass established immunity and spread rapidly around the world. Some of the earliest reports on this flu strain indicated high mortality in people that were infected, raising the possibility of a "perfect storm" like that in 1918—an influenza virus that spread widely and had a high mortality rate. In addition, the swine flu seemed to have a greater effect on healthy young adults than it did on the very old or very young, making it reminiscent of the 1918 influenza. Public health officials monitored the situation closely and recommended that a vaccine against the swine flu be developed at the earliest possible opportunity. Because our global production capacity for influenza vaccine is currently limited, production of the new vaccine had to wait until the seasonal vaccine was completed. In addition, the swine flu virus grew more slowly than anticipated, again delaying completion of the new vaccine. Fortunately, the mortality in infected populations turned out to be much lower than what was originally predicted based on the early reports, but even so, hospital beds were filled to capacity. The vaccine was deployed to people at greatest risk, helping to decrease transmission of and mortality due to the virus. As of August 2010, the WHO has officially declared the H1N1 pandemic to be over. However, our brush with a possible repeat of 1918 taught us several things: we need greater capacity for production of influenza virus, we need to update our technology for production of influenza virus, and we need to increase our capacity to house and treat the sick during a pandemic.

The influenza virus provides an excellent example of how small and large genetic changes in pathogens can lead to increased risk to the human population. Antigenic drift and shift also occur in other pathogens with similar consequences. Many viral enzymes have high mutation rates and can generate antigenic drift in viruses (see Chapter 13). Bacteria may undergo antigenic shift when they acquire genetic information from other bacteria by transformation, transduction, or conjugation (see Chapter 7). These changes may enable a pathogen to evade established immunity as does the influenza virus, or they may lead to the acquisition of new toxins or antibiotic resistance. In either case, the result spells trouble for the human population and can lead to the emergence of an old pathogen with new tricks.

HANTAVIRUS: AN EXAMPLE OF ENVIRONMENTAL CHANGE (NICHE DISRUPTION)

Another way that pathogens can emerge as new and dangerous threats is when they are introduced into a new environment. The role an organism plays in its environment is its **niche**. If something happens to disturb the niche of an organism, it may take on a new role or move into a new environment. One example of how niche disruption can lead to an epidemic occurred in the Four Corners region of the American Southwest during 1993.

In the spring of 1993, several people became sick and died from a mysterious respiratory illness. Scientists attempted to identify the pathogen by screening serum from the victims against antigens from known pathogens. If the victims had been exposed to a known pathogen, then antibodies in their serum should bind to the antigens. These initial attempts at identification of the pathogen ruled out many of the known bacteria and viruses but failed to identify the pathogen.

At this point, epidemiologists began a descriptive study and gathered data about the environment and the behavior of the people who died. During the course of their investigations, they talked with elders from the local Native American tribes and were told of earlier times when high rainfall led to an increase in the mouse population and then mysterious deaths in young people. This information, plus a positive reaction between serum from victims and a

> **REMEMBER**
> Environmental changes can cause the movement of organisms, including microbes, bringing humans into contact with new potential pathogens.

Hantavirus from Scandinavia, led the scientists to the deer mice. Deer mice were trapped and tested for the presence of *Hantavirus*. Based on the sequence of the viral genome, the pathogen was identified as a new member of the *Hantavirus* genus.

The stories about previous outbreaks of the mysterious illness among Native Americans indicate that *Hantavirus* was not new to the Four Corners region. The virus existed in the mouse reservoir and rarely contacted the human population. 1991 and 1992, however, were El Niño years. As a result of El Niño, there was exceptionally heavy rainfall, resulting in a big crop of pinon nuts. The increased food supply caused the mouse population to boom, increasing the possibility of virus transmission in mice and resulting in a high prevalence of *Hantavirus* within the mouse population. The increase in the population may also have caused mice to come closer to human habitation, leaving their droppings in sheds and other areas. When people disturbed these areas, the droppings were stirred up into the air, and the people inhaled the virus. Airborne transmission of the virus resulted in infection, disease, and death.

In the case of *Hantavirus*, when environmental conditions returned to normal, the pathogen retreated to its old niche. In many cases, however, once a pathogen enters a new niche it remains there. This is particularly true when organisms are introduced into new environments. In recent years, we have seen the introduction and gradual spread of the West Nile virus into the United States. Increases in the deer populations on the East Coast of the United States has resulted in more cases of Lyme disease, which is caused by a spirochete that resides in the deer reservoir. These emerging infectious diseases are likely to remain a problem for the immediate future. In addition, increasing global temperatures are causing the migration of species into new areas. Although the movements of larger organisms—like insects, birds, and mammals—are more obvious to us, when larger organisms move, their microbes move with them. Thus, global warming could lead to contact between new pathogens and human populations, resulting in outbreaks of disease.

The Virulence of Epidemics

When a pathogen is first introduced into a population, the results are often devastating. If the population has no established immunity to the pathogen, the pathogen may spread easily and have severe effects. When Europeans first came to the New World, they brought their diseases with them and caused the deaths of countless Native Americans. As West Nile virus spreads into the American bird population, it is killing large numbers of American crows. The European colonists who brought smallpox to the Americas were not affected by this disease nearly as badly as were the Native Americans. Likewise, Middle Eastern birds that carry West Nile virus do not have the same mortality rate as the American crows. It seems that when a pathogen and its host have survived together for a period of time, they may establish a sort of equilibrium with each other.

HERD IMMUNITY

One reason host populations become resistant to pathogens is that as a pathogen infects members of a host population, the survivors acquire immunity to the pathogen. As more members of the population become immune, it is harder for the pathogen to spread from person to person. This phenomenon is called **herd immunity**. The term comes from ecology and has to do with the protection individuals get from being part of a group. For example, if you were a lone zebra on the African savanna and a lion was looking for lunch, you would probably be history. However, if you were a member of a herd of zebras, there would be a good chance that the lion would eat someone else. So, herd immunity refers to the protection you get just by being a member of a group.

> **REMEMBER**
> Vaccination generates herd immunity, protecting not just the vaccinated but also the unvaccinated or immunocompromised people in a population.

The concept of herd immunity is very important to public health, as it is the foundation for vaccination programs. Vaccinating members of a population can generate herd immunity. If only a few people are vaccinated, the immunity won't be sufficient to protect the population because too many people will be able to serve as reservoirs for the pathogen. After a certain threshold is reached, however, the pathogen can be effectively blocked from spreading through the population. Thus, for the good of the whole population, it is very important that people follow the immunization recommendations of their national public health agency. In some areas, questions about the safety of vaccines are causing people to choose to not vaccinate their children for preventable diseases such as pertussis and measles. This often has devastating consequences for the children and, if vaccination rates fall too low, could have consequences for others in the population as well. Common questions about the safety of immunizations are addressed in Chapter 21.

COEVOLUTION OF HOST AND PATHOGEN

Another hypothesis for why pathogens and hosts achieve balance is that, over time, pathogens and their hosts may coevolve with each other. When the pathogen attacks a host population, the most susceptible will be more severely affected and may die. The more resistant members of the population survive and pass on their characteristics to their offspring, leading to a more resistant population over time. Pathogens can also evolve over time. Strains of a pathogen that cause severe disease and kill their hosts may not spread as far as strains that are less harmful (Figure 17.5). If a host is killed quickly, it does not have time to interact with many other hosts and spread the dangerous strain. Hosts that live longer because of less harmful strains may spread those strains farther, resulting in more of the less harmful strains in the pathogen population.

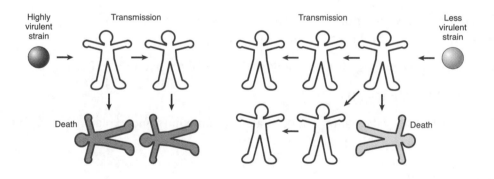

FIGURE 17.5. The coevolution of host and pathogen.

Over time, pathogens and host populations may reach an equilibrium. Highly virulent strains can result in rapid mortality and therefore not spread as widely in the population. Less virulent strains may spread more widely and thus become more common in the population.

REVIEW EXERCISES FOR CHAPTER 17

1. Describe the science of epidemiology, including the three different types of epidemiological analysis.

2. Define the following terms: etiology, incidence, prevalence, sporadic, epidemic, pandemic, endemic, morbidity, and mortality.

3. Describe the role of the CDC in disease surveillance and control.

4. State and explain five factors that contribute to problems with infectious disease in today's society.

5. Describe the structure of the influenza virus.

6. Explain why influenza has a high mutation rate.

7. Describe the processes of antigenic shift and drift and state their consequences.

8. Explain how recombination leads to antigenic shift in influenza virus.

9. Describe the methods used to identify *Hantavirus*.

10. Describe how environmental change led to an epidemic of *Hantavirus* in the Four Corners region of the American Southwest.

11. Explain how pathogens and host populations may evolve to coexist.

12. Define herd immunity.

13. Describe how vaccination can generate herd immunity and explain the benefit of vaccination to public health.

SELF-TEST

1. New influenza epidemics are caused by

   A. genetic change in the virus.
   B. antigenic drift.
   C. antigenic shift.
   D. environmental changes.
   E. Both A and B are correct.

2. Which of the following pairs are **not** correctly matched?

   A. Incidence—the total number of cases of a disease in a population at one time
   B. Mortality—the number of people that died from a disease in a given time
   C. Pandemic—an epidemic that spans more than one continent or population
   D. Etiology—the cause of a disease
   E. Endemic—a relatively constant level of disease in a population

3. True or False. A natural disaster like a flood could lead to a disease epidemic.

4. During his investigation of cholera epidemics in London, John Snow compared the incidence of cholera in people who drank water from the Broad Street pump and in people who drank beer. This represented

   A. a descriptive analysis.
   B. an experimental analysis.
   C. an analytical analysis.
   D. All of the above are correct.
   E. None of the above is correct.

5. True or False. Global travel helps spread pathogens around the world.

6. The CDC does all of the following **except**:

   A. Collects data on nationally notifiable diseases
   B. Releases publications on the incidence and prevalence of disease in the United States
   C. Cooperates with international agencies and governments to monitor emerging infectious diseases
   D. Publishes informative guides on disease for the general public
   E. All of the above are correct.

7. The World Health Organization is attempting to eradicate polio virus from the planet. Humans are the only reservoir for the virus, which can be transmitted in contaminated water. If polio can be prevented from infecting enough humans, the virus will no longer be able to multiply. To achieve this goal, the WHO is working on a global vaccination program. Eradication of polio from the planet is possible because

   A. humans are the only reservoir.
   B. vaccines can generate herd immunity in the human population.
   C. transmission of polio virus is fairly difficult.
   D. Both A and B are correct.
   E. A, B, and C are correct.

8. True or False. Antibodies are molecules that trigger the production of antigens.

9. True or False. If a human pathogen undergoes genetic change, it may become more virulent.

10. True or False. Epidemiologists help prevent the spread of disease.

11. If two flu strains, H2N1 and H7N3 infected a single cell at the same time, what strains might be produced? If H2N1 is a common strain of human flu, but H7N3 is a flu that only infects birds, predict the effect on the human population from each of the strains you predicted.

12. If another country does control the admistration of antibiotics and thus many antibiotic-resistant strains of bacteria exist in that country, should you be concerned? Why or why not?

13. War often leads to epidemics of disease. Explain why this might be so.

Answers

Review Exercises

1. The science of epidemiology tracks the occurrence of disease, seeking to understand its causes and transmission. This enables epidemiologists to make recommendations for the control of disease.

2. Etiology is the study of the cause of disease. Incidence is the number of new cases of a disease in a given time. Prevalence is the total number of cases of a disease during a given time period. Sporadic diseases occur occasionally within a population. An epidemic occurs when the number of cases of a disease exceed expected levels. A pandemic is an epidemic that spreads worldwide. Morbidity is the incidence and prevalence of a disease. Mortality is the number of deaths attributed to a disease in a given time period.

3. The CDC is the national epidemiological agency for the United States. It collects information on the incidence and prevalence of diseases in the United States and analyzes that information for trends. It also participates in global disease surveillance efforts and works with the WHO and the governments of other countries. Based on the data collected, the CDC makes recommendations for the control of disease.

4. Global travel contributes to the worldwide spread of pathogens. Centralized food processing can cause the rapid spread of contaminated foods. Urbanization leads to dense populations in which disease can spread more easily. Suburbanization is resulting in the development of formerly wild lands and bringing people into contact with new pathogens. The misuse of antibiotics contributes to the increasing numbers of antibiotic-resistant bacteria.

5. Influenza is an enveloped RNA virus. Spikes are present on the envelope. The viral RNA is split into eight different pieces.

6. Influenza has a high mutation rate because it copies its RNA with an RNA-dependent RNA polymerase that does not have good proofreading abilities.

7. Antigenic drift results from small genetic changes in pathogens like those that result from point mutations during replication of nucleic acid. The change in the antigens results in evasion of some of the preexisting immunity in the population, and an epidemic can result.

   Antigenic shift results from large genetic changes in pathogens like those that result from recombination. Because the antigens change dramatically, there may be no preexisting immunity in a population. The disease spreads rapidly and can cause a pandemic.

8. When two strains of the influenza virus coinfect a single cell, they may swap pieces of their genetic information as the virions assemble. This leads to influenza strains that have entirely new combinations of antigens (antigenic shift). Some of these combinations may lead to a strain that can infect people but to which there is no preexisting immunity in the population.

9. The *Hantavirus* was identified first by a reaction of serum antibodies from the victims to known *Hantavirus* antigens. Then nucleic acid from the *Hantavirus* was extracted from deer mice and sequenced. This sequence was compared to sequences of other species of *Hantavirus*.

10. An increase in rainfall led to greater production of pinon nuts by the pinon pines. This supported a large increase in the mouse population. Disease transmission among mice was easier, and the prevalence of *Hantavirus* increased in the mouse population. When mice came into contact with human dwellings, they left droppings that dried and became dust in the air. The inhalation of droppings from infected mice led to a fatal respiratory illness.

11. When pathogens are first introduced into a population, there is no preexisting immunity, and the pathogen spreads widely. Very virulent strains of the pathogen may kill hosts more quickly and not be transmitted as often as less virulent strains of the pathogen. Thus, over time, less virulent strains may become more common.

12. Herd immunity is the protection an individual gets by being part of a group.

13. When people are vaccinated, they become resistant to the pathogens they were vaccinated against. As more people in the population become resistant, it is harder for the pathogen to find susceptible hosts and spread through the population. If enough people in a population are vaccinated, a pathogen cannot spread and even people who are not vaccinated will be protected from the pathogen. This is an example of herd immunity. By achieving the right level of vaccination in a population, the population can be protected from disease.

Self-Test

1. E	5. T	9. T
2. A	6. E	10. T
3. T	7. D	
4. C	8. F	

11. Possible strains: H2N1, H7N3, H2N3, and H7N1. H2N1 is already present in the human population, so it would result in epidemics. H7N3 is a bird flu, so it would not affect the human population. Both H2N3 and H7N1, if they could infect human cells, would contain antigens that are new to the human population. These would lead to pandemics.

12. Yes, because global travel allows microbes to travel also.

13. War can disrupt sanitation and force people into new environments where they may encounter pathogens. People who live in refugee camps often do not have clean water and may live in crowded conditions that facilitate the spread of disease.

Innate Immunity

WHAT YOU WILL LEARN

This chapter presents the components of your innate immunity, the physical and chemical defenses of your body that protect you from the wide variety of microbes you encounter every day. As you study this chapter, you will:

- examine how your physical barriers like your skin protect you from infection;
- discover how phagocytes eat up invading bacteria and viruses;
- learn how inflammation helps clear pathogens from your body
- investigate the causes of fever;
- explore your molecular defenses against pathogens.

SECTIONS IN THIS CHAPTER

- Anatomic and Physiologic Barriers to Infection
- Types of White Blood Cells
- Phagocytosis
- Inflammation
- Fever
- Complement
- Interferon

You have intimate encounters with microbes every day. You pick them up on your hands, you inhale them, and you even eat them with your food. Yet, most of the time, you never even knew they were there. This is because most microbes never get past your **innate immunity**, the immunity you have to all types of microbes based on the general structure and chemistry of your body. Because your innate immunity is effective against a wide variety of microbes, it's also sometimes called your **nonspecific defenses**. Most microbes never even get past your first line of defense, which is made up of your intact skin, mucous membranes, and the microbes that normally live on your body. If potentially pathogenic microbes do get past these barriers, your second line of defenses, including phagocytosis and inflammation, then usually blocks them. Very few microbes make it past these defenses and successfully colonize the body. If they do, then they are specifically targeted by the cells of your adaptive immunity (Chapter 19) and usually cleared from the body.

Anatomic and Physiologic Barriers to Infection

Your **skin** is the largest organ in your body, and when it comes to protection from microbes, it is your best friend. Because it is so dry, the surface of your skin is inhospitable to many microbes. It is also slightly acidic because of the fatty acid component of **sebum**, the oily secretion from your sebaceous glands. Skin cells routinely flake off, taking many microbial visitors with them. To penetrate the skin, a microbe would have to be able to get past the **keratin**, a tough protein that reinforces the epithelial cells. Microbes would also have to have a way to penetrate the many layers of closely packed cells that make up the epidermis (Figure 18.1).

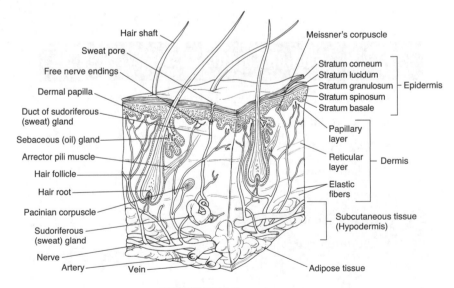

FIGURE 18.1. The skin.

Like the epidermis of the skin, the epithelia of **mucous membranes** are made up of layers of closely packed cells. Mucous membranes line the surfaces of your respiratory tract, digestive system, and genital tract. Although they are moist and not reinforced with keratin, they do have many unique defenses that make them very good barriers to microbes. One of these defenses is **mucus**, a thick substance that traps microbes as they are inhaled or eaten. In addition, your lower respiratory tract is protected from colonization by your **ciliary escalator**, rows of cilia that sweep microbe-laden mucous toward the throat where it can be coughed out of the body.

Microbes are also removed from mucous membranes by secretions of the body. Examples of secretions that perform **mechanical washing** include tears in the eye, saliva in the mouth, urine in the urethra, and vaginal secretions in the vagina. In addition to physical removal of microbes, many secretions also contain the enzyme **lysozyme**, which breaks down peptidoglycan and causes lysis of bacteria. Lysozyme is found in perspiration, tears, saliva, nasal secretions, and tissue fluids. Finally, many microbes cannot survive the temperature or pH of the body and so cannot colonize it. Gastric juice in the stomach creates an extremely acidic environment with a pH of 2, destroying many ingested bacteria that would otherwise be potential human pathogens.

Types of White Blood Cells

Microbes that successfully get past the skin and mucous membranes must next contend with a variety of defensive cells and molecules whose job it is to fight off microbes. Because many different types of cells are involved in both the innate and adaptive defenses of the body, it is generally a good idea to review these cell types and their distinguishing characteristics before examining their role in defense (Table 18.1). With the exception of mast cells, these cells are **leukocytes**, or white blood cells (*leuko* = white; *cyte* = cell). Basophils, neutrophils, and eosinophils are called **granulocytes** because they contain visible granules when stained. **Lymphocytes** are involved in specific immunity.

TABLE 18.1. Distinguishing Characteristic of Cells
Important to Innate and Adaptive Immunity

Type of Cell	Structure/Location	Function
Basophil	Granulocyte, granules stained with basic stain Circulate in blood, can migrate into tissues	Release histamine
Neutrophil	Granulocyte, granules stained with acidic or basic stain Also called polymorphonuclear (PMN) leukocytes owing to the highly lobed nucleus Released into blood, migrate into tissues	Highly phagocytic Numbers increase early in infection (leukocytosis)
Eosinophil	Granulocyte, granules stained with acidic stain Released into blood, migrate into tissues	Somewhat phagocytic Important in defense against parasitic worms
Monocyte	Circulate in blood	Migrate to tissues and differentiate into macrophages
Macrophage	Wandering macrophages migrate through tissues, fixed macrophages remain in a tissue	Highly phagocytic
Lymphocyte	Circulate in blood and lymph, migrate into tissues	Central role in specific immunity
Mast cells	Released into blood, migrate into tissues, especially prevalent in connective tissues of skin, respiratory system, and surrounding blood vessels	Release histamine
Dendritic cells	Circulate in blood and lymph, found in skin, mucous membranes, and organs	Alert lymphocytes by presenting foreign antigen

Phagocytosis

In the vast army of cells that defend your body, the **phagocytes** (*phago* = eat, *cyte* = cell) are like sentries that are constantly on patrol looking for invading bacteria and viruses. They crawl around your body like little amoebae, engulfing and destroying any invaders that they find. In addition to destroying pathogens, phagocytes also alert lymphocytes as to their presence, which is important for the activation of your specific immunity (Chapter 19). Two types of cells are very important in phagocytosis, the **neutrophils** and the **macrophages**. Because neutrophils typically respond first, an increase in their numbers, called **leukocytosis**, is an indicator of a recent infection. This sign of infection can be observed as part of a routine screening of the blood, called a complete blood count (CBC), in which a white blood cell count is performed.

For phagocytosis to occur, the first thing that needs to happen is for the phagocytes to find the pathogens. When microbes enter the body and damage cells, phagocytes are attracted to the scene of the crime by chemicals called **chemoattractants**. Chemoattractants may be released from the damaged cells or be part of the microbes themselves. The chemoattractants diffuse to the phagocytes and bind to receptors on their surface, causing the phagocytes to move toward the chemicals. The movement of the phagocytes toward the chemoattractants is called **chemotaxis**.

After the phagocyte has moved toward the pathogen, it must adhere to it. This **adherence** occurs when receptors on the phagocyte bind to molecules on the microbe. Adherence causes the membrane of the phagocyte to move around the microbe, forming projections called **pseudopods** (Figure 18.2). The pseudopods surround the microbe and fuse together, enclosing the microbe in a membrane-bound sac called the **phagocytic vesicle** or **phagosome** (Figure 18.2).

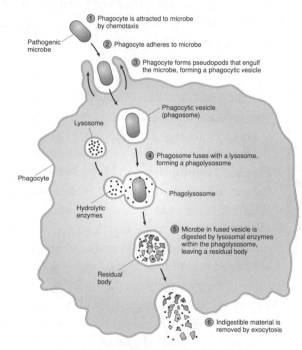

FIGURE 18.2. Phagocytosis.

After the microbe is captured in the phagosome, it can be digested. **Digestion** is triggered when **lysosomes** fuse with the phagosome, forming a **phagolysosome**. Recall that lysosomes are vesicles that contain destructive molecules (Chapter 4), such as hydrolytic enzymes that break down macromolecules and enzymes that can produce toxic forms of oxygen (Chapter 6). When the lysosome fuses with the phagosome, it dumps its destructive molecules onto the microbe. Toxic forms of oxygen are generated in an **oxidative burst** and begin to steal electrons from the molecules that make up the microbe. Hydrolytic enzymes like proteases, nucleases, and lipases break down the macromolecules into smaller components. Macrophages also produce nitric oxide (NO), which kills microbes by inhibiting their ability to produce ATP. The phagocyte recycles usable components and any undigestible material, called the **residual body**, is released from the cell by exocytosis (Figure 18.2).

Inflammation

Invasion of the body by a microbe is often accompanied by **inflammation**. Anyone who has ever had a cut or a scrape is familiar with the signs of inflammation, which are **redness** (erythema), **pain**, **swelling** (edema), and **heat**. In addition to these signs, inflammation may also be accompanied by **loss of function** of the affected area. In addition to being triggered by infection, inflammation can also be triggered by allergies, burns, and injuries like sprains. It is a beneficial response of the body designed to destroy or contain any invading organisms, as well as to repair or replace wounded tissue. **Acute inflammation**, like that in response to injury, develops quickly and doesn't last long. Some diseases, like gum disease or arthritis, can induce **chronic inflammation**, which may be less intense than acute inflammation but which can recur over a long period of time. Recent medical data have suggested that chronic inflammation is harmful to the body and increases the risk of heart attack and certain forms of cancer.

> **REMEMBER**
> Inflammation results from the release of histamine, which increases vasodilation and blood vessel permeability, allowing the delivery of white blood cells and clotting elements to the site of injury.

Inflammation is triggered when cells are damaged, causing them to release chemicals like **histamine** (Table 18.2) that increase **vasodilation** and **blood vessel permeability**. The increased vasodilation allows greater blood flow to the area, which delivers phagocytes and clotting elements such as **fibrinogen**. Greater blood flow also results in the signs of redness and heat. Because blood vessels are more permeable, fluid is able to pass out of the blood vessels into the tissue spaces, causing swelling and delivering clotting elements to the site of injury. Blood clots may form and help to wall off any invading pathogens. Swelling, along with nerve damage or the presence of toxins, can contribute to the pain of inflammation.

The loosening of the blood vessel walls also allows phagocytes to leave the blood and migrate into the damaged tissue. The phagocytes are attracted to the site of damage by microbial compounds, chemicals released from damaged cells, and **kinins** in the plasma (Table 18.2). Inflammatory chemicals cause changes in the cells that line the blood vessels. In response, the phagocytes begin to stick to the blood vessel walls and roll along them in a process called **margination** (Figure 18.3). When the phagocytes find the space between two cells, they squeeze through the capillary walls and move into the tissue. The exit of phagocytes from the blood vessels is called **emigration** (Figure 18.3) or **diapedesis**.

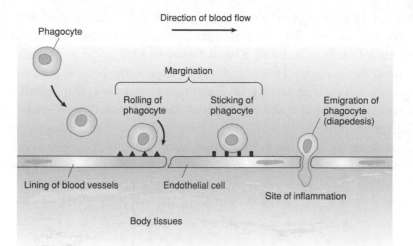

FIGURE 18.3. Margination and emigration.

Phagocytes are attracted to the site of inflammation by chemicals released by damaged cells. Changes in the lining of blood vessels near the site of inflammation cause the phagocytes to roll and stick to the blood vessel wall and then squeeze out and migrate into tissues.

TABLE 18.2. Selected Chemicals Involved in Inflammation

Chemical	Source	Function
Histamine	Mast cells, basophils, platelets	Increases vasodilation and blood vessel permeability
Kinins	Present in blood plasma	Increase vasodilation and blood vessel permeability, chemoattractant for neutrophils, stimulate pain receptors
Prostaglandin	Damaged cells	Helps increase vasodilation and blood vessel permeability, helps phagocytes move through capillary walls
Leukotrienes	Mast cells, basophils	Increase blood vessel permeability, help phagocytes attach to pathogens
Fibrinogen	Present in blood	Converts to insoluble fibrin, forming blood clots

After the phagocytes arrive in the damaged area, they engulf and destroy any invading microbes. Pockets of **pus** may form from body fluids and the remains of damaged cells. In addition to removing any harmful substances, the body will attempt to repair the damaged tissue. The ability to repair the tissue depends on the type of tissue and cells that are involved in the repair. For example, skin repairs easily, whereas nervous tissue does not. Depending on the type of cells involved in the repair, scar tissue may develop. The process of inflammation is not complete until repair is complete.

Sometimes inflammation is only a **local response**, such as the slight inflammation around a small cut or scrape. However, if the damage to the body is severe enough, inflammation can trigger an accompanying **systemic response**. During inflammation, the body increases production and activation of chemicals called **acute-phase proteins**. Many acute-phase proteins are normally present in the blood in an inactive state, becoming activated in response to inflammation or infection. These proteins, which include kinins (Table 18.2), fibrinogen (Table 18.2), and proteins in the **complement** system, have both local and systemic effects. Systemic effects can also be produced by **cytokines** that are released during the inflammatory response. Cytokines are signaling molecules that cells use to communicate with each other.

Fever

Fever is an example of an important systemic response (Figure 18.4). Like inflammation, fever is also a beneficial response of the body that is designed to help protect the body from infection. When phagocytes digest microbes, microbial molecules like the **endotoxin** (lipopolysaccharide) in gram-negative cell walls cause the phagocytes to release fever-inducing cytokines called **pyrogenic cytokines** (*pyro* = fire, *genesis* = give rise to) or endogenous pyrogens. Molecules that enter the body and cause fever, like endotoxin, are called **exogenous pyrogens**. The pyrogenic cytokines that were triggered by the exogenous pyrogens travel through the blood to the hypothalamus.

> **REMEMBER**
> Low-grade fevers are believed to be beneficial in fighting infections.

The hypothalamus receives the cytokine signal and produces **prostaglandin**. The prostaglandin resets the thermostat of the body to a higher set point. As the body temperature rises, people may experience a **chill** during which the skin feels cold and shivering occurs. The skin is cold in response to the constriction of blood vessels that, along with shivering and an increased metabolic rate, helps to raise the body temperature. When the cytokine signal stops, these changes reverse, and **crisis** begins. During crisis, vasodilation and sweating occur, lowering the body temperature.

The importance of fever in fighting human infections is under investigation. On the one hand, scientists have demonstrated that fever is beneficial to survival from infection in a wide variety of organisms. In humans, low-grade fevers have been shown to enhance several of our defenses. The higher temperature may slow the growth rate of some organisms and speed up the rate of repair in the body. The higher temperature also causes cells to release **transferrins**, a type of protein that tightly binds iron and keeps it away from microbes that need it for growth. Also, the IL-1 produced by phagocytes during the fever leads to increased production of T-lymphocytes, which are necessary for the specific immune response (Chapter 19). However, if fever is truly important for fighting infection, then you would predict that lowering fever would have negative outcomes for patients. Surprisingly, scientists who have looked at the effects of lowering fever have found that it has no significant impact on the survival rates of patients. One possibility for these conflicting results is that human innate and adaptive immunity (Chapter 19) are incredibly complex and also redundant. So, perhaps if fever is lowered in humans, other defenses can compensate so that our survival is not compromised. While the research into fever is still ongoing, the general advice of the medical community is to consider low-grade fevers (under 102°F) as beneficial and to allow them to run unchecked. For very young children, three months of age or younger, a medical visit is recommended any time their body temperatures reach 100.4°F, which is the lowest temperature medically recognized as a fever.

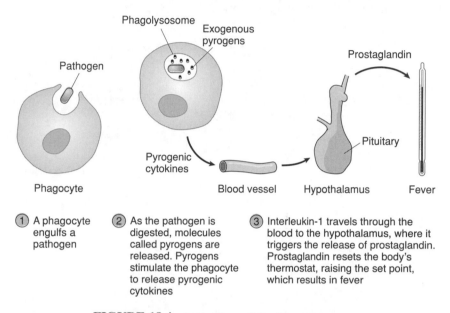

① A phagocyte engulfs a pathogen

② As the pathogen is digested, molecules called pyrogens are released. Pyrogens stimulate the phagocyte to release pyrogenic cytokines

③ Interleukin-1 travels through the blood to the hypothalamus, where it triggers the release of prostaglandin. Prostaglandin resets the body's thermostat, raising the set point, which results in fever

FIGURE 18.4. Activation of the fever response.

Complement

The **complement system** is also involved in systemic responses to infection. This system consists of over thirty different proteins that are produced by the liver and released into the blood. These proteins travel in the blood in an *inactive* form until they are activated by direct or indirect encounters with microbes. Complement activation has many beneficial effects that help clear invading microbes from the system. In response to infection, the liver may increase production of complement proteins so that more of these defensive proteins are available to help destroy the microbe.

There are three different ways for inactive complement proteins to become activated. These include the **classical pathway**, the **alternative pathway**, and the **lectin pathway** of activation. In each case, an encounter with a microbe sets off a chain reaction that leads to the activation of an important complement protein called **C3**. The series of reactions that

REMEMBER
Complement proteins circulate in the blood in an inactive form until they are activated by the presence of microbial molecules.

leads to activation of C3 is different for each of the activation pathways. The details of the activation pathways may seem complicated, but each one is in essence a kind of relay race: One complement protein becomes activated, so it activates the next one in the pathway, which activates the next one, and so on, until C3 is activated.

In the classical pathway of complement activation, antibodies produced by the specific immune response must first bind to the microbe and then bind the complement protein C1 (Figure 18.5). This activates C1 and causes it to interact with C2 and C4. C2 and C4 both split in half, two of their subunits join together, and then activate C3 (Figure 18.5). The defining characteristic of the classical pathway is that it requires the interaction of bound antibody and complement.

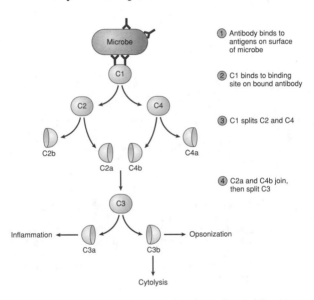

FIGURE 18.5. Classical activation of complement.

In the alternative pathway of complement activation (Figure 18.6), cell wall polysaccharides on the surface of the microbe interact with the complement proteins called Factor B, Factor D, and Factor P. The defining characteristic of the alternative pathway is that it requires the interaction of cell wall polysaccharides and complement.

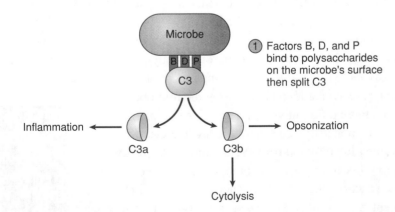

FIGURE 18.6. Alternative activation of complement.

In the lectin pathway of complement activation, lectins, which are proteins produced by the liver, bind to sugars on the surface of the microbe (Figure 18.7). The
bound lectins interact with C2 and C4, causing them to split and join together just as
they do in the classical pathway. The protein that is formed from the joining of C2a
and C4b activates C3. The defining characteristic of the lectin pathway is that it
requires the interaction of bound lectins and complement.

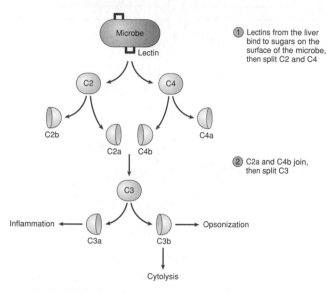

FIGURE 18.7. Lectin activation of complement.

During an infection, any or all of the three activation pathways may be occurring in
the body, resulting in the activation of C3. When C3 is activated, it splits into two subunits, C3a and C3b. This triggers three different beneficial outcomes of complement
activation, all of which help to clear invading microbes from the body (Figure 18.8).
One outcome, called **opsonization**, occurs when C3b binds to the surfaces of
microbes. Phagocytes have receptors for C3b, so this helps them bind to the microbes
and begin the process of phagocytosis.

Another positive outcome of complement activation is
cytolysis, which occurs when complement proteins bind to
the surfaces of microbes and punch holes in their membranes
(Figure 18.8). This occurs through the interaction of several
complement proteins. First, C3b interacts with C5, activating
it and causing it to split into two subunits, C5a and C5b.
C5b binds to C6 and C7, which attach to the surface of the
microbe. Finally, C8 and several C9 molecules attach to
the other complement proteins, forming a circle of proteins
called the **membrane attack complex (MAC)** that extends across the plasma membrane of the microbe (Figure 18.9). Cytoplasm leaks out through the ring of proteins,
causing the microbial cell to lyse.

> **REMEMBER**
> Once activated, complement
> protein C3 activates other
> complement proteins, leading to
> lysis of microbial cells, improved
> phagocytosis, and enhanced
> inflammation.

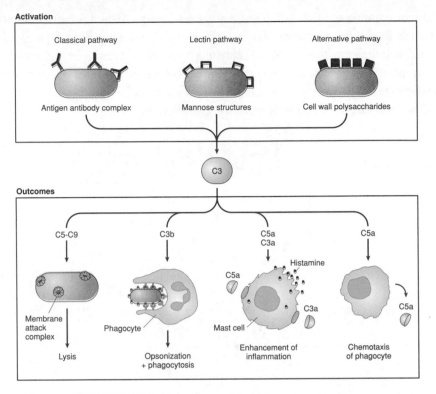

FIGURE 18.8. Outcomes of complement activation.

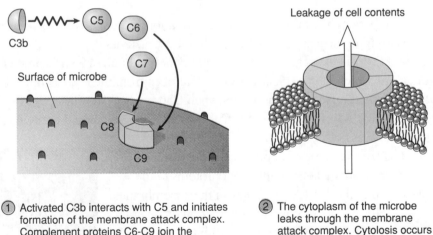

① Activated C3b interacts with C5 and initiates formation of the membrane attack complex. Complement proteins C6-C9 join the complex, forming a ring of proteins that penetrates the membrane of the microbe

② The cytoplasm of the microbe leaks through the membrane attack complex. Cytolosis occurs and the microbe is killed

FIGURE 18.9. The membrane attack complex.

The third positive outcome of complement activation occurs when complement proteins enhance the inflammatory response. Both C3a and C5a can bind to mast cells, basophils, and platelets, causing them to release their histamine. In addition, C5a is a chemoattractant for phagocytes (Figure 18.8).

Because complement proteins travel in the blood in an inactive form, none of these outcomes occur unless there is an infection of the body by a microbe. When there is an infection, complement proteins that are in the same area as the microbes will become activated, helping to trigger localized inflammation, attract phagocytes, and destroy microbial cells. Regulatory proteins in the blood work quickly to deactivate complement proteins, so the response only lasts as long as the infection is present.

Interferon

Another type of defensive proteins, the **interferons (IFNs)**, also helps defend the body against viruses and bacteria. There are three types of interferons, **alpha-interferon**, **beta-interferon**, and **gamma-interferon**. Interferons are produced by fibroblasts in the connective tissue, as well as by leukocytes. Alpha- and beta-interferon help defend the body against viral infections. When a cell is infected with a virus, it begins to produce alpha- and beta-interferons (Figure 18.10). The interferons leave the infected cell and travel to neighboring cells where they bind to receptors on the cell surface. The interferon signals the neighboring cell to produce **antiviral proteins (AVPs)**, such as nucleases that can degrade viral nucleic acid when it enters the cell. Interferons also activate proteins that

> **REMEMBER**
> Alpha- and beta-interferon are released by virally infected cells, then they travel to neighboring cells and signal them to activate their antiviral defenses.

are involved in apoptosis, making neighboring cells more likely to commit suicide if they are infected. The alpha- and beta-interferons do not help the originally infected cell, but these interferons warn the neighboring cells to arm themselves against viral infection. Gamma-interferon is produced by activated lymphocytes that have encountered microbes. It signals phagocytes, causing them to kill bacteria.

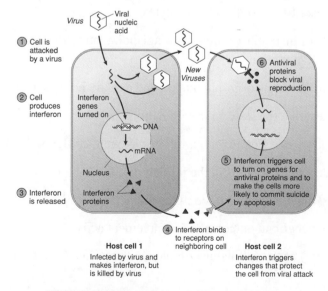

FIGURE 18.10. Alpha- and beta-interferon.

Because viruses reproduce inside of host cells and have few proteins of their own, viral infections are difficult to treat with drugs. The discovery of alpha- and beta-interferons raised great hopes for their use as natural antiviral substances. Recombinant DNA technology was used to insert the human genes for interferons into bacteria and induce the bacteria to make the human protein for pharmaceutical use. Unfortunately, interferons have only had limited success as antiviral drugs. Alpha-interferon is used in the treatment of herpes, hepatitis B, and hepatitis C. It has also had some success in the treatment of Karposi's sarcoma and hairy cell leukemia. Beta-interferon is used in the treatment of multiple sclerosis.

REVIEW EXERCISES FOR CHAPTER 18

1. Describe the anatomic barriers of the skin.

2. Describe the anatomic barriers of the mucous membranes.

3. Describe how temperature and pH can restrict microbial growth in the body.

4. Describe the function of lysozyme and what locations in the body produce lysozyme.

5. State the function and location in the body of the following cell types: neutrophil, basophil, eosinophil, monocyte, macrophage, lymphocyte, mast cell, and dendritic cell.

6. Name and describe the steps in phagocytosis.

7. State the role of histamine, kinins, prostaglandin, leukotrienes, and acute-phase proteins in inflammation.

8. State the cardinal signs of inflammation and explain how each is caused.

9. Name and describe the steps in inflammation.

10. Define cytokine, margination, emigration, pyrogen, chill, and crisis.

11. Name and describe the steps in the initiation of fever.

12. Describe the three pathways of complement activation.

13. Describe the beneficial outcomes of complement activation.

14. State the role(s) of the following complement proteins: C3a, C5a, C3b, and C5–C9.

15. Describe how alpha- and beta-interferons trigger cells to protect themselves against viral infection.

SELF-TEST

1. Which of the following cell types are phagocytic?

   A. Basophil
   B. Neutrophil
   C. Lymphocyte
   D. Monocyte
   E. None of the above is correct.

2. Which of the following is a characteristic of the skin that helps prevent infection?

   A. Dryness
   B. Keratin
   C. Many layers of cells
   D. pH
   E. All of the above are correct.

3. If you arranged the following steps of phagocytosis in order, which event would be fourth?

   A. Fusion of phagosome and lysosome
   B. Exocytosis of residual body
   C. Adhesion of phagocyte to pathogen
   D. Digestion of pathogen
   E. Chemotaxis of phagocyte

4. Which of the following complement proteins acts as a chemoattractant for phagocytes?

   A. C3b
   B. C3a
   C. C5a
   D. C8
   E. C9

5. Which of the following types of complement activation results from antigen–antibody reactions?

   A. Classical pathway
   B. Alternative pathway
   C. Lectin pathway
   D. Antibody pathway
   E. Antigen pathway

6. True or False. After a cell produces alpha-interferon, it is protected from viral attack.

7. During inflammation, redness is a direct result of

   A. increased vasodilation.
   B. increased blood vessel permeability.
   C. margination.
   D. emigration.
   E. None of the above is correct.

8. When phagocytes squeeze out through capillary walls, it is termed

   A. margination.
   B. emigration (diapedesis).
   C. inflammation.
   D. edema.
   E. complement activation.

9. True or False. An inherited genetic defect that prevented the production of complement protein C3 would be expected to result in greater numbers of microbial infections.

10. During fever, the body releases iron-binding proteins called

    A. interleukins.
    B. cytokines.
    C. prostaglandins.
    D. transferrins.
    E. siderophores.

11. Vaccinia viruses, including the smallpox virus, have receptors that bind Interleukin-1 and prevent its activity in the body. Therefore, infection by these viruses would be expected to

    A. cause fever.
    B. block fever.

12. The drug ibuprofin blocks the production of prostaglandin. What effect does ibuprofin have on the body and why?

13. To be a successful parasite, a microbe must avoid destruction by complement. For each of the following examples, explain how the strategy enables the microbe to avoid destruction by complement.

    a. Group A streptococci release an enzyme that degrades C5a.

    b. Herpesvirus internalizes antigen–antibody complexes.

Answers

Review Exercises

1. The epidermis consists of multiple layers of closely packed cells. This helps prevent microbes from getting past this barrier. In addition, the upper layer of cells contains a tough protein called keratin that reinforces the skin. The skin is dry and has a low pH owing to the breakdown of sebum. All of these conditions slow down the growth of many microbes. Finally, skin cells are continuously being shed along with the microbes that are on them.

2. The epithelium of all mucous membranes consists of many layers of closely packed cells that help prevent microbes from penetrating this layer. The thick, sticky mucus, which is produced by certain cells, does an excellent job of trapping microbes. In the respiratory tract, the ciliary escalator moves this mucus into position to be coughed out of the body.

3. Many microbes cannot grow at 37° Celsius and so are not potential human pathogens. In certain areas of the body, pH can restrict microbial growth. For example, the acidic gastric juice of the stomach can denature proteins, resulting in the death of microbes that are ingested. The breakdown of sebum on the skin makes this a slightly acidic environment which will also prevent the growth of some microbes.

4. Lysozyme breaks down peptidoglycan, thus weakening bacterial cell walls. It is present in tears, saliva, sweat, nasal secretions, and tissue fluids.

5. Neutrophils, which are found in the blood and in tissues, are highly phagocytic. Basophils, which are found in the blood and in tissues, release histamine. Eosinophils, which are found in the blood and in tissues, are mildly phagocytic. Monocytes are found in the blood; they also migrate to tissues where they differentiate into macrophages. Macrophages are found in the tissues and are highly phagocytic. Fixed populations of macrophages remain in certain tissues; others wander the body. Lymphocytes are found in the blood, lymph, and tissues. They are involved in specific immunity. Mast cells, which are found in blood and connective tissue, release histamine. Dendritic cells are found in the blood, lymph, and tissues. They present antigen to lymphocytes.

6. Phagocytes are attracted to the site of infection by molecules released by damaged cells or by molecules on the microbes themselves. This is called chemotaxis. Adhesion of the phagocyte to the microbe occurs when receptors on the phagocyte bind to the microbe. Pseudopods wrap around the microbe, forming a phagocytic vesicle or phagosome, and the microbe is taken into the cell by endocytosis. Digestion occurs when the phagosome fuses with the lysosome, releasing hydrolytic enzymes and other toxic molecules onto the microbe. After the microbe is broken down, undigested material called the residual body is released by exocytosis.

7. Histamine and kinins both increase vasodilation and blood vessel permeability. In addition, kinins are also chemoattractants and can stimulate pain receptors. Prostaglandins increase the effects of the histamines and kinins and help phagocytes move through capillary walls. Leukotrienes increase the permeability of blood vessel walls and help phagocytes attach to pathogens. Acute-phase proteins have a variety of affects, including facilitating clearance of the pathogen (complement) and blood clotting (fibrinogen).

8. Redness and heat are both caused by the increased blood flow to the area owing to increased vasodilation. Swelling is caused by the increased blood vessel permeability, which allows fluid and cells to migrate into tissue. Pain is caused by swelling pressing on nerves and also by the release of molecules like kinins that stimulate pain receptors.

9. Damaged cells release chemicals such as prostaglandins, kinins, leukotrienes, and histamine. These chemicals trigger vasodilation and increased blood vessel permeability. Host molecules such as kinins and molecules on microbes act as chemoattractants for phagocytes. Changes in the cells that line the blood vessels near the site of infection cause the phagocytes to begin to stick to the walls. The phagocytes roll along the walls in a process called margination and then squeeze out into the tissues in a process called emigration. Clotting factors delivered by the blood help wall off any infection, and phagocytes eliminate the pathogens. Damaged tissue is repaired.

10. A cytokine is a molecule that communicates between cells. Margination is the rolling of phagocytes along blood vessel walls. Emigration is the passage of phagocytes out of blood vessels and into tissues. A pyrogen is a molecule that can induce fever. Chill occurs as body temperature is rising and is marked by cold skin and shivering. Crisis occurs as body temperature falls and is marked by warm skin and sweating.

11. As phagocytes digest pathogens, molecules called pyrogens that are part of pathogens can trigger the phagocyte to release interleukin-1. Interleukin-1 travels to the hypothalamus where it triggers the release of prostaglandin. Prostaglandin resets the body's thermostat.

12. Classical activation occurs by the interaction of complement proteins with antibody–antigen complexes. When antibodies are bound to microbial antigen, they display binding sites for these complement proteins. Alternative activation occurs by the interaction of complement proteins with cell wall polysaccharides. Lectin activation occurs when proteins called lectins bind to sugars on the surfaces of microbes. The bound lectins can interact with complement proteins. All three pathways lead to the activation of C3.

13. After complement is activated, complement proteins help to clear pathogens from the body. C3b binds to the surfaces of microbes, facilitating phagocyte binding. This process is called opsonization. C3a and C5a bind to mast cells and basophils,

causing them to release histamine. This increases the inflammatory response. In addition, C5a is a chemoattractant for phagocytes. C5 through C9 interact with each other and the surfaces of microbes to form the membrane attack complex. This complex punches holes in the surfaces of microbes and causes them to lyse.

14. C3a and C5a bind to mast cells and basophils and cause them to release histamine. C5a is also a chemoattractant for phagocytes. C3b binds to the surfaces of microbes, facilitating phagocytosis (opsonization). C5–C9 form the membrane attack complex.

15. Alpha- and beta-interferons are produced by virally infected cells. They are released by these cells and then bind to receptors on neighboring cells. This signal causes the neighboring cells to produce antiviral proteins such as nucleases and proteases that will help protect them from viral attack.

Self-Test

1. B	5. A	9. T
2. E	6. F	10. D
3. D	7. A	11. B
4. C	8. B	

12. Ibuprofen decreases inflammation and pain. Prostaglandin enhances the effects of histamines and kinins, thus increasing blood vessel permeability and vasodilation. By blocking prostaglandin production, ibuprofin lessens these effects, resulting in less swelling, inflammation, and associated pain.

13. a. Enhancement of inflammation is blocked because C5a cannot bind to mast cells and basophils, nor can it act as a chemoattractant for phagocytes.
    b. Activation of complement by the classical pathway is blocked because there are not any antigen–antibody complexes present on the surface of the virus.

Adaptive Immunity

WHAT YOU WILL LEARN

This chapter explores the workings of the adaptive immune system, the defenses that target specific molecules like those found on pathogens. As you study this chapter, you will:

- review the structure and function of the immune system;
- compare antigens and antibodies;
- discover the role of pattern-recognition receptors in immune system activation;
- investigate activation of T and B cells;
- explore how antibodies help clear infections from the body;
- compare the primary and secondary immune responses;
- examine why the immune system normally doesn't attack self molecules.

SECTIONS IN THIS CHAPTER

- Types of Immunity
- The Immune System
- Antigens
- Antigen Presentation and T-cell Activation
- B-cell Activation
- Antibodies
- The Secondary Immune Response
- Self-Tolerance

How many times have you been sick in your life? How many times have you had a cold, the flu, a strep throat, or maybe something more serious? You probably took antibiotics to help get rid of some of the illnesses caused by bacteria, but for others you probably just waited until it was over. Every time you got sick, it meant a pathogen got around your innate immunity (Chapter 18). But every time you got well on your own, it was because your **specific immune response** came to your rescue.

The specific immune response has certain characteristics that make it extremely effective. When it is activated, it responds *specifically* to the pathogens that are invading the body. Unlike your innate defenses that were effective against many types of pathogens, the specific immune response is tailored to each particular pathogen when it attacks. Although the immune response is specific, it is *diverse* and has the ability to respond to a wide variety of pathogens. For many of these pathogens, it retains an immunologic *memory* that protects you from further attacks. In a healthy person, it also has *self-tolerance*, the ability to distinguish between self and nonself. In other words, your immune system does not normally attack your own cells. These features of the immune system are a result of a complex system of cells, particularly **T** and **B** **lymphocytes**, which can recognize and respond to foreign molecules. These cells use signaling molecules to communicate with each other and the cells involved in your innate immunity, creating a web of defenses that is very effective at clearing pathogens from your body.

Types of Immunity

Our innate defenses (Chapter 18) give us a certain amount of **innate immunity** to pathogens, an ability to resist diseases by virtue of our structure and physiology. In addition to this, the immune response enables us to develop **acquired immunity**, or resistance to specific pathogens that we develop during our lifetimes. Any time our immune system responds to a pathogen, we generate **active immunity**, which is typically long-lasting. If active immunity is triggered in response to an infection, it is called **naturally acquired active immunity**. If it is triggered by a vaccine, it is called **artificially acquired active immunity**. It is also possible to acquire short-term **passive immunity** from pathogens by receiving defensive proteins called antibodies from someone else. Breast-fed babies receive antibodies from their mothers, providing **naturally acquired passive immunity**. If you are exposed to some pathogens, such as the hepatitis virus, it is possible to receive an injection of antibodies to help clear the virus from your system. This is an example of **artificially acquired passive immunity**.

The Immune System

The ability to acquire immunity depends on a functional **immune system**, which is composed of the **lymphatic system** and certain organs and tissues (Figure 19.1A). **Lymphatic vessels** collect fluid, called **lymph**, from the tissues and transport it around the body, eventually returning it back to the circulatory system. As the lymph circulates, it is passes through lymphatic organs called **lymph nodes** (Figure 19.1B). The lymph nodes act as filters, using a cleanup crew of phagocytes and lymphocytes to screen out microbes and harmful materials from the lymph before it is returned to the blood. Lymphatic tissue is also associated with various organs of the body, including the spleen, tonsils, and appendix. Additionally, patches of lymphatic tissue underlie the mucosa (mucosa-associated lymphoid tissue or MALT), the gut (gut-associated lymphoid tissue or GALT), and the skin (skin-associated lymphoid tissue or SALT).

> **REMEMBER**
> The immune system is like a giant filtration system, collecting material from the tissues and giving white blood cells a chance to clean up any pathogens before the fluid is returned to the blood.

The cells of the immune system begin as stem cells in the **bone marrow**. The stem cells develop into different types of precursor cells that give rise to phagocytes and lymphocytes (Figure 19.2). **T lymphocytes** must also travel to the **thymus** as part of their normal development. Lymphocytes communicate with each other and with phagocytes by means of special communicating molecules called **cytokines**. **Interleukins** are a particular type of cytokines that are often used to communicate between white blood cells. The effectiveness of the immune system depends upon this complex communication (Figure 19.2).

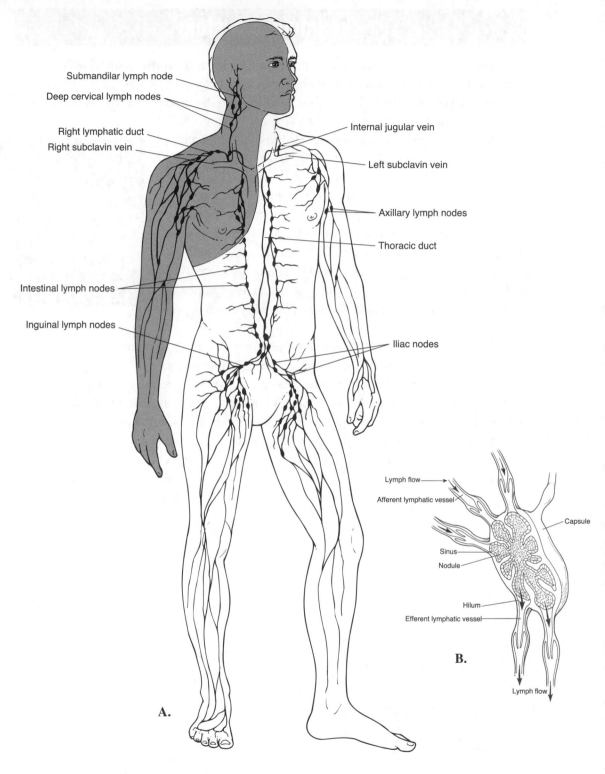

FIGURE 19.1. A. Lymphatic vessels and nodes. B. A lymph node.

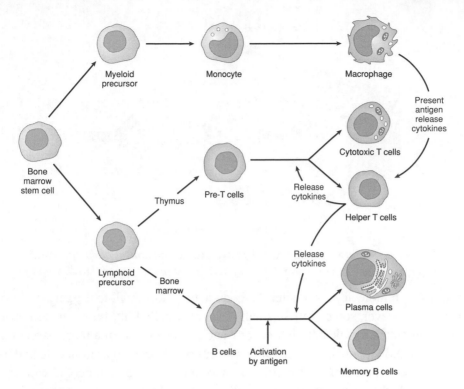

FIGURE 19.2. Overview of the cells involved in the immune response.

Macrophages, T cells, and B cells all arise from stem cells in the bone marrow, but each has a unique developmental pathway. Macrophages present antigen to helper T (T_H) cells, activating them. T_H cells release cytokines that stimulate T_H cells, cytotoxic (T_C) cells, and B cells.

ANTIGENS

The immune response is activated when the cells of the immune system recognize that something foreign has entered the body. What they are actually responding to are molecules called **antigens**. Antigens are molecules that trigger the production of defensive proteins called **antibodies**. Because all cells are made up of molecules, a microbe entering the body is basically a little ball of antigens. Antigens are typically large molecules like proteins and polysaccharides. For a white blood cell to respond to an antigen, it must have a receptor on its surface that can bind to it. Each receptor is very specific and can only bind to a particular subunit of the antigen called an **epitope** or **antigenic determinant** (Figure 19.3). Because antigens are large molecules, a single antigen may have several distinct epitopes, and each epitope is capable of triggering the production of a unique population of antibodies. So, when a microbe enters the body, many different populations of antibodies can be generated, each one responding to a particular epitope on a particular antigen.

> **REMEMBER**
> Antigens are typically the large molecules that make up cells, and they have the potential to trigger the production of defensive proteins called antibodies.

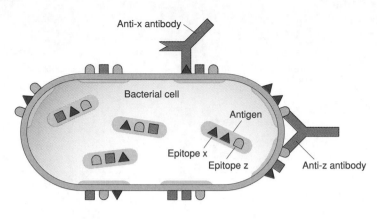

FIGURE 19.3. Antigens.

Antigens are large molecules. They have distinct subunits, called epitopes.
Each epitope can trigger a population of antibodies that is specific to the epitope.

Small foreign molecules that enter the body may be too small to trigger antibody production. However, some of these small molecules, called **haptens**, can combine with existing **carrier molecules** in the body and become a part of a larger structure (Figure 19.4). Now the hapten functions as an epitope of the larger molecule and can generate antibody production. Penicillin is an example of a hapten. People who are not allergic to penicillin do not make a molecule that can act as a carrier for penicillin. When penicillin enters their bodies, it simply goes unnoticed by the immune system and does not cause an allergic reaction. In people who are allergic, however, penicillin attaches to a carrier molecule and triggers an immune response.

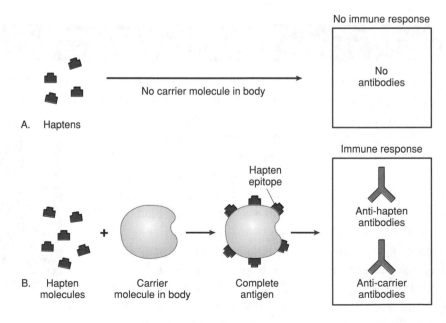

FIGURE 19.4. Haptens.

Haptens are small molecules that do not trigger antibody
production unless they are attached to larger molecules.

Antigen Presentation and T-Cell Activation

For antigens to trigger an immune response, they must be recognized by lymphocytes, particularly **T lymphocytes**, or **T cells**. T cells arise from stem cells in the bone marrow and are then processed in the thymus, after which they can differentiate into several different types of T cells. **Helper T cells (T_H cells)** play a central role in the coordination of the immune response, releasing cytokines that activate phagocytes and other lymphocytes. **Cytotoxic T cells (T_C cells)** recognize and destroy body cells that are infected or cancerous. All T cells recognize and bind to antigen with a receptor called the **T-cell receptor**. Different types of T cells can be distinguished by another surface protein, which has different forms. T_H cells have a surface protein called CD4; T_C cells have a surface protein called CD8.

The activation of T_H cells by antigen is of crucial importance to the immune response. In order for T_H cells to bind antigen with their T-cell receptor, the antigen must be presented to them by an **antigen-presenting cell (APC)**, such as a phagocyte, dendritic cell, or B cell. Phagocytes and dendritic cells are like sentries, patrolling the body for microbes. When they encounter microbes, they take them in and break them down. As the microbe is being broken down into its component parts, fragments of microbial antigen are then displayed on the surface of the antigen-presenting cell where they can be recognized by T_H cells. The antigen-presenting cells are like sentries who discover an enemy and then show the evidence to their commanding officer, the T cell.

When antigen is displayed on the surface of an antigen-presenting cell, it must be displayed in a particular way. T cells only recognize antigen that is displayed within a set of surface proteins called the **major histocompatibility complex (MHC)**. This complex of proteins is essential in the recognition of self versus non-self and gets its name from its importance in matching organ donors and recipients (*histo* = tissue). There are two types of major histocompatibility complexes found on human cells. Major histocompatibility complex type I (MHCI) is found on all nucleated cells. Major histocompatibility complex type II (MHCII) is found on antigen-presenting cells. When antigen-presenting cells are digesting microbes, they attach microbial antigen to MHCII proteins that are on their way to the cell surface. The microbial antigen fits into a cleft on the MHCII just like a hot dog in a bun (Figure 19.5). Antigen-presenting cells that are displaying foreign antigen within their MHCII can activate T_H cells. A T_H cell with a T-cell receptor that can recognize the foreign antigen binds to the MHCII–antigen complex (Figure 19.6).

> **REMEMBER**
> Antigen-presenting cells capture antigen and display it within cell surface proteins called the major histocompatibility complex, type II (MHCII).

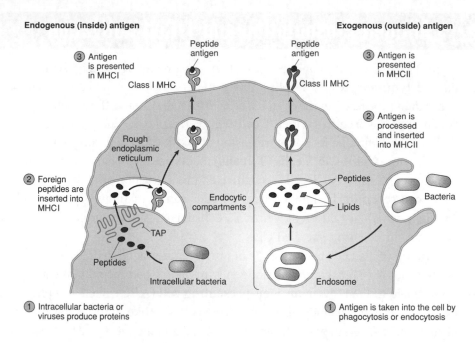

FIGURE 19.5. Antigen presentation.

Endogenous antigens are those made inside the cell. This includes molecules made by intracellular pathogens as well as abnormal proteins in cancer cells. Exogenous antigens are those found outside the host cell and include those found on circulating bacteria and viruses.

In addition to recognizing the MHC-antigen complex, antigen-presenting cells must give T_H cells a second signal in order to activate them. These signals, called **co-stimulatory signals**, are proteins that antigen-presenting cells produce on their surface in response to the presence of molecules that signal danger to the body. Certain molecules, like peptidoglycan, flagellin, and lipopolysaccharide, for example, aren't part of your cells at all. So, if they're in your body, it's because of a microbe. Molecules like these that are part of microbes, but not our cells, are called **pathogen-associated molecular patterns (PAMPs)**. And, it turns out that our antigen-presenting cells are very good at detecting these sorts of molecules due to the presence of receptors called **pattern-recognition receptors (PRRs)**. If an antigen-presenting cell detects a foreign molecule (a PAMP) with one of its pattern-recognition receptors, the antigen-presenting cell will make the co-stimulatory proteins and put them on its surface. So, to use the analogy of a sentinel reporting to the general again, you can think of the sentinel showing the antigen to the general in the MHC complex and saying, "Look what I found," and then putting up the co-stimulatory proteins on its surface and saying, "And by the way, it looks dangerous." The T cell binds to the MHC-antigen complex with its T-cell receptor and then binds to the co-stimulatory

> **REMEMBER**
> Antigen-presenting cells must display two signals—presented antigen plus co-stimulatory signals—in order activate helper T cells.

proteins with matching protein on its surface. The combination of the two signals—antigen recognition plus co-stimulatory signals—activates the T_H cell. An activated T_H cell produces cytokines that stimulate immune system cells, including itself. One cytokine, called **Interleuken-2 (IL-2)**, binds to receptors on the T_H cells and stimulates it to divide, producing many copies of itself. This process of selecting and copying one particular T_H cell, called **clonal selection**, and then copying it, called **clonal expansion**, creates a population of activated T_H cells that are specific to the displayed antigen. These activated T_H cells release IL-2 and other cytokines that activate **B cells** and T_C **cells** that are also specific to this antigen. Again, we can think of the T_H cell as a commanding officer that starts to send out messages to the troops, getting them ready to fight the invading microbe.

> **REMEMBER**
> Activated T_H cells produce Interleuken-2 (IL-2), which self-stimulates the activated T_H cell to divide and make many copies of itself. IL-2 also stimulates other white blood cells in the immune system.

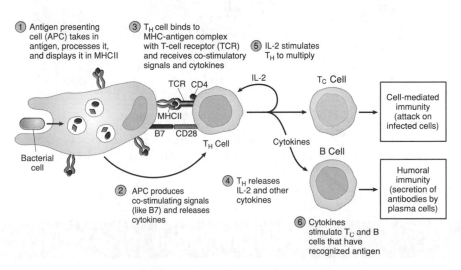

FIGURE 19.6. Helper T-cell activation.

When T_C cells receive cytokines from T_H cells, they are stimulated to divide and mature. Like T_H cells, T_C cells have a T-cell receptor and can only recognize antigen if it is presented within a major histocompatibility complex. Unlike T_H cells, however, T_C cells recognize and bind to antigen that is displayed in **MHCI** (Figure 19.5), which is present on the surface of almost all cells in the body. Typically, if a body cell is infected by a pathogen, fragments of the foreign antigen will get displayed in the MHCI on the surface of the cell. This acts like a little sign to the T_C cell, telling it that the cell has been invaded. When a T_C cell binds a body cell that is displaying foreign antigen in its MHCI, the T_C cell is stimulated to release **perforin**, a hollow protein that can insert itself into cell

> **REMEMBER**
> T_H cells detect antigen that's presented in MHCII, whereas T_C cells detect antigen that's presented in MHCI.

membranes (Figure 19.7). Perforin creates pores that allow destructive enzymes to enter the cell, resulting in **cytolysis**. The action of T_C cells, which is called **cell-mediated immunity**, is very important in the control of viral infections and cancer, as well as bacterial infections.

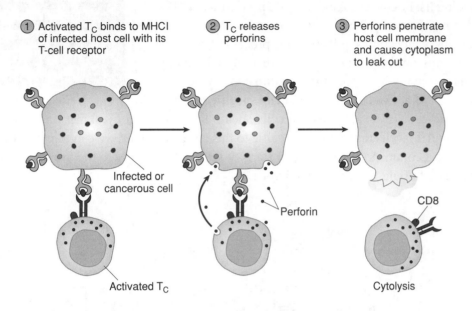

① Activated T_C binds to MHCI of infected host cell with its T-cell receptor

② T_C releases perforins

③ Perforins penetrate host cell membrane and cause cytoplasm to leak out

Infected or cancerous cell

Perforin

CD8

Activated T_C

Cytolysis

FIGURE 19.7. Cell-mediated immunity.

Cytotoxic T cells recognize abnormal antigen displayed in major histocompatibility complex type I and destroy the infected or cancerous cell.

B-Cell Activation

Cytokines from T_H cells also activate the other major branch of immunity, **humoral immunity**. Humoral immunity refers to the production of antibodies by **B cells**. B cells arise from stem cells in the bone marrow and then mature and migrate to lymphoid tissues, where they produce antibodies. Each B cell has surface receptors specific to one epitope and is capable of producing antibody specific to that epitope. For antibody production to occur, B cells must be activated by antigen.

There are two categories of antigens that activate B cells, **T-dependent antigens** and **T-independent antigens** (Figure 19.8). As the name implies, activation by T-dependent antigens requires cytokines from T_H cells. Activation by T-independent antigens does not require cytokines from T_H cells. Although both types of activation result in antibody production, activation by T-dependent antigens gives a stronger response, resulting in the development of immunologic memory. Activation by T-independent antigens is weaker and does not result in immunologic memory.

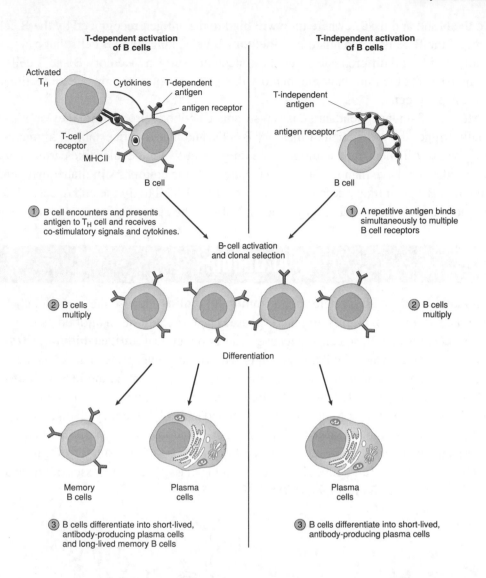

FIGURE 19.8. B-cell activation.

In order for a B cell to be activated, it must first encounter antigen. T-dependent antigens are processed by the B cell and presented to T_H cells within the MHCII on the surface of the B cell (Figure 19.8). A T_H cell that recognizes this antigen binds to the MHCII–antigen complex on the B cell, receives a co-stimulatory signal, and releases cytokines that stimulate the B cell to divide and differentiate. This **clonal selection** and **expansion** of the B cell results in the proliferation of B cells that can respond to this particular antigen. Some of the B cells differentiate into antibody-producing cells called **plasma cells** (Figure 19.8), which are relatively short-lived. Antibodies produced by the plasma cells are released

> **REMEMBER**
> T-dependent activation of a B cell results in the production of plasma cells, which make antibodies, and memory B cells, which help generate immunologic memory.

into the blood and tissues where they will bind to the antigen recognized by the B cell. Some of the B cells differentiate into **memory B cells,** long-lived B cells that can rapidly divide and differentiate when they encounter antigen. Memory B and T cells are the basis for the immunologic memory that protects you from repeated infections by some pathogens.

Although T-independent antigens can stimulate antibody production, they do not usually trigger the production of memory B cells, and thus do not establish immuno-logic memory. T-independent antigens are long, repetitive molecules like certain poly-saccharides that have many copies of one epitope. If the epitopes simultaneously bind to enough B-cell antigen receptors, it can trigger the B cell to divide and differentiate into plasma cells that produce antibodies specific to the epitope (Figure 19.8).

Antibodies

Antibodies are a type of protein called **immunoglobulins** because they are globular proteins that function in the immune response. The basic unit of all antibodies is **bivalent** (Figure 19.9), so called because it has two identical **antigen-binding sites** (*bi* = two). Each bivalent antibody is composed of four polypeptide chains that are held together by covalent bonds. There are two long chains, called the **heavy chains,** and two short chains, called the **light chains** (Figure 19.9). The specificity of an anti-body is due to the precise folding of the combined ends of one heavy chain and one light chain to form the **variable region,** which can bind to only one specific epitope of an antigen (Figure 19.9). The rest of the antibody consists of the **constant region,** which is the same for all antibodies within a single class. The five classes of antibodies are summarized in Figure 19.9 and Table 19.1.

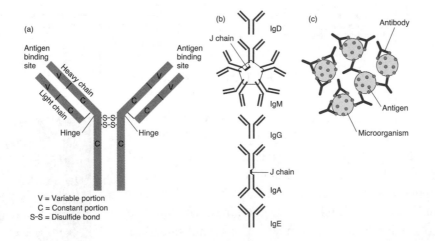

FIGURE 19.9. Antibodies.

A. The structure of an antibody molecule showing the four chains that make up the molecule. B. The structures of five different types of antibodies. C. The reaction between antibody molecules and antigens on the surface of a microorganism. Antibody molecules bind the microorganisms together and assist phagocytosis.

TABLE 19.1. The Five Classes of Antibodies

Class of Antibody	Structure	Location in Body	Major Functions
IgG	Monomer	Circulates in blood and lymph, present in intestine, can cross into tissues including the placenta	Defends against circulating bacteria and viruses, protects fetus
IgM	Pentamer (5 bivalents)	In blood and lymph, monomer (bivalent) is surface receptor on B cells	Produced first in response to infection, very effective at agglutination
IgA	Dimer (two bivalents attached by secretory component or J chain)	Secretions such as mucous and breast milk, blood, lymph	Naturally acquired passive immunity to newborns, defends against pathogens on mucosal surfaces
IgD	Monomer	Surface of B cells, blood, lymph	Antigen receptor
IgE	Monomer	Bound to surface of basophils and mast cells	Antigen receptor

After B cells are activated and antibodies are produced, the antibodies circulate around the body binding specifically to epitopes they recognize. The binding of antibody to antigen has a number of positive outcomes that help the immune system clear pathogens from the body (Figure 19.10). One beneficial outcome is that phagocytes can more easily bind to pathogens that are coated with antibody. This enhancement of phagocytosis, called **opsonization**, occurs because phagocytes have receptors for the stem, or F_C region, of antibodies (Figure 19.10). Because antibodies have more than one binding site for antigen, they can bind more than one antigen at one time. This clumping of antigens is called **agglutination** (Figure 19.10). Agglutination helps prevent the spread of pathogens and makes it easier for phagocytes to clean them up. When antibodies bind to the surfaces of pathogens, it also results in **neutralization** of the pathogen (Figure 19.10). Bacteria and viruses both need to bind to host cell receptors, which is prevented (neutralized) by the presence of antibody on their surface. Neutralization can also block toxins from binding to host cell receptors, thus preventing the effect of the toxin.

> **REMEMBER**
> IgM is produced first in response to an infection, followed by IgG.

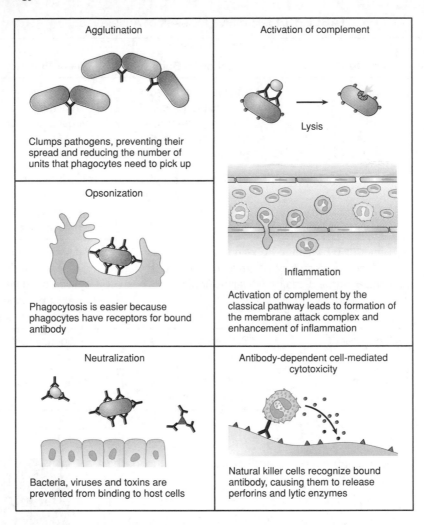

FIGURE 19.10. The outcomes of antibody–antigen reactions in the body.

REMEMBER
Antibodies bind to antigens, helping to clear them from the body by agglutination, opsonization, neutralization, activation of complement, and antibody-dependent cell-mediated cytotoxicity (ADCC).

The binding of antibody to antigen can also indirectly cause beneficial outcomes by the activation of complement. Recall that bound antibody can interact with complement protein C1, resulting in **classical activation** of the complement system. Thus, antibody production also triggers all of the beneficial effects of the complement system: opsonization by complement, enhancement of inflammation including chemoattraction of phagocytes, and formation of the membrane attack complex (Chapter 18). In fact, the complement system of proteins was originally named because of their interaction with bound antibodies. The effects of these proteins were seen to *complement* the action of antibodies.

In addition to triggering complement, the presence of bound antibody also stimulates a population of cells that can assist in the removal of pathogens. These cells, called **natural killer cells**, are derived from the precursors to lymphocytes, but they are *not* T cells. They do not have T-cell receptors and they are not MHC-dependent. Instead, they have receptors for the F_C region of antibodies and for complement. If they encounter a pathogen that has bound antibody or complement, they react by releasing perforins and other lytic enzymes that destroy the pathogen (Figure 19.11). This response is particularly useful against pathogens like parasitic worms that are too large to be engulfed by phagocytosis. In addition to the natural killer cells, other cells can display *natural killer activity*. In other words, these cells act like natural killer cells and release destructive compounds in response to stimulation by bound complement or bound antibody. Macrophages, eosinophils, and neutrophils can all display natural killer activity. The destruction of pathogens by natural killer cells or cells displaying natural killer activity is called **antibody-dependent cell-mediated cytotoxicity (ADCC)**. This is an enormous term, but it literally describes the response. Cells are destroyed ("cytotoxicity") by natural killer cells ("cell-mediated") that recognize bound antibody ("antibody-dependent").

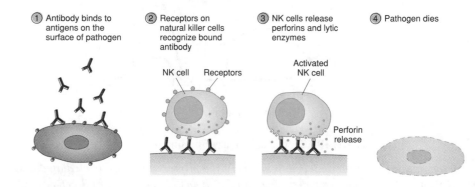

FIGURE 19.11. Antibody-dependent cell-mediated cytotoxicity.

The Secondary Immune Response

The production of antibodies by activated B cells is a beneficial short-term response that leads to the immediate clearance of the pathogen by opsonization, agglutination, neutralization, activation of complement, and antibody-dependent cell-mediated cytotoxicity. The activation of B cells by T-dependent antigens also generates a beneficial long-term response called the **secondary immune response** (also called **immunologic memory** or the **anamnastic response**). This response, which is mediated by memory B cells and memory T cells, leads to very rapid clearance of the pathogen upon a second introduction to the body.

The **primary response** of the body to the first introduction of antigen is very different from that of the **secondary response** (Figure 19.12). When the body first encounters an antigen and the immune response is triggered, the first class of antibody to be produced is IgM. These large antibodies are very good at agglutinating the pathogen and helping to restrict its spread.

After a few days, the B cells switch into production of IgG antibodies. IgG antibodies circulate through the body, helping to clear the pathogen. During the primary response, it can take more than a week before there is strong production of IgG antibody.

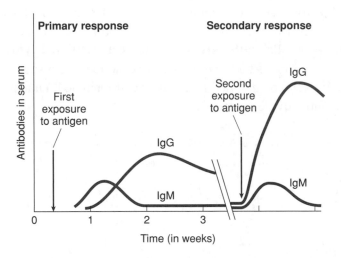

FIGURE 19.12. The secondary immune response.

When the antigen is first introduced to the body, the primary response is mediated first by IgM then by IgG. Upon second exposure to the same antigen, memory B cells quickly differentiate into plasma cells that produce large amounts of IgG antibody. This IgG antibody circulates and blocks the pathogen from establishing itself in the body.

Because memory B cells are created during the primary response and are descended from activated B cells, the secondary response to the antigen is quite different. When the body encounters an antigen for the second time, memory B cells rapidly divide and differentiate into plasma cells and begin producing both IgG and IgM antibodies. IgG production is particularly strong and far exceeds that of the primary response. This large quantity of IgG antibody circulates throughout the body, tagging the pathogen for destruction. The pathogen is rapidly cleared from the system, and no signs or symptoms of illness result.

Self-Tolerance

Antibody production against an antigen is obviously a very effective way of triggering destruction of that antigen. Unfortunately, some people with **autoimmune disorders** produce antibodies to their own antigens, causing destruction of their own cells. For example, the disease myasthenia gravis is triggered by antibodies that bind to acetylcholine receptors at the junctions between nerves and muscles. Normal communication between nerves and muscles is interrupted, and muscle responses become progressively weaker. Eventually, the muscles that control the diaphragm may fail to respond, and death occurs as a result of respiratory failure. To understand autoimmune disorders fully, we must understand the basis for **self-tolerance**. In other words, we must understand why a healthy body does not attack its own antigens. This is an area of science that is still being explored, and the answers are not fully known.

The ability of the immune system to tolerate self depends on its power to know when to react and when to stay calm. In other words, to protect you without harming you, your immune system needs to attack pathogens but leave your cells alone. How exactly it does that is still a matter of debate, but two main hypotheses have proved useful in explaining scientists' observations. One hypothesis, sometimes called the **stranger hypothesis**, proposes that the functioning of the immune system fundamentally depends on its ability to recognize self molecules versus molecules found on pathogens. According to the stranger hypothesis, dendritic cells activate when they recognize molecules unique to pathogens using specific receptors (**pattern-recognition receptors**), then alert T-cells to the problem. Thus, an immune response is initiated in response to the presence of a molecule that's foreign to the body. The other hypothesis, sometimes called the **danger hypothesis**, proposes that the immune system activates when it detects molecules indicating that cellular damage has occurred. According to the danger hypothesis, damaged cells release materials that are not normally seen outside the cell, and these molecules can activate dendritic cells, which in turn activate T cells. Scientists conducting research on the immune system have found evidence that supports both of these hypotheses, creating the possibility that the actual basis of tolerance may be quite complicated and contain elements of both hypotheses.

One process that seems to contribute to self-tolerance is **clonal deletion**, a process that eliminates T and B cells that would react to self-antigens. Humans have vast populations of T cells and B cells, each of which recognizes a specific epitope. T cells recognize antigens with T-cell receptors; B cells use antibodies on their surfaces for receptors. During fetal development, populations of T cells and B cells are generated. These populations can make every variety of antigen receptor that is possible given the genetic make-up of the individual. This includes cells that would have the ability to react with self-antigen. These populations of T and B cells are then tested for their reaction against self-antigens, and those that react are destroyed.

REVIEW EXERCISES FOR CHAPTER 19

1. Distinguish between naturally and artificially acquired immunity, as well as passive and actively acquired immunity.

2. Define antigen, antibody, epitope, and hapten.

3. State the role of the major histocompatibility complex in the immune response.

4. Describe antigen presentation and how it is connected to T-cell activation.

5. State the events of T_H-cell activation and the order in which they occur.

6. State the consequences of T_H-cell activation.

7. Distinguish between T cells and B cells.

8. Distinguish between T_H (CD4) and T_C (CD8).

9. Describe the steps of cell-mediated cytotoxicity.

10. Describe the steps of T-dependent B-cell activation and the order in which they occur.

11. Differentiate between T-dependent and T-independent B cell activation.

12. Define clonal selection and explain its relevance to both T cells and B cells.

13. Describe the structure of antibodies.

14. State the five classes of antibodies, their locations in the body, and their functions.

15. Describe the beneficial outcomes of antibody production (antigen–antibody reactions).

16. Distinguish natural killer cells from cytotoxic T cells.

17. Describe the process of antibody-dependent cell-mediated cytotoxicity (ADCC).

18. Describe the secondary immune response (memory response).

19. Define self-tolerance.

SELF-TEST

1. Which class of antibodies is secreted in breast milk?

   A. IgG
   B. IgA
   C. IgM
   D. IgE
   E. IgD

2. If the following steps of T_H-cell activation are put in order, which one would be third?

   A. Antigen-presenting cell produces Interleukin-1 (IL-1)
   B. TH binds to MHC–antigen complex
   C. TH produces Interleukin-2 (IL-2)
   D. Antigen-presenting cell places antigen in MHCII
   E. Antigen-presenting cell encounters antigen

3. A person who had a recent infection with a pathogen would have high amounts of which type of antibody in the blood?

   A. IgA
   B. IgM
   C. IgG
   D. IgD
   E. IgE

4. Which of the following is **not** triggered by antigen–antibody reactions?

   A. Opsonization
   B. Neutralization
   C. Agglutination
   D. Destruction of pathogens by T_C cells
   E. Activation of complement

5. True or False. Natural killer cells produce perforin.

6. The secondary immune response relies upon which type of cell?

   A. Plasma cells
   B. T_C cells
   C. Memory B cells
   D. Natural killer cells
   E. None of the above is correct.

7. The subunits of antigens that trigger unique populations of antibodies are called

   A. haptens.
   B. variable regions.
   C. F_C regions.
   D. epitopes.
   E. MHCII.

8. True or False. T-dependent antigens are small molecules that must combine with larger molecules to trigger antibody production.

9. Which of the following confers artificially acquired active immunity?

   A. Getting sick
   B. Getting a vaccine
   C. Getting a shot of antibodies (gamma globulin)
   D. Breastfeeding
   E. None of the above is correct.

10. Which outcome of antigen–antibody reactions clumps pathogens?

    A. Neutralization
    B. Agglutination
    C. Opsonization
    D. Complement activation
    E. Antibody-dependent cell-mediated cytotoxicity

11. HIV attacks T_H cells. Explain why HIV infection leads to acquired immunodeficiency syndrome (AIDS), an inability to fight off disease.

12. Which would make a better vaccine, a T-dependent antigen or a T-independent antigen? Why?

Answers

Review Exercises

1. Naturally acquired immunity occurs as a result of exposure to pathogens or through natural biological events like breastfeeding. Artifically acquired immunity occurs through medical intervention. When the body is challenged by antigen and produces its own immune response, this is called actively acquired immunity. If the body receives temporary protection by the acquisition of antibodies from an outside source like breastmilk, this is called passively acquired immunity.

2. An antigen is a molecule that triggers antibody production and can be bound by those antibodies. An antibody is a defensive protein that binds specifically to antigens. Epitopes are subunits of antigens that are each capable of triggering a unique population of antibodies. Haptens are small molecules that by themselves

do not trigger antibodies but that can trigger antibodies when they are bound to another molecule called a carrier.

3. The major histocompatibility complex is a large protein complex present on the surface of human cells. As antigens are processed inside cells, fragments of antigen are displayed within major histocompatibility complexes.

4. When antigen-presenting cells engulf antigen, they display fragments in MHCII molecules. These antigens, within the MHCII complex, are assessed by helper T cells. If a helper T cell is capable of binding to the antigen–MHC complex and receives a signal from the antigen-presenting cell, the helper T cell will be activated and release cytokines that stimulate B cells and cytotoxic T cells. In addition, all cells display fragments of intracellular antigens within MHCI molecules. These antigens, when displayed within the MHCI complex, can be assessed by cytotoxic T cells, which will react by destroying infected cells.

5. Antigen-presenting cells engulf antigen and display antigenic fragments within MHCII. The antigen-presenting cells also release the cytokine, Interleukin-1. Helper T cells bind to the MHCII–antigen complex with their T-cell receptor and receive the IL-1 signal. This stimulates the helper T cell to release Interluekin-2 and other cytokines that induce helper T-cell proliferation and stimulate both B cells and cytotoxic T cells.

6. After helper T cells are activated, they release cytokines. The cytokine IL-2 self-stimulates the helper T cells to proliferate, creating a larger population of helper T cells that react to the specific antigen presented by the antigen-presenting cell. These cytokines also contribute to the activation of cytotoxic T cells and B cells. Cytotoxic T cells will circulate through the body, destroying any cells that are displaying abnormal antigen in their MHCI complexes. B cells will divide and differentiate into plasma cells and memory B cells. Plasma cells produce antibodies that help clear the body of infection. Memory B cells are long-lived and help protect the body against future infection by the same pathogen.

7. T cells and B cells are both lymphocytes that arise from stem cells in the bone marrow. T-cell precursors travel to the thymus, where T-cell differentiation is completed. There are several different types of T cells. Two of the most important are the helper T cells and cytotoxic T cells. Helper T cells are very important in communication and coordination of the immune response. They release cytokines that activate other immune system cells. Cytotoxic T cells destroy infected and cancerous cells. B-cell precursors can differentiate into B cells within the bone marrow. B cells develop into plasma cells and memory B cells. Plasma cells produce antibodies that help clear the body of infection. Memory B cells are long-lived and help protect the body against future infection by the same pathogen.

8. Helper T cells have a protein on their surface called CD4. They bind to MHCII complexes on the surfaces of antigen-presenting cells. If they bind antigen within an MHCII complex and receive signals from the antigen-presenting cell, they release cytokines that activate other immune system cells. Cytotoxic T cells have a protein on their surface called CD8. They bind to MHCI complexes and, if they detect abnormal antigen within the complex, they will destroy the cell.

9. Activated cytotoxic T cells bind to MHCI complexes with their T-cell receptors. If they detect abnormal antigen within the MHCI complex, the T_C cells release perforins. Perforins are proteins that introduce channels into plasma membranes. The perforins penetrate the membrane of the infected or cancerous cell, causing cell lysis.

10. A B cell that has encountered T-dependent antigen displays the antigen within MHCII complexes. The B cell binds with an activated helper T cell and receives cytokines from the T_H. This causes the B cell to proliferate. Some cells differentiate into antibody-producing plasma cells, others become long-lived memory B cells.

11. T-dependent antigens will only activate B cells if the B cells also receive signals from a helper T cell. T-independent antigens are repetitive molecules that can simultaneously bind to several antigen-receptors on the surface of B cells, thus activating the B cell.

12. There are many different T cells and B cells within the body, each capable of recognizing a specific epitope. When a pathogen enters the body, its antigens react with those T cells and B cells that can specifically bind the antigen. After these T cells and B cells are activated, they multiply, forming many copies of themselves. This is called clonal selection. It is important because it increases the number of lymphocytes that have the capability of responding to the antigen that is challenging the body.

13. Antibodies are proteins that are made up of four polypeptide chains. Two of the chains are longer and are called the heavy chains. The two shorter chains are called the light chains. The four chains are held together by covalent bonds with the two longer chains toward the inside of the antibody. At one end of each of the chains is the variable region, the region that is unique for each antibody. The variable regions of the four chains fold together to create a precisely shaped pocket that is specific to one epitope. This pocket is the antigen-binding site. Each antibody has two antigen-binding sites and is called a bivalent. Outside of the variable regions of the four polypeptide chains are the constant regions. These regions are the same for any one class of antibody.

14. IgG antibodies are the smallest. They circulate in the blood and the lymph and are capable of passing into tissues and across the placenta. They defend against circulating bacteria and viruses. IgM antibodies are the largest. They are made of five

bivalents and so have ten antigen-binding sites. They are found in the blood and the lymph. They are produced early in reponse to an infection and are particularly good at agglutinating antigens. IgA antibodies consist of two bivalents joined by a short peptide called the J chain or secretory component. IgA antibodies are produced by B cells associated with the mucosa and help defend against pathogens that enter the mucosa. They are also released into breast milk. IgD antibodies, along with IgM monomers, serve as antigen receptors on the surfaces of B cells. IgE antibodies are antigen receptors on the surfaces of mast cells and basophils.

15. Because antibodies have more than one antigen binding site, a single antibody can bind to more than one antigen at one time, causing antigens to clump. This is called agglutination. It helps prevent the spread of pathogens and facilitates phagocytosis by reducing the number of units a phagocyte needs to pick up. Because phagocytes have receptors for bound antigen, any pathogen that is coated with antigen is also more easily phagocytosed. This is called opsonization. Pathogens that are covered with antibody may no longer be able to bind to host cells, preventing them from becoming established. This is called neutralization. Natural killer cells also have receptors for bound antibody and will release perforins and hydrolytic enzymes on pathogens that have bound antibody on them. This destroys the pathogens and is called antibody-dependent cell-mediated cytotoxicity. Finally, complement proteins are also capable of interacting with bound antibody. This activates complement via the classical pathway and leads to the enhancement of inflammation and formation of the membrane attack complex.

16. Although both natural killer cells and cytotoxic T cells can release perforins and trigger cell lysis, their targets and how they recognize these targets are different. Natural killer cells have receptors for bound antibody and will nonspecifically target any foreign cell that has bound antibody on it. Cytotoxic T cells use their T cell receptor to react specifically to antigens displayed in MHCI complexes. Each T_C cell can only react to a specific epitope. They target infected or cancerous body cells.

17. Natural killer cells have receptors for bound antibody and complement. When their receptors bind to bound antibody or complement, it triggers the natural killer cells to release perforins, hydrolytic enzymes, and toxic forms of oxygen that can destroy the pathogen.

18. During the first exposure to a pathogen, the body may develop memory B and T cells, long-lived cells that are capable of responding quickly if challenged with antigen. When the body is exposed to the same antigen a second time, these cells divide rapidly. The memory B cells give rise to plasma cells that begin producing large quantities of IgG antibody. The memory T cells release cytokines. The IgG antibodies circulate, quickly blocking establishment of the pathogen before illness can result.

19. Self-tolerance is the ability of the immune system to distinguish between self anti-gen and foreign antigen.

Self-Test

1. B	5. T	9. B
2. A	6. C	10. B
3. B	7. D	
4. D	8. F	

11. Helper T cells are critical for the coordination of the immune response. After they are activated by antigen-presenting cells, T_H cells release cytokines that activate other immune system cells, including cytotoxic T cells and B cells. Because HIV destroys T_H cells, it lowers the ability of the body to respond to both pathogens and cancer. Without the T_H cells, B cells do not become activated and produce antibodies that help clear pathogens from the body. Also, T_C cells do not become activated to deal with infected or cancerous cells.

12. T-dependent antigens make better vaccines because they lead to the production of plasma cells and memory B cells. T-independent antigens do not trigger memory B-cell production and thus do not establish immunologic memory.

Pathogenicity of Microorganisms

20

WHAT YOU WILL LEARN

This chapter takes a look at the various tools and tricks pathogens use in order to colonize the body and get past your defenses. As you study this chapter, you will:

- examine how pathogens enter the body;
- discover the importance of attachment;
- learn how pathogens evade defenses like complement and phagocytosis;
- explore how pathogens damage host tissues with toxins.

SECTIONS IN THIS CHAPTER

- Ability to Invade Tissues
- Evasion of Host Defenses
- Damage to the Host

The human body has amazing defenses against invasion by microorganisms. Most microbes we encounter are prevented from establishing an infection by our innate immunity (Chapter 18). Those that do establish an infection are usually defeated by our adaptive immunity (Chapter 19). So why is it that we get sick at all? How do some microbes manage to get around our defenses? Why does their presence in our bodies cause disease?

The ability of a microorganism to cause disease is its **pathogenicity**. A more familiar term might be **virulence**, which refers to the degree of pathogenicity. For example, an organism that very easily causes disease would be considered highly virulent. Microorganisms that are highly virulent have the ability to colonize the host easily and disrupt the host physiology. Any molecules that a microorganism makes and that help them to cause disease are called **virulence factors**. In other words, virulence factors are like the tool kit of a pathogen. What makes one pathogen different from another is what they have in their tool kit.

> **REMEMBER**
> A virulence factor is any molecule that helps a pathogen cause disease.

The ability of a pathogen to cause disease depends on three main issues. First, the pathogen must be able to invade the host. Second, it must be able to evade the host defenses. And finally, it may produce enzymes or toxins that damage host tissues and contribute to the signs and symptoms of disease. The different strategies that pathogens use to invade the human body are truly amazing, and understanding the details of these mechanisms can help us figure out new ways to prevent disease.

Ability to Invade Tissues

MULTIPLICITY OF INFECTION

One factor that determines whether an encounter with a pathogen will lead to disease is the **multiplicity of infection**, or number of pathogens encountered. Typically, the greater the number of pathogens encountered, the more likely it is that infection and disease will result. A common measure that is used to compare the effect of numbers on virulence is the ID_{50}. The ID_{50} is the number of pathogens (cells or virions) that individuals would have to encounter in order for 50 percent of the population to become sick. For example, *E. coli* O157:H7 and *Salmonella enteriditis* are both capable of causing diarrheal disease in humans. The ID_{50} for *E. coli* O157:H7 is ten cells, for *S. enteriditis* it is 10,000 cells. Thus, *E. coli* O157:H7 is a more virulent pathogen because fewer invading cells are required to cause disease.

One reason the number of invading organisms may make a difference in the establishment of disease is that some pathogenic bacteria turn on the genes for certain virulence factors only when they are in a dense population. The ability to sense the presence of other bacteria is called **quorum sensing**. Bacteria that do quorum sensing have the ability to sense molecules made by other bacterial cells. The presence of these molecules at a certain concentration signals that the population is of a certain density, and individual bacteria may respond by turning on genes. The products of these genes may help the bacteria invade the human body, making the bacteria more virulent.

PORTALS OF ENTRY

The routes by which pathogens enter the body are the **portals of entry**. The most common portal of entry is through the mucous membranes lining the respiratory tract. This occurs when pathogens are inhaled and results in respiratory diseases such as the common cold, influenza, measles, and tuberculosis. Many pathogens, such as those that cause diarrhea, cholera, hepatitis A, and polio, enter through the mucous membranes of the gastrointestinal tract. These pathogens may be present in contaminated water or food. Pathogens that enter within food particles may be protected from the stomach acid. The pathogens that cause sexually transmitted diseases like chlamydia, AIDS, genital herpes, and genital warts, enter through the mucous membranes of the genitourinary tract.

Although it is generally much harder to penetrate the skin than the mucous membranes, the skin is a portal of entry for some pathogens. Some fungi can colonize the skin because they produce **keratinase**, which breaks down the protein keratin that toughens the skin. Other pathogens may enter through pores or the ducts of sweat glands. If a pathogen penetrates through the skin as a result of a break in the skin, such as those caused by cuts, insect bites, or punctures, this is referred to as **parenteral route** of entry. Microbes may also penetrate mucous membranes by the parenteral route through breaks or abrasions.

> **REMEMBER**
> The most common portal of entry for human infections is through the respiratory tract.

The portal of entry a pathogen uses to access the body may affect the ID_{50} for that pathogen. For example, *Bacillus anthracis*, the causative agent of anthrax, can enter the body as endospores through three different portals of entry. The ID_{50} for entry through the skin by the parenteral route is ten to fifty endospores. This results in cutaneous anthrax and the development of a black lesion on the skin. The ID_{50} for entry through the respiratory tract is 10,000 to 20,000 endospores. Inhalation anthrax is the most lethal form of anthrax because it progresses very quickly and leads to lesions in the lungs and bleeding in the brain. The ID_{50} for entry through the gastrointestinal tract is 250,000 to 1,000,000 endospores. Gastrointestinal anthrax leads to severe vomiting and diarrhea and may result in death. *Bacillus anthracis* is an example of a very versatile pathogen that may cause disease through several portals of entry. However, not all pathogens are this versatile. Some organisms may be pathogenic if they enter through one portal but do not cause disease if they enter through a different portal.

ATTACHMENT (ADHESION)

After pathogens enter the body, most must attach to host cells to cause disease. The attachment of viral pathogens has already been discussed in Chapter 13. The attachment of bacterial pathogens follows the same principle: Molecules on the surface of the bacterium are used to attach to receptors on host cells (Figure 20.1). The molecules that bacteria use to attach are called **adhesins**, and they are often found on structures used for attachment like fimbriae and the glycocalyx. For example, *Neisseria gonorrhoeae* must attach to cells of the genitourinary tract with its fimbrial adhesins to cause

REMEMBER
An adhesin is a molecule that bacteria use to attach to host cells or structures. Adhesins are often part of pili, fimbriae, capsules, and slime layers.

the disease gonorrhea. After *Streptococcus mutans* attaches to proteins on the surface of teeth with its glycocalyx, acids produced by its metabolism contribute to tooth decay. Enteropathogenic *E. coli* first attaches by fimbrial adhesins to epithelial cells in the gastrointestinal tract and then actually deposits additional receptors for itself into the membrane of the host cells!

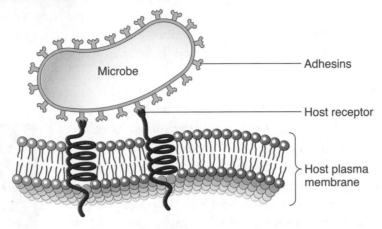

FIGURE 20.1. Adhesins.

Proteins on the surfaces of bacteria attach to receptors on host cells.

The secretion of sticky molecules for attachment by some pathogens leads to dense communities of microbes called **biofilms**. A biofilm consists of one or more species of microbes embedded in a matrix of sticky molecules. We have all experienced biofilms when we get up in the morning and feel a slick coating on our teeth or when we have stepped onto a rock coated with slippery substances at the beach. The matrix not only helps the members of the biofilm community to attach, it also protects them from drying out and from the action of antimicrobial substances. This can make organisms that live in biofilms difficult to kill. For example, in the lungs of patients with cystic fibrosis, *Pseudomonas aeruginosa* forms a biofilm community that is resistant to antibiotics. Biofilm communities can also be difficult to eradicate from ventilation systems and other moist areas in hospitals.

INVASION OF HOST TISSUES

Once established, the growth and spread of pathogens may be facilitated by the production of enzymes that help them destroy tissue or obtain nutrients. Enzymes like **collagenase** and **hyaluronidase** help break down molecules like collagen and hyaluronic acid that support cells and hold them together. This allows pathogens to penetrate into host tissues. The enzyme **fibrinolysin** breaks down the fibrin that holds blood clots together and may enable a bacterium to move beyond a clot and continue its spread through the body. Some bacteria also produce **hemolysins** that enable them to lyse red blood cells, which contain nutrients like iron that may be essential for bacterial growth. Enzymes called **invasins** allow pathogens like *Salmonella* to rearrange the cytoskeleton of the host cell. This can allow the pathogens to enter the cell and even use the cytoskeleton to move through the cell and then infect neighboring cells.

Evasion of Host Defenses

After pathogens have entered the body and begun to establish themselves, they must still escape the host defenses if they are to multiply. Thus, successful pathogens have various strategies to escape the action of complement (Chapter 18), phagocytosis (Chapter 18), and even the adaptive immune response (Chapter 19).

COMPLEMENT EVASION

Complement activation (Chapter 18) or its effects may be avoided if complement proteins cannot bind to molecules on the surface of bacteria. For example, *Streptococcus pyogenes* avoids opsonization by complement because it has a surface protein, called **M protein**, to which C3b will not bind. Some pathogens avoid destruction by complement by causing complement activation to occur at a safe distance away from the surface of the cell. For example, some gram-negative bacteria have very long polysaccharides that trail off their surfaces, while others may actually shed their polysaccharides.

EVASION OF PHAGOCYTOSIS

Just as pathogens can evade complement, they have many strategies for avoiding destruction by phagocytes (Chapter 18). Some pathogens avoid phagocytes by colonizing glands that are not patrolled by phagocytes or by multiplying inside of host cells. Others release compounds that inhibit chemotaxis of phagocytes. Some may even disguise themselves from phagocytes by hiding under host molecules. For example, *Staphylococcus aureus* uses the enzyme **coagulase** to convert host fibrinogen to fibrin, the material that is the basis for blood clots. *S. aureus* that is located under a pile of fibrin molecules is effectively hidden from phagocytes.

   Even if phagocytes are capable of locating pathogens, certain pathogens have ways of preventing the phagocytes from destroying them. The presence of a glycocalyx or proteins on the surface of a pathogen can prevent phagocytes from being able to bind. For example, strains of *Streptococcus pneumoniae* that make a capsule are virulent because they resist phagocytosis; strains that do not make a capsule are not virulent because they do not survive in the body. Pathogens like *Rickettsia*, which causes Rocky Mountain spotted fever, escape from the phagosome into the cytoplasm of the phagocyte. Once in the cytoplasm, these bacteria are safe from destruction by the lysosome and can multiply. Some strains of *Salmonella* escape destruction by the lysosome by preventing it from fusing with the phagosome. Other bacteria neutralize the destructive compounds of the lysosome and so can survive within the phagolysosome. Examples of this include *Bacillus anthracis*,

> **REMEMBER**
> Bacteria that produce a glycocalyx can avoid capture by phagocytes.

which produces enzymes to neutralize the oxidative burst, and *Mycobacterium tuberculosis*, which is protected by its waxy wall. Still other bacteria produce proteins called **leukocidins** that kill phagocytes.

EVASION OF ADAPTIVE IMMUNITY

It is more challenging to avoid our adaptive immune response, and not many pathogens can do so for very long. Some bacteria produce **IgA protease**, which destroys the IgA antibodies that are common on the surfaces of mucous membranes. Bacteria and viruses that multiply inside of host cells may successfully avoid the specific immune response, particularly if they prevent their own antigens from being displayed on the cell surface. (If viral or bacterial antigens are displayed in MHCI complexes, then the infected cells would be subject to destruction by T_C cells.) Other bacteria turn genes on and off to change their antigens. This **antigenic variation** prevents the immune system from making the correct antibodies to target the pathogen. After B-cell activation has occurred and antibodies are produced, the pathogen has changed, and the antibodies will not bind to the new antigens.

Damage to the Host

As pathogens multiply within the body, the signs and symptoms of disease are caused by destruction of host tissues and by the response of the body to the infection. As bacteria multiply in the body, they may compete for nutrients necessary for host cell function. For example, most bacteria require iron to grow and may produce proteins called **siderophores** that are capable of taking iron away from host proteins like transferrins that carry iron around in the body. Bacteria and viruses that multiply inside of host cells may directly destroy host cells. Even if host cells are not immediately destroyed by intracellular pathogens, cell function is impaired by the use of cellular resources by the pathogen. The negative effects on virally infected host cells are called **cytopathic effects.** Even bacteria that don't reproduce inside of cells may still directly damage host cells by production of enzymes such as hemolysins and leukocidins.

PRODUCTION OF TOXINS (TOXIGENESIS)

During bacterial infections, significant damage can be done to the host by the production of bacterial toxins. There are two main categories of toxins found in bacteria, **exotoxins** and **endotoxins**. Both gram-negative and gram-positive bacteria produce exotoxins. Exotoxins are proteins that are made inside the cell and then released. Endotoxins are the lipid A portion of the lipopolysaccharide found in gram-negative cell walls, which are released from the wall as these bacteria die.

EXOTOXINS

The fact that exotoxins are proteins has many implications. Like all proteins, their activity is specific: Each exotoxin has a unique function. Also, heat or chemicals can denature them. When they are denatured, they are called **toxoids**. Because they can be denatured, and because they are highly immunogenic, effective vaccines against exotoxins can be produced. For example, both the diphtheria and tetanus vaccines are vaccines against exotoxins.

The genes for exotoxins may be located on plasmids and phages; this location facilitates their transfer between bacteria. Recall that when a phage carrying a toxin gene enters a bacterial cell it may insert its DNA into the DNA of the bacterium (see Chapter 13). This occurs in the lysogenic cycle of a bacteriophage and results in a dormant phage called a **prophage**. The presence of the prophage may give the bacterium the ability to produce that toxin. The acquisition of a new characteristic caused by the presence of a prophage is called a **lysogenic conversion**, and it may contribute to the pathogenicity of the organism.

> **REMEMBER**
> Exotoxins are protein toxins that are produced and released by both gram-negative and gram-positive bacteria. Because they are proteins, exotoxins can be denatured.

Some exotoxins are extremely deadly, requiring very small quantities of toxin to kill people. A number that is often used to compare the potency of toxins is the LD_{50}. The LD_{50} represents the dose of toxin that would be required to kill 50 percent of the test population. Botulinum toxin, produced by *Clostridium botulinum*, is one of the most deadly exotoxins known. The LD_{50} of ingested botulinum toxin for a 70-kilogram adult human is estimated to be as little as 10 micrograms. In contrast, the lethal doses for poisons such as strychnine or cyanide are much higher. The lethal dose of strychnine for adults ranges from 30,000 to 120,000 micrograms; that of potassium cyanide is 200,000 to 300,000 micrograms.

Although each exotoxin is specific in its effects, there are three main categories of exotoxin based on structure and activity. These are A–B toxins, superantigens, and membrane-disrupting toxins (Table 20.1). A–B toxins, also called Type III toxins, all have a similar structure (Figure 20.2). These toxins are made up of two types of subunits, A and B. A is for the active subunit, and B is for the binding subunit. When an A–B exotoxin is synthesized inside the bacterial cell, it is inactive. It is released from the cell, and the B subunit of the toxin facilitates binding and uptake of the exotoxin into the host cell. After the toxin is inside the host cell, the A subunit becomes activated and carries out the specific function of the toxin.

There are many specific functions of A–B exotoxins. Some, like diphtheria toxin, block protein synthesis in host cells, preventing them from functioning. Other A–B exotoxins act as **neurotoxins** and inhibit neurotransmission. The botulinum toxin produced by *Clostridium botulinum* blocks communication between nerves and muscles, blocking muscular contraction and resulting in a flaccid paralysis. It is the cause of the dangerous food poisoning known as botulism. Recently, the use of this toxin for cosmetic purposes has become more common. If it is introduced under the skin of the face in very minute quantities, it can prevent contraction of facial muscles and thus prevent wrinkles. The tetanus toxin, tetanospasmin, is produced by *Clostridium tetani* and is another example of a neurotoxin. It prevents the inhibition of skeletal muscle and thus leads to a spastic paralysis. This can lead to lockjaw or more serious paralysis and even death. A–B exotoxins can also act as **enterotoxins** that bind to intestinal cells and cause them to release fluids and electrolytes. For example, cholera toxin induces intestinal cells to pump sodium ions into the lumen of the gastrointestinal tract. Chloride ions follow the sodium ions, creating a hypertonic environment in the lumen. This causes movement of water into the lumen. Too much water in the gastrointestinal tract leads to the diarrhea and loss of electrolytes that are associated with cholera. Other types of A–B exotoxins include **cardiotoxins** that affect heart cells, **hepatotoxins** that affect liver cells, and **leukotoxins** that attack white blood cells.

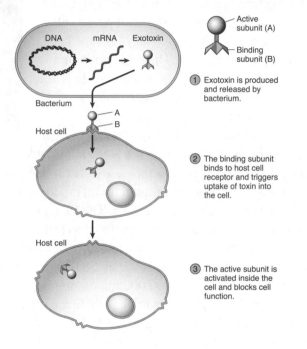

FIGURE 20.2. A–B exotoxins.

A–B exotoxins are proteins that can have various negative effects on cells. They are very specific and can block function of nerve cells (neurotoxins), intestinal cells (enterotoxins), cardiac muscle cells (cardiotoxins), and liver cells (hepatotoxins).

TABLE 20.1. Exotoxins

Category	Function
A–B toxins	
Neurotoxins	Inhibit neurotransmission
Enterotoxins	Cause intestinal cells to release fluid and electrolytes
Cardiotoxins	Inhibit functioning of heart cells
Hepatotoxins	Inhibit functioning of liver cells
Leukotoxins	Kill white blood cells
Superantigens	Overstimulate the immune system
Membrane-disrupting toxins	Disrupt membranes and cause cell lysis

Another group of exotoxins are the **superantigens**, or type I exotoxins. These exotoxins cause nonspecific activation of T_H cells (Figure 20.3). Recall that normally only the T_H cells, with T-cell receptors that can recognize specific microbial antigens within the MHC–antigen complexes, will become activated during an infection. This can be as little as 0.01 percent of your entire T_H-cell population. In contrast, the presence of superantigens may result in the activation of as much as 25 percent of your T_H cells. This causes a massive overproduction of cytokines, resulting in fever, nausea and vomiting, diarrhea, an excessive drop in blood pressure (**shock**), and even death. For example, the symptoms of toxic shock syndrome caused by *Staphylococcus aureus* are attributable to a superantigen. Shock that is induced by a bacterium is referred to as **septic shock**.

> **REMEMBER**
> Superantigens overstimulate the immune system by activating T_H cells that have not recognized antigen. Activation of too many T cells results in overproduction of cytokines, which can lead to septic shock.

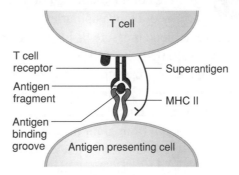

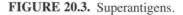

FIGURE 20.3. Superantigens.

Superantigens are toxins that cause an abnormally strong immune response. They do this by binding between helper T cells and antigen-presenting cells. This stimulates activation of T_H cells that are not specific to the antigen. Too many T_H cells are activated, resulting in overproduction of cytokines. This can lead to overly increased vasodilation and blood vessel permeability and thus shock and even death.

The third group of exotoxins is the **membrane-disrupting toxins**, or type II exotoxins. These toxins disrupt membranes by forming protein channels in the membrane or by affecting the phospholipids. Some membrane-disrupting toxins disrupt plasma membranes; for example, hemolysins disrupt the plasma membrane of red blood cells by forming protein channels. Others may disrupt the membranes of vesicles within the cell such as those that enable pathogens to escape phagosomes and move into the host cytoplasm. Leukocidins kill white blood cells by forming protein channels in lysosomal membranes, causing the lysosomes to release their destructive compounds into the cytoplasm of the white blood cell.

ENDOTOXINS

Endotoxins are different from exotoxins in both structure and function. Because endotoxins are lipid, they are not inactivated by heat. Also, although they trigger an immune response, it is not sufficient to protect fully against the toxin. Because of these two factors, effective vaccines against endotoxin cannot be produced.

Endotoxins are not as specific as exotoxins in that all endotoxins from a wide variety of gram-negative bacteria essentially function in the same way. Because gram-negative bacteria are processed by antigen-presenting cells, they simulate the APCs to overproduce cytokines such as interleuken-1 (IL-1) and tumor necrosis factor alpha (TNF-α). IL-1 travels to the hypothalamus and induces the release of prostaglandin (see Chapter 18). The prostaglandin resets the body's thermostat producing fever and increases blood vessel permeability and vasodilation, leading to movement of fluid out of the blood and into tissue spaces. The permeability of blood capillaries is also increased by TNF-α, resulting in a further decrease in blood pressure. If blood pressure decreases too dramatically, septic shock may result. IL-1 and TNF-α can also trigger activation of blood-clotting proteins, leading to the formation of blood clots. These clots can block blood flow through capillaries and result in damage to tissues and organs.

> **REMEMBER**
> Endotoxin refers to lipid A, which is part of the lipopolysaccharide in the outer membrane of gram-negative bacteria. Because endotoxin is a lipid, it is not denatured by heat.

The combination of fever, shock, and tissue damage from blood clots can result in severe consequences, and even death, from gram-negative infections. Because endotoxins released as gram-negative cells are destroyed, antibiotic therapy to treat gram-negative infections may actually result in an initial increase in these signs and symptoms. In addition, because endotoxins are not destroyed by heat, endotoxin may be present even after sterilization of equipment or fluid in which gram-negative bacteria were present. If this equipment or fluid comes into contact with people, the endotoxins can cause damage to the body even if no living bacteria are present.

As we have seen, pathogens use their virulence factors, such as adhesins, enzymes, and toxins, to gain entry into host cells and avoid host defenses. Toxin production and competition for resources result in damage to host cells and tissues. This damage, combined with the host responses, such as cytokine production, inflammation, and fever, lead to the signs and symptoms of disease. As pathogens reproduce inside the host, they can exit the host and be transferred to new hosts, restarting the disease process. The **portal of exit** for most pathogens is the same as the portal of entry; that is, most pathogens leave the body the same way they entered it.

REVIEW EXERCISES FOR CHAPTER 20

1. Define pathogenicity, virulence, virulence factor, quorum sensing, ID_{50}, and LD_{50}.

2. Describe the portals of entry for microbes.

3. Define adhesin and biofilm and indicate their importance to disease.

4. Describe the ways in which microbes evade host defenses, including complement, phagocytosis, and the specific immune response.

5. Name and describe the enzymes used by bacteria to spread through host tissues.

6. Explain why antigenic variation can increase virulence of a pathogen.

7. Explain why penetration into the host cell cytoskeleton can increase virulence of a pathogen.

8. Compare and contrast the mechanisms by which bacteria and viruses damage host cells.

9. Define siderophore.

10. Compare and contrast exotoxins and endotoxins.

11. Explain why exotoxins are specific in their activity.

12. Explain how vaccination can be used to generate immunity against exotoxins.

13. Define lysogenic conversion.

14. Describe the steps in the attachment and entry of A–B toxins.

15. Describe the possible effects of A–B toxins, superantigens, and membrane-disrupting toxins.

16. Describe the effects of endotoxins.

17. Explain why vaccination is not useful against endotoxins.

SELF-TEST

1. Some strains of *E. coli* cause urinary tract infections (UTIs). To cause UTIs, these bacteria must avoid being flushed out by the flow of urine. Which of the following virulence factors might help them do this?

   A. Hemolysin
   B. Endotoxin
   C. Adhesin
   D. Leukocidin
   E. Neurotoxin

2. If a bacterium is introduced into the body by a bug bite on the arm, this represents which portal of entry?

   A. Mucous membranes
   B. Gastroinestinal
   C. Genitourinary
   D. Skin
   E. Parenteral route

3. Possible defenses against phagocytes include all of the following **except**:

   A. Produce leukocidins
   B. Prevent fusion of phagosome with lysosome
   C. Release chemoattractants
   D. Synthesize a capsule
   E. Produce superoxide dismutase and catalase

4. True or False. The tetanus vaccine is actually an injection of denatured tetanus toxin.

5. Superantigens

   A. disrupt host membranes.
   B. have an active and a binding subunit.
   C. are not denatured by heat.
   D. cause an overproduction of cytokines.
   E. All of the above are correct.

6. True or False. Both bacteria and viruses can cause lysis of host cells.

7. True or False. A bacterium that is infected with a virus may be more virulent as a result.

8. True or False. Invasins help pathogens to invade the body by breaking down connective tissue.

9. All of the following are true of siderophores **except**:

    A. They are produced by bacteria.
    B. They bind iron.
    C. They help bacteria compete for nutrients with host cells.
    D. They are also called transferrins.
    E. All of the above are correct.

10. True or False. One bacterium may have many different virulence factors.

11. If you were to drink water contaminated with the three diarrheal pathogens listed below, which one would be most likely to establish an infection?

Pathogen	ID_{50}
Salmonella gastroenteritis	$>10^3$ cells
Giardia lembli	10 oocysts
Cryptosporidium parvum	1042 oocysts

12. A turkey was thawed improperly, which allowed multiplication of *Salmonella*, a gram-negative, rod-shaped bacterium. Then the turkey was roasted, which killed all the *Salmonella*. People who ate the turkey, however, developed fever, chills, and hypotension. Explain why the people showed each of these reactions.

13. Examples of virulence factors of two pathogens follow. For each, describe how the virulence factor contributes to the pathogenicity of the organism.

    Bordetella pertussis: Filamentous hemagglutinin adhesin (FHA) binds to leukocyte receptor and triggers uptake into macrophages without triggering an oxidative burst.

    Legionella pneumophila: Prevents fusion of phagosome with lysosome.

Answers

Review Exercises

1. Pathogenicity is the ability of an organism to cause disease. Virulence is the degree of pathogenicity. A virulence factor is any molecule a pathogen has that contributes to its ability to cause disease. Quorum sensing is the ability of some microbes to sense the density of their population. The ID_{50} is the number of cells of a particular pathogen it would take to cause disease in 50 percent of a test population. The LD_{50} is the dose of a toxin that it would take to kill 50 percent of a test population.

2. Microbes commonly enter through the respiratory tract by being breathed in through the mouth or the nose. They may also gain entry to the gastrointestinal tract when contaminated food or water are consumed. Access to the genitourinary tract may result from direct contact with an infected individual. If entry is through a break in the skin or mucous membranes, this represents the parenteral route of entry.

3. An adhesin is a protein that a bacterium uses to attach to host cells. Attachment is necessary for most bacteria to cause disease. Biofilms are populations of microbes embedded in layers of sticky molecules. Biofilms are also used for attachment and may protect the cells within them from antimicrobial substances.

4. Some gram-negative bacteria evade complement by modifying their surface polysaccharides to keep complement away from the cell surface. Phagocytosis may be evaded in a number of ways. Some microbes colonize areas of the body where there are no phagocytes. Others release molecules that block chemotaxis of phagocytes. Still others prevent phagocytes from binding to them by the production of a capsule or by the presence of certain surface proteins. Some microbes can escape from the phagosome before lysosomal fusion and others can survive inside of the phagolysosome. A specific immune response may be avoided if the pathogen is an intracellular parasite and blocks or varies the display of its molecules on the cell surface. Also, some bacteria produce proteases that are capable of breaking down antibodies.

5. Collagenase breaks down collagen in connective tissue. Hyaluronidase breaks down hyaluronic acid in tissue cement. Hemolysins lyse red blood cells. Fibrinolysin breaks down blood clots and can enable a pathogen to spread. Invasins allow pathogens to rearrange the cytoskeleton so they can move through cells.

6. If a pathogen is capable of varying its antigens during an infection, it makes it harder for the specific immune response to target the pathogen. By the time antibody production occurs, the pathogen may have changed its surface antigens, making the antibodies obsolete.

7. If a pathogen can rearrange the cytoskeleton and use it to penetrate and move though cells, it may have two possible advantages. One is that it may be able to hide from the specific immune system while it is inside the cell. The other is that it may be able to use the cytoskeleton like railroad tracks to move through the cells and into new ones.

8. Both bacteria and viruses can compete with host cells for nutrients. Also, both can lead to cell lysis, although bacteria do not always do so. Bacteria often produce toxins that affect cell function.

9. Siderophores are molecules produced by bacteria that have a strong affinity for iron. They can take iron away from host iron-binding proteins.

10. Exotoxins are proteins, whereas endotoxins are lipid A. Because exotoxins are proteins, they are specific and can be denatured by heat. Because they can be denatured, inactive toxins, called toxoids, can be used for vaccines. This is also possible because exotoxins are highly immunogenic. Because endotoxins are lipid, they are resistant to heat, and toxoids cannot be made. Also, they do not trigger a strong immune response. Thus, there are no vaccinations for endotoxins.

11. Exotoxins are specific because they are proteins. Proteins perform specific functions.

12. Denatured exotoxins called toxoids are injected into the body. This triggers an immune response, including the ability to produce antibodies to the toxin rapidly.

13. A lysogenic conversion is when a bacterium acquires a new trait because it is infected with a lysogenic virus.

14. A–B exotoxins are synthesized inside of the bacterial cell and released. The B subunit binds to the host cell and triggers uptake of the toxin into the cell. Once inside, the A subunit becomes active and performs a specific function in the cell.

15. A–B toxins can block protein synthesis, prevent neurons from functioning properly, cause intestinal cells to release electrolytes and fluids into the gastrointestinal tract, target heart cells, and target liver cells. Superantigens cause the nonspecific activation of helper T cells. This results in overproduction of cytokines, which can lead to too much blood vessel permeability and vasodilation, resulting in shock and even death. Membrane-disrupting toxins destroy membranes. This can result in cell lysis.

16. Endotoxins stimulate macrophages to release IL-1 and TNF-α. These cytokines increase blood vessel permeability and vasodilation, which can result in shock. In addtion, IL-1 triggers fever. Activation of blood-clotting proteins can also occur leading to obstruction of capillaries and tissue death.

17. Endotoxins cannot be denatured, so a safe vaccine cannot be made. In addition, it does not trigger a strong immune response.

Self-Test

1. C	5. D	9. D
2. E	6. T	10. T
3. C	7. T	
4. T	8. F	

11. *Giardia lembli* has the lowest ID_{50}. This means it takes less of this pathogen to cause disease in 50 percent of the population.

12. Because *Salmonella* is gram-negative, it has endotoxin, which is not destroyed by heat. Even though roasting killed the bacterium, the endotoxin remained. Endotoxin stimulates macrophages to release IL-1 and TNF-α. These cytokines

increase blood vessel permeability and vasodilation, which can result in shock (hypotension). In addition, IL-1 triggers fever. As the body temperature rises, a person might feel a chill. Activation of blood-clotting proteins can also occur leading to obstruction of capillaries and tissue death.

13. For *Bordatella pertussis,* the oxidative burst is the release of free radicals, which will oxidize and destroy the microbe. Without the oxidative burst, *Bordatella* may survive within the macrophage. For *Legionella pneumophila:* By preventing fusion of the phagosome with the lysosome, *Legionella* protects itself from digestion by the enzymes in the lysosome.

21

Practical Applications of Immunology

WHAT YOU WILL LEARN

This chapter introduces the ways in which people have harnessed the power of the immune system in order to diagnose and prevent disease. As you study this chapter, you will:

- take a detailed look at how vaccines really work;
- consider the benefits and risks of vaccination to human populations;
- examine some of the ways antibodies are used to diagnose disease.

SECTIONS IN THIS CHAPTER

- History of Vaccination
- Type of Immunity Triggered by Vaccines
- Types of Vaccines
- The Effect of Immunization on Populations
- Diagnostic Immunology

Our understanding of the human immune system has led to many advances in health care. First among these is the ability to *vaccinate* and protect people against infectious disease. Specific populations of antibodies are used to identify microorganisms, detect pregnancy, and administer tests for drug use. An assessment of *titer*, the amount of antibody in the serum, can provide valuable information about a person's health. Likewise, *differential white blood cell counts* assess the percentage of each type of white blood cell in a sample of 100 white blood cells. An increase or decrease in certain types of white blood cells can provide clues for the identification of disease. An assessment of *complement levels* (CH_{50} or CH_{100}) can determine whether deficiencies in the complement system are contributing to diseases like autoimmune diseases. New therapies based on the immune system are also being developed. For example, interferon is used in the treatment of several viral diseases, multiple sclerosis, and some types of cancer. As we learn more about the immune system, this knowledge will undoubtedly be used to develop more tools for the identification and treatment of disease.

History of Vaccination

The development of vaccines began many years ago. The first known attempts at protection were for the disease smallpox. Unlike the relatively harmless chicken pox we see today, smallpox was a very dangerous and often deadly disease. Smallpox epidemics occurred frequently in Europe and killed most of the Native Americans on the East Coast of the United States after Europeans introduced the disease. People who survived smallpox could not get the disease again, but were left with ugly scars on their skin.

The observation that previously infected people became immune to smallpox was the basis for the first efforts at immunization. Chinese physicians would collect the scabs from people with smallpox, grind them into a powder, and then insert that powder into the nose of the person to be immunized. Mary Montague, the wife of the

> **REMEMBER**
> The first vaccine used cowpox virus in order to prevent smallpox infection.

English ambassador to Turkey, observed a similar practice in the early 1700s in Turkey. She reported that old women would dip a needle into smallpox scabs then stick the needle into a vein of the person to be immunized. Lady Montague was so impressed by her observations that she chose to her have her own children treated. This type of immunization, called *variolation*, became common in Europe. However, variolation was a dangerous form of immunization. Although most people experienced a period of mild illness and then became immune, some people would develop a serious and sometimes deadly case of smallpox. In addition, even people who experienced a mild form of the illness could spread the illness to others, in whom it could be much more severe.

The English physician, Edward Jenner, developed a safer form of immunization in the late 1700s. Jenner was intrigued when a dairymaid told him she could never get smallpox, because she had already had cowpox. Cowpox, a mild disease that causes lesions on cow's udders, would get transferred to the hands of dairymaids as they milked the cows. Although it caused scarring on the hands of the dairymaids, it did no other lasting damage. In fact, dairymaids were envied for their unscarred complexions. Based on the local belief that cowpox somehow protected people from smallpox, a farmer named Benjamin Jesty deliberately infected his family with cowpox in 1774. His family never contracted smallpox for the rest of their lives.

The idea that cowpox somehow protected people from smallpox, combined with his knowledge of variolation, led Jenner to begin experiments in 1796 on the development of a smallpox immunization that was based on cowpox. In May, Jenner attempted to immunize an 8-year-old boy with fluid taken from a cowpox lesion on the finger of a dairymaid. Then, in July, he deliberately exposed the boy to smallpox. Fortunately, the immunization worked, and the boy did not contract smallpox. Smallpox vaccination became widespread, leading to the complete eradication of the disease. Not a single case of smallpox has been reported since October of 1977. The type of immunization that Jenner developed was named **vaccination** (*vacca* = cow). A **vaccine** is a suspension of bacteria or viruses that have been weakened or killed and are used to induce immunity. (In the case of smallpox, a less harmful virus was used.)

Type of Immunity Triggered By Vaccines

Vaccination is possible because the immune system has the ability to develop *immunologic memory*. Recall that initial exposure to an antigen causes a *primary immune response*, including antibody production as well as the creation of long-lived **memory B cells** (Chapter 19). Upon second exposure to an antigen, a *secondary immune response* occurs, causing memory B cells to divide rapidly and to differentiate into plasma cells that release large quantities of IgG antibodies (Figure 19.12). The IgG antibodies circulate around the body binding to the antigen and triggering the destruction of the pathogen (Figure 19.10). Because vaccines contain microbial antigen, when they enter your body, they trigger a primary immune response, including the formation of memory B cells. In essence, they educate your immune system about what this particular pathogen looks like, so your immune system can be ready if the real thing shows up. Then, if you do get an infection, your immune system responds with a secondary immune response, and your body clears the pathogen before it can establish itself in your body and before you ever knew it was there. Because vaccines trigger an immune response, they confer *artificially acquired active immunity*.

Types of Vaccines

WHOLE AGENT VACCINES

Although all vaccines contain microbes or parts of microbes, the particular components of vaccines can be different. Many of the vaccines currently in use (Table 21.1) are **whole agent** vaccines, which contain entire bacterial cells or viral particles.

> **REMEMBER**
>
> Whole agent vaccines that contain attenuated pathogens generate the most robust immunity because the immune system is introduced to many different antigens as the pathogen goes through its normal life cycle.

Obviously, it wouldn't be a good idea to inject the body with a pathogen that was fully functional, so the microbes in whole agent vaccines are either **inactivated** or **attenuated**. Inactivated microbes have been killed by treatment with heat or chemicals. Attenuated microbes are still capable of reproduction but have been weakened in some way, usually by long-term growth in abnormal culture conditions. When the attenuated microbe enters the body, it reproduces itself, resulting in a relatively long-term exposure to the immune system and triggering a very effective immune response. However, one disadvantage of attenuated vaccines is that they occasionally mutate back to the pathogenic form and have the potential to cause epidemics. For example, for every 2.4 million doses of oral polio vaccine that are administered, one case of vaccine-associated paralytic poliomyelitis (VAPP) occurs. In some cases, this risk has been eliminated by the use of genetic engineering to permanently remove genes for virulence factors (Chapter 20) that are necessary for disease.

TABLE 21.1. Types of Common Vaccines Used to Prevent Disease in Humans

Disease	Type of Vaccine
Bacterial diseases	
Diphtheria	Toxoid (inactivated toxin)
Tetanus	Toxoid (inactivated toxin)
Pertussis (whooping cough)	Acellular proteins or killed bacteria
Bacterial pneumonia (pneumococcus)	Purified antigen (capsular polysaccharides)
Haemophilus influenzae meningitis	Conjugated vaccine (capsular polysaccharides bound to protein)
Viral diseases	
Measles	Attenuated virus
Mumps	Attenuated virus
Rubella (German measles)	Attenuated virus
Polio	Inactivated virus (Salk, injected) or attenuated virus (Sabin, oral)
Influenza	Inactivated virus (injected) or attenuated virus (nasal)
Hepatitis B	Purified antigen made by recombinant DNA technology
Varicella (chicken pox)	Attenuated virus

The advantages and disadvantages of inactivated and attenuated vaccines are compared in Table 21.2. One potential disadvantage to all whole agent vaccines is that they occasionally trigger complications that are associated with the disease; for example, a small percentage of people who receive the measles vaccine develop post-vaccine encephalitis. However, the risk of developing encephalitis as a result of the vaccine is much less than that of developing encephalitis in response to an actual measles infection.

TABLE 21.2. Comparison Between Inactivated and Attenuated Vaccines

Characteristic	Inactivated Vaccine	Attenuated Vaccine
Components	Pathogen that has been inactivated (killed) by irradiation or treatment with chemicals	Pathogen that has been weakened by growth in adverse conditions
Type of immunity	Primarily humoral immunity is induced	Humoral and cell-mediated immunity is induced
Requirement for boosters	Requires multiple boosters to be effective	Generally requires only a single booster to be effective
Relative stability	More stable	Less stable
Ability to revert to virulent pathogen	None	May revert

PURIFIED ANTIGEN VACCINES

Vaccines that consist of specific, purified antigens can overcome some of the disadvantages of whole agent vaccines. **Toxoid vaccines**, for example, consist of denatured exotoxin from bacterial pathogens. When toxoids enter the body, they enable the production of antitoxin antibodies that can bind to and neutralize the exotoxin. The vaccines for diphtheria and tetanus are both toxoid vaccines. Many pathogens produce polysaccharide capsules that help protect them from phagocytosis. If **polysaccharide vaccines** consisting of the capsular polysaccharides are used, these vaccines can trigger the production of antibodies specific to the capsule. These antibodies bind to the capsule, resulting in opsonization (Chapter 19) and phagocytosis. One disadvantage of polysaccharide vaccines, however, is that polysaccharides activate B cells in a T-independent fashion and thus do not lead to the formation of memory B cells (Chapter 19). One solution to this problem is to attach the polysaccharides to protein molecules. Because of the protein, this **conjugated vaccine** activates B cells in a T-dependent fashion, resulting in both antibodies to the polysaccharides and the production of memory B cells. For

> **REMEMBER**
> Bacterial exotoxins, like the pertussin and tetanus toxins, can be denatured and used in toxoid vaccines.

example, the vaccine for *Haemophilus influenzae* type b (Hib) is a conjugated vaccine of capsular polysaccharides to a protein carrier.

Recombinant DNA technology is being used to generate new approaches to vaccine development. To obtain sufficient quantities of an antigen for a vaccine, the genes for that antigen can be taken from the pathogen and introduced into organisms that are easy to culture. The recombinant organism produces the antigen, which is then purified and used for a **recombinant antigen vaccine**. Alternatively, harmless microbes can be engineered to act as carriers, or vectors, of antigen and then introduced into the body as **recombinant vector vaccines**. For example, the cowpox virus could be genetically modified to contain genes from a particular pathogen. When the recombinant cowpox virus is introduced into cells, it produces the antigen, triggering an immune response.

DNA VACCINES

Another new strategy that holds great promise is the development of **DNA vaccines**. If DNA from a microbe is injected into animal muscle tissue, animal cells transcribe and translate the DNA, producing the microbial antigen. The microbial antigen is displayed in the major histocompatibility complexes of the animal cells (Chapter 19), triggering an immune response. Tests of DNA vaccines in animals have shown they are effective in establishing immunity for a number of pathogens, including the influenza virus. Human trials for several DNA vaccines are currently underway.

The Effect of Immunization on Populations

In countries where immunization is readily available, children today lead a relatively disease-free life as compared to the children of just one hundred years ago. Diseases such as diphtheria, measles, mumps, rubella (German measles), pertussis (whooping cough), and poliomyelitis that once caused serious complications and even death can now be prevented in children who receive their immunizations (Table 21.3). In fact, polio is the target of the world's largest public health initiative ever—the Global Polio Eradication Initiative. The World Health Organization (WHO), the Centers for Disease Control and Prevention (CDC), Rotary International, and the United Nations Children's Fund (UNICEF) have teamed together under this initiative to try to eliminate polio from the human population forever. Since the initiative's launch in 1988, great progress has been made, and the initiative continues to work to eliminate polio.

TABLE 21.3. Recommended Childhood and Adolescent Immunization Schedule—United States, 2010*

Vaccine	Birth	1 mo	2 mo	4 mo	6 mo	12 mo	15 mo	18 mo	24 mo	4–6 yr	11–12 yr	13–18 yr
			Range of Recommended Ages								Preadolescent Assessment	
Hepatitis B	HepB #1	HepB #2			HepB #3							
Diphtheria, tetanus, pertussis			DTaP	DTaP	DTaP		DTaP			DTaP	Td	
Haemophilus influenzae type b			Hib	Hib	Hib	Hib						
Inactivated polio			IPV	IPV		IPV				IPV		
Measles, mumps, rubella						MMR #1				MMR #2		
Varicella							Varicella					
Pneumococcal			PCV	PCV	PCV	PCV						
Influenza						Influenza (yearly)						
Vaccines below this line are for selected populations												
Hepatitis A											HepA (2 doses)	
Human papillomavirus											HPV (3 doses)	
Meningococcal											MCV	

*Data from CDC.

HERD IMMUNITY

The effect of immunization on the incidence of a disease is dramatically illustrated by what happened after the measles vaccine was licensed (Figure 21.1). This situation contrasts sharply with that of countries in which vaccination is not readily available: For example, measles causes 10 percent of deaths in children younger than age 5 worldwide (approximately 1 million deaths annually). Overall, the World Health Organization estimates that vaccine-preventable diseases cause 20–35 percent of infant deaths worldwide.

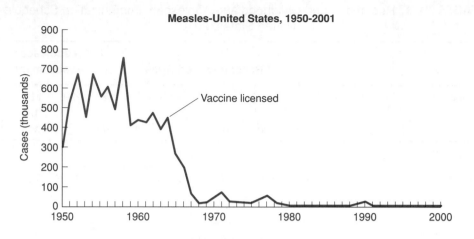

FIGURE 21.1. Measles incidence in the United States
before and after licensing of the measles vaccine (data from CDC).

Vaccination reduces the incidence of disease by creating **herd immunity** in the population (Chapter 17). As more and more people become vaccinated, it becomes harder for a pathogen to spread through the population because there are fewer susceptible hosts. Because the pathogen cannot spread easily, even people who are not vaccinated will be less likely to get the disease. However, vaccination only protects a population if enough people get vaccinated. Otherwise, herd immunity is not established, and the population is vulnerable to epidemics.

RISKS ASSOCIATED WITH VACCINES

Although vaccines are available to most children in the United States, not all American children are vaccinated. The percentage of children vaccinated in the United States varies considerably by state, with some states having as few as 60 percent of 19- to 35-month-olds current on their shots. This has led to epidemics of vaccine-preventable diseases such as whooping cough and measles. Many parents who choose not to vaccinate their children do so because they fear the side effects associated with the vaccine. Almost all vaccines have some risk of side effects. However, these side effects are generally much less dangerous than the risk of actually catching the disease. The differences in risk of side effects and risk of disease is illustrated by the old form of the pertussis vaccine, which was a whole agent vaccine that had a fairly high rate of severe side effects. One in every 100 doses caused persistent, inconsolable crying; one in every 330 doses caused a fever of greater than 105°F, and one in every 1750 doses caused seizures with fever. Because of these side effects, many people chose not to vaccinate their children. In fact, in 1975,

> **REMEMBER**
> If enough people are vaccinated, a population develops herd immunity to the pathogen, which protects all members of the population, including infants that are too young to develop good immunity on their own, and the elderly, whose immunity may be fading.

the Japanese government decided to discontinue the vaccine. In the 3 years *before* pertussis vaccination was halted, there were 400 cases of pertussis with ten cases resulting in death. In the three years *after* pertussis vaccination was halted, there were 13,000 cases of pertussis with 113 cases resulting in death. The Japanese government quickly reinstated the pertussis vaccine (data from the Vaccine Education Center of The Children's Hospital of Philadelphia).

The example of the pertussis vaccine in Japan helps illustrate the difference between *perceived risk* and *real risk*. If something tragic happens to a child, and the story is related in the national news, everyone hears about it, and the *perceived risk* of the threat is great. Based on probability, the chances of the same tragic occurrence happening to another child may be one in a million. That represents the *real risk*. Many people choose not to vaccinate their children because of the *perceived risk* of the vaccine. They may have heard stories of the side effects and potential complications of a vaccine, and they have a very real fear for their children's health. This perceived risk may seem stronger than a bunch of dry statistics that represent the real risk of side effects. And, because vaccinations have reduced the incidence of many harmful diseases, people have no first-hand knowledge of the severity of these diseases. The perceived risk of contracting the disease is small. In a way, vaccination is a victim of its own success: Because vaccination has decreased the risk of dangerous diseases, people no longer perceive its value.

> **REMEMBER**
> *Perceived risk* of a danger is the risk a person feels, and this is often inflated by sensational news stories and rumors. *Real risk* of a danger is based on a large sample of people and gives a better estimate on the likelihood of the danger becoming real.

If vaccination rates decline and herd immunity falls below safe levels, a population may learn a hard lesson like that of pertussis in Japan. To counteract this trend in the United States, there are many efforts to increase education about vaccines and to encourage parents to get their children vaccinated. Answers to some common questions about vaccines are presented in Table 21.4. More information on these topics can be found on the Centers for Disease Control and Prevention web site (*www.cdc.gov*) under vaccines. Vaccination rates may also be improved by strategies that would make vaccination easier. This could include use of more combined vaccines, so that fewer shots are necessary, or alternative methods of vaccine delivery, such as nasal sprays or skin patches. More effective vaccines that do not require boosters would also improve the situation. And perhaps most importantly, access to affordable health care is critical to the success of any vaccination program.

TABLE 21.4. Answers to Some Concerns About Vaccinations

Concern	Answer
Safety of vaccines	
Are vaccines safe?	No vaccine is 100 percent harmless to all people. Some vaccines can cause side effects, which range from slight inflammation to more severe reactions. However, for a vaccine to be licensed in the United States, the number of people affected by side effects must be *very small* compared to the number of people who would be affected by the disease. In addition, the side effects are generally much less severe than the disease symptoms themselves.
Does the MMR vaccine cause autism?	Origin of the concern: A 1998 study of twelve British children with intestinal disorders revealed that eight children developed autism shortly after receiving the MMR shot. (Note: Since the publication of these results, the study was found to be flawed in its design, and it has been officially retracted from the British medical journal *Lancet*. The doctor who wrote the study was found to have falsified his data, and was barred from practicing medicine in Britain.) Evidence: Many scientific studies comparing vaccinated and unvaccinated children, including a study of more than 500,000 Danish children, have found no connection between the MMR vaccine and autism.
Does the Hepatitis B vaccine cause MS?	Origin of the concern: Individual adults have reported that they developed MS or other autoimmune disorders after receiving the HepB vaccine. Evidence: Multiple studies comparing incidence of MS among vaccinated and unvaccinated people have found no link between the HepB vaccine and MS.
Do vaccines contain mercury and does this contribute to autism, ADHD, and learning disabilities?	Origin of the concern: Mercury is known to impair brain development. Mercury is found in thimerosal, a preservative that was used in vaccines. Evidence: Preservatives are required in vaccines to prevent the growth of contaminating pathogens. The level of thimerosal that was used in vaccines was determined to be safe by the U.S. Food and Drug Administration, and no harmful reactions resulting from thimerosal in vaccines have been reported except for minor inflammation at the site of injection. However, because of concerns about thimerosal, its use in vaccines for children in the United States has been stopped. Thimerosal is still used in influenza vaccines.

TABLE 21.4. Answers to Some Concerns About Vaccinations

Concern	Answer
Vaccines and the immune system	
Do vaccines overwhelm the immune system?	The human immune system has the capacity to respond to billions of different antigens. When children in the United States receive all of the recommended immunizations, the total number of antigens introduced into their systems is fewer than 150. Also, the immune system has the ability to replenish the cells that respond to these antigens constantly. Thus, immunizations have essentially no negative impact on the capacity of the immune system to respond to other pathogens.
Do vaccines weaken the immune system?	Studies have shown that children who receive vaccines are no more susceptible to other infections than are children who are not vaccinated. In fact, they may be less susceptible, especially to secondary infections that can follow vaccine-preventable diseases.

For more concerns and more detailed information, see Vaccine Beliefs and Concerns at *www.cdc.gov*; data also taken from the Vaccine Education Center of The Children's Hospital of Philadelphia, *www.chop.edu*.)

FUTURE DIRECTIONS FOR VACCINES

Many efforts are currently underway to develop new vaccines and to make vaccines available in areas of the world that do not have good health care systems. Recombinant DNA technology is providing new ways to manufacture and deliver vaccines. For example, vaccines may soon be delivered to children in poorer countries as genetically modified foods. These foods come from plants that have been genetically engineered to express microbial antigens. It is hoped that when the foods are eaten, the microbial antigens will interact with the immune system cells in the intestine, triggering immunity. Not only are these vaccines easy to administer, but also many do not require refrigeration. Several plant-derived vaccines are currently undergoing clinical trials.

One area in which vaccines are badly needed is for diarrheal pathogens. Diarrhea is one of the leading infectious killers of children worldwide, causing approximately 2 million deaths per year. Many cases of infant diarrhea are caused by rotavirus. Recently, a successful rotavirus vaccine was developed using genetic engineering. Albert Kapikian modified a rotavirus from monkeys, which was not pathogenic to humans, with genes for the major antigen from four strains of human rotavirus. The oral vaccine was successful in clinical trials and approved by the U.S. Food and Drug Administration (FDA) in 1997. Unfortunately, about one in 10,000 children in the United States who were given the vaccine developed a serious complication in which one section of the small intestine folds into another one. Although only one in 1 million children died from this complication, as compared to five to ten in 1 million who might ordinarily die from rotavirus infection in the United States, the use of this vaccine in the United States was put on hold. However, the data are currently being reviewed, and the vaccine is again in development.

The development of new vaccines is expensive and requires a great deal of research and development. After a vaccine is developed in the United States, the FDA must license it. The vaccine is first tested in animals. If it is safe, it is then tested in adults and finally children. These trials can take years or even decades to be completed. After a vaccine is licensed, it is then considered by the CDC, the American Academy of Pediatrics (AAP), and the American Academy of Family Physicians (AAFP). These organizations compare the risk and benefit of the vaccine, as well as the financial cost of vaccinating a significant portion of the population. If they decide the benefits outweigh the risks and the cost, they will vote to recommend the vaccine. At that point, the vaccine is finally offered to the population.

Because vaccine development is so costly and the approval process so lengthy, vaccines are not the most profitable ventures for pharmaceutical companies. Even if successful, vaccines are typically only administered once or a few times during a lifetime. When this is contrasted with drugs like allergy medications, arthritis remedies, and antacids that are taken all the time, the profit margin of vaccines starts looking pretty small. Although the high cost of vaccines is a limiting factor in their development, the recent rise in antibiotic-resistant bacteria (Chapter 23) has created renewed interest in vaccines, and many vaccines, including ones for malaria and HIV infection, are currently in development.

DIAGNOSTIC IMMUNOLOGY

MONOCLONAL AND POLYCLONAL ANTIBODIES

In addition to developing vaccines to fight disease, our knowledge of the immune system has enabled us to develop many tests for the diagnosis of disease and identification of specific antigens. To identify a specific antigen, you can use an antibody or other molecule that specifically binds the antigen. In other words, you need to have molecules that will bind only to that antigen and no other. When a microbe enters the human body, it contains many antigens, each of which contains several epitopes. Each epitope triggers the production of unique populations of antibodies made by many different B cells. The resulting populations of antibodies are called **polyclonal antibodies** because they are made by many different B cell clones.

> **REMEMBER**
> Monoclonal antibodies are highly specific because they recognize a single epitope.

To have a very specific population of antibodies that recognize only one epitope, you want a population of **monoclonal antibodies** that is produced by just one B cell. Unfortunately, normal B cells do not reproduce for very long in cell culture, so it is not possible to isolate one and grow it in a dish to make antibodies. This problem was solved when scientists developed a protocol for fusing normal mouse B cells to cancerous mouse B cells called myelomas, which reproduce indefinitely in cell culture. When fused together, the B cell and myeloma form a *hybridoma* that both grows well in cell culture and produces the same antibody as that of the normal B cell (Figure 21.2). By first injecting mice with a particular antigen, then isolating B cells and creating hybridomas, it is possible to generate monoclonal antibodies that are specific to the injected antigen.

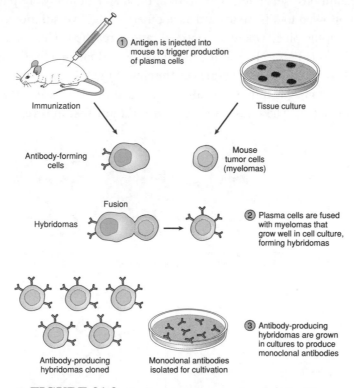

① Antigen is injected into mouse to trigger production of plasma cells

Immunization

Tissue culture

Antibody-forming cells

Mouse tumor cells (myelomas)

Fusion

Hybridomas

② Plasma cells are fused with myelomas that grow well in cell culture, forming hybridomas

Antibody-producing hybridomas cloned

Monoclonal antibodies isolated for cultivation

③ Antibody-producing hybridomas are grown in cultures to produce monoclonal antibodies

FIGURE 21.2. Production of monoclonal antibodies.

Monoclonal antibodies are very useful as diagnostic tools for the identification of antigen and in the treatment of disease. In medicine, they are used to identify pathogens, type tissues for transplantation, identify tumors, and determine pregnancy. They are also used to test for the presence of drugs such as cocaine. In basic biological research, they are used to identify the presence and location of particular proteins within cells. Recently, monoclonal antibodies have been used in the treatment of disease. For example, toxins can be attached to monoclonal antibodies that are specific to antigens found only on cancer cells. This will deliver the toxin specifically to those cells, killing them while sparing noncancerous cells.

One problem with the use of monoclonal antibodies to identify antigen is that antibodies are too small to see. Therefore, there must be some indirect method of determining whether they have found and bound to their antigen. There are several ways of determining whether antigen–antibody reactions have occurred, including the use of physical reactions, chemical reactions, and fluorescence.

ANTIGEN–ANTIBODY REACTIONS

Precipitation reactions are physical reactions between *soluble* antigens, or antigens that dissolve in water, and IgG or IgM antibodies. When there is just the right balance of antigen and antibody, called the *optimal ratio*, the antigens and antibodies will form a large lattice that is visible to the eye as a cloudy white line. To get the right balance of antigen and antibody, separate solutions of antigen and antibody are placed next to each other and allowed to diffuse toward each other. As the two solutions combine, the lattice will form at the place where the optimal ratio is reached. This place is called the *zone of equivalence* (Figure 21.3). If the precipitation is done in a broth, the lattice becomes visible as a band called the **precipitin ring** (Figure 21.4). If the antigen and antibody are placed in agar, a line of visible precipitate forms between them when the optimal ratio is reached. These tests are called **immunodiffusion tests**.

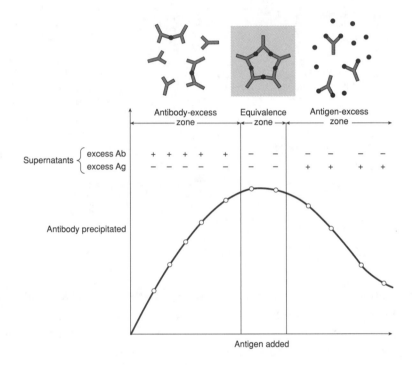

FIGURE 21.3. A precipitation curve.

The greatest amount of precipitation occurs where the ratio of antibody (Ab) to antigen (Ag) is equivalent.

The presence of antibodies to *particulate* antigens, such as antigens on cells, can be detected by **agglutination reactions**. Agglutination reactions are physical reactions that result in the linking of antigens by antibodies. In **direct agglutination tests**, a series of small containers called microtiter wells are filled with the same concentration of known antigen. Serum that is to be tested for the presence of antibodies is added to each of the wells at increasingly dilute concentrations (Figure 21.5). If there is enough antibody present in a well to cause agglutination, the antigen and antibody will link together and form an even mat on the bottom of the well. If there is not enough antibody present in a well to link the antigen, the antigen will roll down the sides of the well and form a pellet in the bottom (Figure 21.5). The most dilute concentration of serum that produces a positive reaction is called the antibody **titer**. If there is no reaction, then antibodies to the antigen were not present in the serum. Direct agglutination tests can be used to determine infection by the causative agents of brucellosis and tularemia. Agglutination of red blood cells is called **hemagglutination** and is the basis for the determination of blood type.

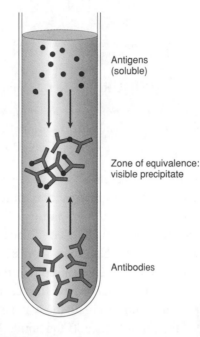

Antigens (soluble)

Zone of equivalence: visible precipitate

Antibodies

FIGURE 21.4. The precipitin ring test.

Two solutions, one containing antibody and the other containing antigen, are added one at a time to a test tube. As the antibodies and antigen diffuse toward each other, visible precipitation will occur at the zone of equivalence. This is visible as a white line of precipitate across the test tube.

Agglutination reactions can also be used to detect antibodies to *soluble* antigens if the antigens are first attached to particles such as tiny latex beads. In these **indirect agglutination tests**, the antigen-coated beads are placed in microtiter wells, and the test proceeds exactly as it does in direct agglutination tests. If antibody is present in the serum, agglutination will cause the antigen-coated beads to form a mat on the bottom of the well. Indirect agglutination tests can also be used to detect antigen by attaching antibodies to the latex beads. If antigen is present in the sample, agglutination will occur and be visible. This test is commonly used to identify the causative agent of plague, *Yersinia pestis*.

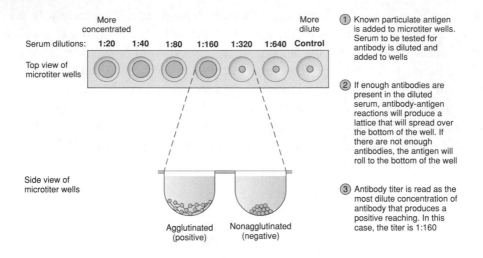

FIGURE 21.5. Agglutination.

A direct agglutination test is shown. It tests serum for the presence of antibody to a known particulate antigen.

Fluorescent-antibody (FA) techniques can be used to make antigen–antibody reactions visible on a special type of microscope, called a fluorescence microscope. In **direct FA tests**, a sample to be tested for a particular antigen is attached, or fixed, to a slide. Known antibodies that have an attached molecule of fluorescent dye are washed over the surface of the sample. If the antigen is present in the sample, the antibodies will bind to it. The surface of the slide is rinsed to remove any unbound antibody. The slide is viewed on the fluorescence microscope, which has special filters to create the wavelength of light needed to activate the fluorescence. If the antibodies attached to the sample, they will be visible as little glowing dots (Figure 21.6). If the antigen was not present in the sample, all of the antibodies will have washed away.

> **REMEMBER**
> Antibodies can reveal the presence of certain antigens by a variety of techniques, including observation of precipitation, agglutination, and fluorescent markers.

Indirect FA tests can detect the presence of particular antibodies in a sample. In these tests, known antigen is fixed to the slide. The serum to be tested for antibodies to the antigen is washed over the surface of the slide. If antibodies are present to the antigen, they will bind to the sample. The slide is then rinsed to remove any unbound antibody. To detect any bound antibody that remains, another antibody is used. This antibody, called anti-human immune serum globulin (anti-HISG), will recognize and bind to any human antibodies. So, if bound antibody is present on the slide, the anti-HISG will attach to them and not be rinsed away (Figure 21.6). The anti-HISG has a fluorescent molecule attached to it so that it can be seen on a fluorescence microscope. If antibody was present in the serum to be tested, the antibody-antigen complexes will be visible as small glowing dots on the slide.

Direct fluorescent antibody test

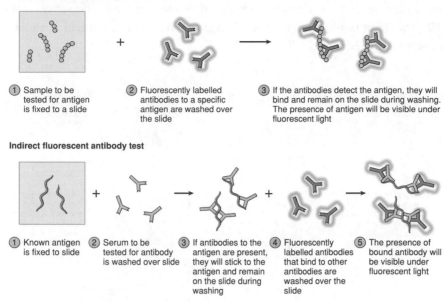

① Sample to be tested for antigen is fixed to a slide

② Fluorescently labelled antibodies to a specific antigen are washed over the slide

③ If the antibodies detect the antigen, they will bind and remain on the slide during washing. The presence of antigen will be visible under fluorescent light

Indirect fluorescent antibody test

① Known antigen is fixed to slide

② Serum to be tested for antibody is washed over slide

③ If antibodies to the antigen are present, they will stick to the antigen and remain on the slide during washing

④ Fluorescently labelled antibodies that bind to other antibodies are washed over the slide

⑤ The presence of bound antibody will be visible under fluorescent light

FIGURE 21.6. Fluorescent-antibody techniques.

Enzyme-linked immunosorbent assays (ELISA) can also be used to detect either antigen or antibody. When **ELISA** are used to detect antigens (Figure 21.7), antibodies are attached to the bottoms of microtiter wells. Next, the sample to be tested is added to the well. If the antigen is present, it will bind to the antibody on the bottom of the well. The well is then rinsed so that any antigens not recognized by the antibody are washed away. More antibodies specific for the antigen that is being tested are added to the well. If the antigen is present, these antibodies will bind to it, sandwiching the antigen between the two sets of antibodies. The second set of antibodies is *enzyme-linked*; that is, they have an enzyme attached to them. The well is rinsed, and any unbound antibody is washed away. Finally, a chemical that reacts with the enzyme attached to the second set of antibodies is added. If these antibodies are present, a visible chemical reaction occurs. For example, the well might turn blue. If no antigen is present in the test sample, then the second set of antibodies will wash away, and no color reaction will occur (Figure 21.7).

ELISAs can also test for the presence of antibody. These tests begin by attaching known antigen to the bottom of microtiter wells. Next, the serum to be tested is washed over the antigen. If antibody to the antigen is present, it will attach and not be washed away when the well is rinsed. Next, enzyme-linked anti-HISG is added to the well. This antibody will bind to any human antibodies that are present and has an attached enzyme. If antibody was present in the serum and is attached to the well, the anti-HISG will bind to it. The well is again rinsed to wash away any unbound antibody. Finally, the

> **REMEMBER**
> In a direct ELISA, the enzyme is linked to the primary antibody. In an indirect ELISA, the enzyme is linked to a secondary antibody.

substrate for the enzyme is added. If anti-HISG is present, a visible chemical reaction will occur (Figure 21.7). If there is no antibody to the antigen in the serum, then the anti-HISG will not bind and will be washed away. When the chemical is added, there will be no reaction (Figure 21.7).

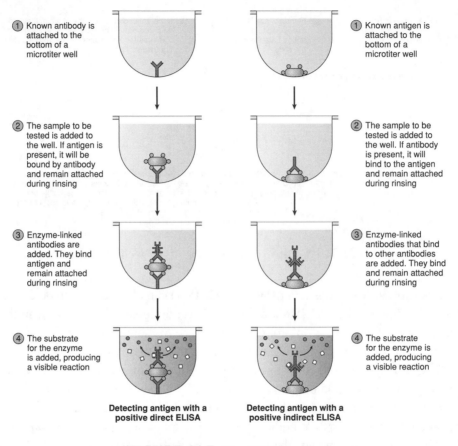

① Known antibody is attached to the bottom of a microtiter well

② The sample to be tested is added to the well. If antigen is present, it will be bound by antibody and remain attached during rinsing

③ Enzyme-linked antibodies are added. They bind antigen and remain attached during rinsing

④ The substrate for the enzyme is added, producing a visible reaction

Detecting antigen with a positive direct ELISA

① Known antigen is attached to the bottom of a microtiter well

② The sample to be tested is added to the well. If antibody is present, it will bind to the antigen and remain attached during rinsing

③ Enzyme-linked antibodies that bind to other antibodies are added. They bind and remain attached during rinsing

④ The substrate for the enzyme is added, producing a visible reaction

Detecting antigen with a positive indirect ELISA

FIGURE 21.7. The ELISA method.

The availability of monoclonal antibodies has greatly increased the possibilities for the diagnosis of human pathogens. Increasingly, pathogens are identified by rapid diagnostic immunologic tests that can take as little as 10 minutes! Because there is antigenic variation among pathogens, however, traditional culture and identification methods must sometimes be used. In the coming years, we will undoubtedly see more advances in immunologic testing.

REVIEW EXERCISES FOR CHAPTER 21

1. How do vaccines trigger immunity?

2. Compare the advantages and disadvantages of attenuated and inactivated whole agent vaccines.

3. Copy the following table and fill it in.

Types of Antigens Used in Vaccines	Example Vaccine

4. How are DNA vaccines different from purified antigen vaccines?

5. What is the benefit of vaccination to human populations?

6. What is the difference between real risk and perceived risk?

7. What are the real risks associated with vaccines? How significant are these risks?

8. What is the process involved in making a vaccine available to the public in the United States?

9. Compare and contrast monoclonal and polyclonal antibodies.

10. Describe three types of diagnostic tests that rely upon antigen–antibody reactions.

11. Describe both the direct and indirect ELISA assays.

SELF-TEST

1. Which type of vaccine could cause the disease it is designed to prevent?

   A. Inactivated whole agent vaccine
   B. Toxoid vaccine
   C. Conjugated vaccine
   D. Attenuated whole agent vaccine
   E. DNA vaccine

2. True or False. Vaccines prevent disease because they block processes essential to pathogen reproduction such as the synthesis of peptidoglycan.

3. True or False. Herd immunity benefits everyone in a population, even those who are not vaccinated.

4. True or False. If an unvaccinated person got polio today, they would get a very mild form of the disease because vaccination has prevented major polio epidemics in the United States for about 40 years.

5. A father reads on an Internet site about a woman who claims her child developed autism after receiving the MMR vaccine. This makes the father afraid to have his child vaccinated. Then, the father heard about the result of a study that examined all children born in Denmark from January 1991 through December 1998. There were a total of 537,303 children in the study; 440,655 of the children were vaccinated with MMR, and 96,648 were not. The researchers did not find a higher risk of autism in the vaccinated group of children than in the unvaccinated group. The father made the decision to have his child vaccinated.

   The initial fear the parent felt about MMR vaccination was the result of _____1_____, while the statistical results from the study on Danish children represented _____2_____.

   A. 1. real risk; 2. perceived risk
   B. 1. perceived risk; 2. real risk
   C. 1. real risk; 2. real risk
   D. 1. perceived risk; 2. perceived risk
   E. None of the above is correct.

6. Which of the following tests would yield a positive result that looked like a cloudy white line?

   A. Agglutination
   B. Precipitation
   C. ELISA
   D. Fluorescent-antibody test
   E. None of the above is correct.

7. Which of the following could be used to detect antigen in a patient's serum?

   A. Direct agglutination
   B. Indirect ELISA
   C. Direct ELISA
   D. Indirect fluorescent-antibody
   E. All of the above.

8. If you put the following steps of an indirect ELISA that is used to screen for antigen in order, which would occur second?

   A. Add enzyme-linked antibody
   B. Add primary antibody
   C. Attach antibody to bottom of microtiter well
   D. Add sample to be tested
   E. Add substrate for enzyme-linked antibody

9. Which of the following tests would yield a positive result that looked like cell clumping?

   A. Agglutination
   B. Precipitation
   C. ELISA
   D. Fluorescent-antibody test
   E. None of the above is correct.

10. Which of the following tests for the presence of a free soluble antigen?

    A. Agglutination
    B. Precipitation
    C. ELISA
    D. Fluorescent-antibody test
    E. None of the above is correct.

Answers

Review Exercises

1. Vaccines introduce weakened or inactivated pathogens or parts of pathogens into the body. This causes the body to respond with a primary immune response, which generates production of antibody and memory B cells. Thus, when the body encounters the real pathogen, a secondary immune response is triggered. This response involves rapid division and proliferation of memory B cells, generating plasma cells that produce large amounts of IgG antibody. Circulating IgG antibody prevents establishment of the pathogen and protects the body from disease.

2. Attenuated whole agent vaccines will actually reproduce in the body, resulting in a long-term exposure of antigen to the immune response. This generates a very robust form of immunity. Inactivated whole agent vaccines do not reproduce and often do not generate as strong of an immune response. However, although it is rare, attenuated vaccines occasionally cause disease in people, whereas inactivated vaccines cannot.

3.

Types of Antigens Used in Vaccines	Example of Vaccine
Toxoid	Diphtheria
Polysaccharide	Pneumococcus
Conjugated	*Haemophilus influenzae* type B (Hib)
Recombinant	Hepatitis B

4. Purified antigen vaccines introduce antigens directly into the body. DNA vaccines introduce the DNA for antigens into muscle cells. The muscle cells transcribe and translate the DNA to produce the antigen themselves.

5. Vaccination generates herd immunity in populations. As more people in the population become immune, it becomes harder for a pathogen to spread from person to person. By blocking the spread of the pathogen, even unvaccinated people have increased protection.

6. Real risk is based on an assessment of probability. It typically involves the comparison of statistics from studies that involve large groups of subjects. Perceived risk is an individual's or population's feeling about how much risk there is in a situation. It may be based on anecdotal evidence from the reports of a few individuals.

7. Attenuated whole agent vaccines do cause disease in a very small percentage of people who receive the vaccine. Some vaccines cause side effects, which can be very unpleasant, in a very small percentage of people who receive the vaccine. Although these risks are real, they are generally very much smaller than the real risks associated with actually contracting the disease or having an epidemic of a disease in a population.

8. First, researchers in laboratories test molecules or strains of attenuated pathogens for their ability to generate immunity. After a promising vaccine has been developed, the FDA must license it. This involves testing in animals, then adults, and finally children. After it is licensed, the data is examined by leading medical organizations like the Centers for Disease Control and Prevention, the American Academy of Pediatricians, and the American Academy of Family Physicians. They compare the risks and benefits of the vaccine and make a decision about whether to offer the vaccine to the public.

9. Monoclonal antibodies are populations of antibodies that all bind specifically to the same epitope. Polyclonal antibodies are populations of antibodies that bind to more than one epitope.

10. Precipitation reactions detect soluble-free antigen. Solutions to be tested for antigen are placed next to solutions containing specific antibody. The two solutions are allowed to diffuse toward each other. If the antigen is present, antigen and antibody will combine to form a lattice when the concentrations of each reach the zone of equivalence. The lattice is visible as a cloudy white line.

   Agglutination reactions can be used to detect antibodies to antigens that are particulate (direct agglutination) or soluble (indirect agglutination). In these tests, antigen is placed in the bottom of wells. The serum to be tested for antibodies to the antigen is diluted into various concentrations that are then placed into the wells with antigen. If enough antibody is present in the diluted serum, antigen–antibody reactions will occur, forming a lattice that spreads the antigen over the bottom of the well. If insufficient antibody is present, then lattices do not form, and the antigen pellets in the bottom of the well.

Fluorescent antibody reactions can be used to test for antigen (direct FA) or antibody (indirect FA). To test for antigen, a sample is fixed to a slide. Known antibodies to the antigen that is being looked for are washed over the surface of the slide. The slide is rinsed to remove any unbound antibody. If the antigen was present in the sample, the known antibodies will have stuck to the slide. These antibodies are marked with a chemical tag that is visible under fluorescent light. To test for the presence of antibody to a particular antigen, known antigen is fixed to the slide. The serum to be tested is washed over the slide. The slide is rinsed to remove any unbound antibody. Then another solution of fluorescently labeled antibodies that bind to other antibodies is washed over the slide. The slide is rinsed. If antibodies were present in the serum to be tested, they will have stuck to the known antigen. The antibodies that bind to other antibodies will also have stuck, and their fluorescent tag will be visible.

11. A direct ELISA is used to detect antigen. Monoclonal antibodies to the antigen that is being sought are fixed to the bottom of a microtiter well. The solution to be tested is passed over the antibodies. The wells are then washed to remove any unbound antigen. Additional antibodies specific to the antigen are added to the wells. These antibodies are linked to an enzyme. The wells are washed again to remove any unbound antibody. The substrate for the enzyme is added to the well. If the antigen was present in the test solution, then a sort of sandwich of antibody–antigen–enzyme linked antibody will have stuck to the bottom of the well. The enzyme will act on the substrate, and an observable reaction will occur, indicating a positive test.

An indirect ELISA tests for the presence of antibody to a particular antigen. The antigen in question is fixed to the bottom of a microtiter well. The serum that is to be tested for antibody is added to the well. The well is washed to remove any unbound antibody. A second set of enzyme-linked antibodies that are capable of binding to other antibodies is added to the well. The well is washed to remove any unbound antibody. If antibody was present in the serum to be tested, a stack of antigen–antibody from serum–enzyme linked antibody will be stuck to the bottom of the well. The substrate for the enzyme is added, and an observable reaction occurs.

Self-Test

1. D	5. B	9. A
2. F	6. B	10. B
3. T	7. E	
4. F	8. D	

22

Control of Microbial Growth

WHAT YOU WILL LEARN

This chapter provides an overview of the physical and chemical methods that people use to control microbial growth. As you study this chapter, you will:

- explore the history of microbial control;
- learn some of the terminology used to describe microbial control;
- investigate the factors that affect the success of efforts to control microbes;
- compare the physical methods used to control microbes;
- consider the chemical methods used to control microbes;
- examine the reasons that some microbes are more resistant to control efforts than others.

SECTIONS IN THIS CHAPTER

- History of the Control of Microorganisms
- Terminology
- Factors Affecting the Success of Microbial Control Agents
- Physical Methods of Microbial Control
- Chemical Control of Microorganisms
- Resistance to Control Efforts

We do things everyday, consciously and unconsciously, to control the growth of microbes. At home, we put our food in the refrigerator, wash our dishes, and wash our hands. Some of us make jellies and jams to preserve fruits or make jerky to preserve meats. We might use cleaners that are supposed to *fight mold and mildew* or *kill bacteria dead,* and we put chlorine in our swimming pools. If we find something that is slimy or hairy growing somewhere in our houses, we say "Yuck!" and throw it out!

History of the Control of Microorganisms

We are not unique in our efforts to keep our microbial friends and enemies at bay. As long as there have been people, people have probably tried to keep microbes from eating their food and growing in places where they weren't wanted. The Ancient Egyptians were masters at the control of microbial growth, as evidenced by their mummies that still resist decay today. The Ancient Greeks burned sulfur in their homes to purify the air, and the Ancient Hebrews would burn the garments of people with leprosy to prevent its spread. In modern times, we have seen the development of a vast array of techniques to control microbes, including refrigeration, water treatment, disinfectants, food preservatives, and autoclaves. Among all of these amazing advances, we might tend to overlook one of our simplest, but most effective, advances in the control of microbes: the use of soap and water to wash our hands.

Although it may seem simple today, handwashing was not a standard practice in many human societies before the 1800s. Even surgeons did not typically clean their hands or their tools between patients. It wasn't until 1846, when a doctor named Ignaz Semmelweis began making careful observations of a childbirth ward in Vienna General Hospital, that the idea of handwashing was born. Women who delivered babies in the childbirth ward that was attended by surgical students were much more likely to contract **puerperal sepsis**, or childbed fever, than were women who were attended by midwives. Semmelweis noted that the surgeons would come to attend the women directly after dissecting cadavers. As was typical of the time, they would not change clothes or wash their hands first. In 1847, Semmelweis ordered the surgical students to wash their hands before attending to the women in the childbirth ward. The incidence of puerperal sepsis dropped dramatically as a result. Unfortunately, when news of Semmelweis's reforms spread, the hospital administration felt that they were being criticized and made Semmelweis's life very unpleasant. He returned to his native Hungary, joined the staff of a new hospital, and again instituted life-saving reforms. Again, instead of praising his efforts, the medical establishment resisted his changes. Semmelweis suffered a mental breakdown and died in a sanitarium.

REMEMBER
Dr. Ignaz Semmelweis first demonstrated that handwashing can reduce disease transmission.

Because of the lack of sanitary practices in the 1800s, a person who underwent surgery was very likely to die from an infection contracted during the surgery. This situation improved in the 1860s owing to the work of an English surgeon named Joseph Lister. Lister had heard of Semmelweis's handwashing reforms and decided to apply similar principles to surgery. He made a spray of carbolic acid (phenol), which he knew killed bacteria, and began spraying his solution on surgical wounds during surgery. Lister's new practice was very successful in preventing infection, and other surgeons quickly adopted it.

In addition to advances in the control of microbes in medical environments, the 1800s also saw progress in the control of food spoilage organisms. In 1864, Louis Pasteur was asked by the Emperor Napoleon III to help solve a problem in the French wine industry. For reasons the vintners did not understand, batches of developing wine would suddenly sour and turn into vinegar. Pasteur examined batches of good and bad wine under the microscope and observed that the organisms he found in each were different: The good batches contained abundant yeast cells, while the bad batches were overrun with rod-shaped bacteria. To solve the problem, Pasteur realized that the vintners needed a way to control the composition of the microbial population in the wine without damaging the wine's flavor. He developed the process of **pasteurization**, a brief heating of the wine that would kill the spoilage bacteria without ruining the wine.

Terminology

Depending on the organisms present and the method used, there is a range of effectiveness of control measures. Pasteurization, for example, merely reduces the numbers of microbes to safe levels. Complete killing of all organisms, including spores and viruses, is called **sterilization**, which is commonly achieved by using heat combined with pressure. If only vegetative cells are killed, and spores and viruses remain, the process is called **disinfection**. Disinfection is often achieved by the use of chemicals called disinfectants. The term disinfection is used to describe the removal of microbes from inanimate objects. When living tissue is disinfected, the process is called **antisepsis** (*sepsis* = decay), and the chemical used is an antiseptic. **Aseptic technique** is used during surgery and in microbiology labs to prevent microbial contamination. When living tissue is merely wiped clean, for example by an alcohol wipe before an injection, this is called **degerming**. Degerming physically removes some microbes and dirt from an area but does not kill significant numbers of microbes. In the food industry, the numbers of microbes on eating utensils are lowered to safe levels for human consumption by **sanitization** in a dishwasher or chlorine solution.

> **REMEMBER**
> Many common procedures, such as pasteurization, disinfection, and sanitization, lower the number of microbes but don't completely eliminate them.

The ability of many chemical products to control microbes is indicated on the label. The suffix *–cide* indicates killing; the suffix *–static* indicates inhibition of growth. These suffixes are paired with prefixes indicating the type of organism affected. For example, a **bacteriocide** kills bacteria, and a **fungicide** kills fungi. A **germicide** kills microbes, although it may not actually kill resistant structures like endospores. Something that is **bacteriostatic** merely inhibits bacterial growth. Other prefixes that are used include *algi-* for algae, *spori-* for spores, *viru-* for viruses, and *tuberculo-* for mycobacteria.

Factors Affecting the Success of Microbial Control Agents

The success of control measures on a population of microorganisms depends on a number of factors. One of these is the *number* of microorganisms present in the population. The greater the number of microbes, the longer it takes to kill them (Figure 22.1). Thus, *duration* of exposure to the control agent is important. The longer the exposure, the greater the reduction in the microbial population. The *composition* of the population is also significant. If resistant organisms, like endospore-producers or unenveloped viruses, are present, it will be more difficult to achieve sterilization. When chemical agents are being used, the *concentration* of the chemical will affect its success. For many chemicals, the greater the concentration, the greater the antimicrobial activity. This is not always true, however. Alcohols, for example, work better if some water is present, and many are most effective in a 70 percent solution of alcohol in water.

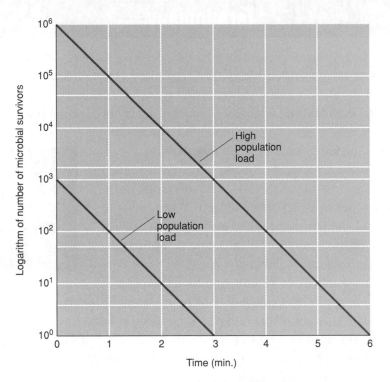

FIGURE 22.1. A microbial death curve.

When microbial populations are controlled with heat or chemicals, the cells die at a constant rate. The greater the number of cells in the population, the longer it will take to kill all of the cells.

The *environment* to be cleaned also affects the success of antimicrobial substances. If organic matter, such as dirt, food, blood, or feces, is present, they can attract and bind the active ingredients in antimicrobial chemicals and reduce their effectiveness. Thus, it is better to clean a wound before applying antiseptic, or to wipe down a table or countertop before using disinfectant. The temperature of the environment can also make a difference. Many chemical disinfectants work better at slightly elevated temperatures, so it is generally a good idea to use hot water when cleaning. An exception to this is the use of chlorine in sanitization of eating utensils. Because chlorine evaporates rapidly at higher temperatures, sanitization using chlorine is done in a cold-water solution.

> **REMEMBER**
> It's important to clean a surface or a wound before you apply a chemical disinfectant or antiseptic. If you don't clean first, the chemical will be absorbed by the dirt or blood and won't actually reach the surface where you're trying to kill microbes.

The goal of all control measures is to kill microbial cells. Cells can be killed by disrupting their plasma membranes and causing them to lyse, by denaturing their proteins, or by damaging their DNA. Cells can also be killed if their molecules are disrupted by oxidation. Thus, almost all control measures do one or more of these things. Some control measures use physical means, such as heat or radiation, to kill cells; others use chemicals.

Physical Methods of Microbial Control

Heat is probably the most widely used method to control microorganisms. Both moist heat, such as boiling or steam, and dry heat, such as naked flame, can achieve sterilization. Moist heat denatures microbial proteins, whereas dry heat oxidizes various microbial components. The boiling temperature of water is 100°C. At this temperature, vegetative cells and eukaryotic spores die within 10 minutes, viruses die within 30 minutes, and prokaryotic spores die within 20 hours. To achieve more rapid sterilization, an **autoclave** (Figure 22.2) may use pressure to raise the temperature of steam above boiling. Complete sterilization can be achieved after 15 minutes in an autoclave at a pressure of 121 pounds per square inch (psi) and a temperature of 121°C. Boiling water or steam is often used to sanitize dishes or other equipment. Autoclaves are used to sterilize biological growth media and other solutions, as well as medical instruments and equipment.

> **REMEMBER**
> An autoclave is essentially a big pressure cooker that uses heat and pressure to sterilize liquids and equipment.

Pasteurization also uses moist heat to control microbial growth in solutions. Although it was initially developed by Pasteur to control the growth of microbes in wine, today it is also used to preserve milk, juice, beer, yogurt, and ice cream. The classic method for pasteurization of milk is a relatively mild increase in temperature to 63°C for 30 minutes. This destroys pathogens in the milk and reduces the numbers of other microbes, increasing the potential storage time of the milk. More recently, **flash pasteurization**, also known as **high-temperature short-time (HTST) pasteurization** has been used to give the same effect. In flash pasteurization, the milk is raised to a higher temperature for a much shorter length of time, for example 72°C for 15 seconds. **Ultra-high-temperature (UHT) treatment** of milk results in sterilization of milk, enabling it to be stored without refrigeration. In UHT treatment, which takes less than 5 seconds, milk is poured in a thin stream through a chamber of superheated steam, raising its temperature to 140°C for less than a second, and then cooled in a vacuum chamber.

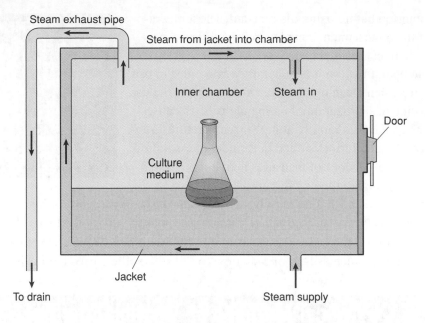

FIGURE 22.2. An autoclave.

*Steam flows into the inner chamber, raising the internal pressure.
The increased pressure raises the temperature of the steam above that
of boiling water. Because of the increased temperature, sterilization can be
achieved in relatively short periods of time.*

Dry heat created by **direct flaming** or ovens can also be used to sterilize. Hospitals use large ovens to achieve **hot-air sterilization** of glassware and instruments. Treatment at 170°C for 2 hours is sufficient to destroy all organic material, including endospores. **Incineration**, or burning, is also used by hospitals to sterilize medical waste such as contaminated dressings and wipes, as well as routine garbage such as paper cups. In the microbiology laboratory, a naked flame is used to sterilize the inoculating tools used to transfer microorganisms to new media. Because the heat in water is more easily transferred to substances, moist heat is generally more efficient than dry heat. For example, the endospores of *Clostridium botulinum* are destroyed in 5 minutes at 121°C in a moist environment, but 2 hours at 170°C is required to destroy endospores in a dry oven.

Cold temperatures are also used very commonly to control the growth of microbes. Low temperatures like those found in household refrigerators lower the metabolic rate of bacteria and prevent most pathogens from growing. Psychrotrophs are more resistant to the effects of low temperatures and can continue to grow. Freezing typically stops all microbes from growing and can kill most eukaryotes and some bacteria.

Because water is necessary for microbial growth, some control measures make water unavailable to microbes. Water can be removed by **desiccation**, or drying. Drying can preserve all fish, meat, and fruit. Water can also be made unavailable to microbes by creating an environment with a high **osmotic pressure**. Salt or sugar can be used to create a hypertonic environment that will make it difficult for microbes to obtain the water they need. For example, fish and meat are sometimes preserved by salting, whereas fruits are preserved as jellies or jams that contain high concentrations of sugar. Drying, salting, and sugaring all restrict microbial growth by preventing the microbes from obtaining water. They do not, however, kill microbes.

> **REMEMBER**
> Filtration is a useful way to sterilize materials that would be damaged by heat, such as prescription drugs.

Some items are sensitive to temperature changes and cannot be sterilized using heat. **Filtration** can be used to sterilize heat-sensitive solutions such as drugs, vitamins, vaccines, and some media. Filtration physically screens microbes out of the solution to be sterilized. Frequently, this is done by the use of **membrane filters**, flexible pieces of thin plastic that have very small holes in them, holes so small that not even viruses can pass through. The filter is placed between the solution and a sterile container (Figure 22.3). A vacuum is applied to draw the solution through the filter. As the solution passes through the filter, all viruses and bacteria are trapped on the filter, creating a sterile solution in the bottle.

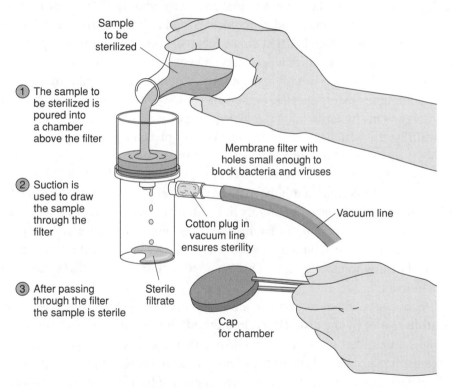

FIGURE 22.3. Filter sterilization.

Temperature-sensitive items can also be sterilized by the use of radiation. There are two types of radiation that can kill microbes, **ionizing radiation** and **ultraviolet (UV) radiation**. Ionizing radiation, which includes X-rays and gamma rays, energizes the electrons in water molecules and causes water to ionize, forming hydrogen ions (H^+) and highly reactive hydroxyl radicals (OH^-). Hydroxyl radicals oxidize the molecules of microbial cells, resulting in microbial death. DNA absorbs ultraviolet radiation, causing adjacent thymines to form covalent bonds with each other (Figure 7.16). These **thymine dimers** distort the DNA and cause errors during DNA replication. Because ultraviolet radiation does not penetrate materials well, it is only useful for the sterilization of surfaces. UV lamps, called germicidal lamps, are often found in hospital rooms where it is used to control microorganisms in the air. UV lamps are also used to sterilize the surfaces in **laminar flow hoods**, special closed environments that are used for particularly challenging aseptic work. Ionizing radiation is used to sterilize drugs and disposable medical supplies such as plastic syringes and gloves.

Chemical Control of Microorganisms

Chemicals can also be used to control the growth of microorganisms, although relatively few chemicals are used to actually sterilize things. Instead, chemicals are generally used as disinfectants and antiseptics to keep the levels of microorganisms at safe levels. Chemicals work in the same way that physical agents do: They disrupt plasma membranes, denature proteins, oxidize molecules, and damage DNA. Depending on their *active ingredient*, which is listed on the label, disinfectants and antiseptics will have a particular effect on microbial cells and may target a particular group of microorganisms. The labels usually indicate which organisms are affected by the chemical, sometimes listing specific organisms by name, and other times indicating groups of organisms by terms ending in the suffixes *-static* and *-cidal*. By reading labels carefully, the best disinfectant for a particular job can be chosen.

The labels on disinfectants also have instructions on how the disinfectant should be used. Because chemical concentration and time of exposure both affect the success of disinfectants, it is important to use the disinfectant properly. For example, if bacteria are spilled in the microbiology laboratory, it is often necessary to leave a bacteriocide on the spill for a certain period of time in order to ensure it is safely cleaned up. Because organic matter can bind to the active ingredients in disinfectants and prevent them from attacking microbial cells, it is also important to clean surfaces before applying the disinfectant.

The effectiveness of a disinfectant against a particular organism can be tested by the **use–dilution test** or by the **disk diffusion method**. For the standard use–dilution test, disinfectants are diluted according to the manufacturer's directions and tested against three organisms: the enteric bacterium *Salmonella choleraesius*, the gram-positive *Staphylococcus aureus*, and the resistant gram-negative bacterium *Pseudomonas aeruginosa*. During the test, the bacteria are inoculated onto different metal rings, placed in the disinfectant solution for 10 minutes, and then transferred into fresh media to see if any bacteria are still alive and can grow. This test can also be used to test disinfectants against other organisms, viruses, or spores. In the disk-diffusion method, the bacteria to be tested are spread across the entire surface of an agar plate to form a **lawn**. Sterile paper disks are then soaked in the test chemical and placed on top of the plate. The plates are incubated to allow the bacteria to grow. If the chemical inhibits their growth, a clear area will be seen around the paper disk (Figure 22.4).

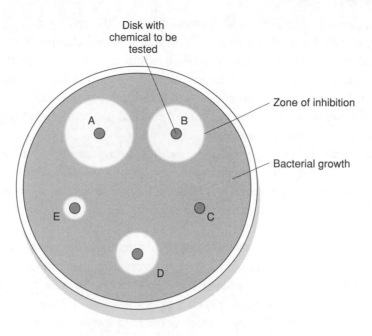

FIGURE 22.4. The disk diffusion method.

Paper disks that contain the chemical to be tested are placed on a petri plate that was inoculated with a lawn of bacteria. If the chemical inhibits bacterial growth, clear zones called zones of inhibition will appear around the disk as the lawn grows. The larger the zone, the greater the inhibition.

Phenols (Figure 22.5) denature proteins and disrupt plasma membranes and can be bacteriocidal if used at high enough concentrations. Although first used by Joseph Lister as an antiseptic, they are now generally considered too toxic to the skin for antiseptic use. At very low concentrations, they have a numbing effect and can be found in some lip balms and throat lozenges. One advantage of phenols is that they are relatively stable compounds that remain active in organic matter such as pus, saliva, and feces. Also, because they disrupt lipids, they are effective against the waxy wall of the mycobacteria. The active ingredient in the common disinfectant Lysol is the phenol *o*-phenylphenol (Figure 22.5).

> **REMEMBER**
> At high concentrations, phenols can be bacteriocidal.

Derivatives of phenols are also used as disinfectants. **Phenolics** are phenols that have been chemically modified to either reduce their toxicity or to increase their effectiveness. **Bisphenols** are molecules that contain two phenol groups connected by a chemical group. The bisphenol hexachlorophene (Figure 22.5), an active ingredient in pHisoHex, is used in medical environments because it is very effective against the gram-positive staphylococci and streptococci. Many of the antibacterial products that have emerged recently, from antibacterial soaps to plastics such as cutting boards and children's toys, contain the bisphenol triclosan. Triclosan is effective against gram-positive bacteria, many gram-negative bacteria, and fungi. Ironically, the recent explosion of antibacterial products containing triclosan may soon make it ineffective as a microbial control agent. The more a product is used, the greater the chance that populations of resistant microbes will emerge, which is already happening for triclosan.

Biguanides (Figure 22.5) disrupt plasma membranes and are effective at killing most bacteria and fungi. The biguanide chlorhexidine is used to control microbial growth on skin and mucous membranes. It is the active ingredient in many surgical hand scrubs and is used to clean the skin of surgical patients before surgery. It is also used in some periodontal rinses.

The **halogens** (Figure 22.5) chlorine, iodine, bromine, and fluorine are a group of elements that are very **electronegative**, or have a strong ability to attract electrons. Although their exact mechanism of action on microbial cells is not known, they have the ability to oxidize molecules, stealing electrons and disrupting cellular components. Iodine (I_2), which has a long history of use as an antiseptic, is thought to denature proteins by adding iodine atoms to the amino acid tyrosine. *Tincture of iodine*, 2 percent iodine in a 50 percent solution of alcohol, is a very effective antiseptic, killing bacteria, various fungi, some viruses, and even some endospores. However, because it stains and can cause allergic reactions, it is no longer widely used. Now, the more common form of iodine is as an *iodophore*, which consists of iodine bound to an organic molecule. The organic molecule releases the iodine gradually over time. Iodophores are not as effective as tinctures of iodine, but they do not stain or cause allergic reactions. They are commonly found in many products for topical use such as Betadine, Wescodyne, Virac, Prepodyne, Iosan, and Surgidyne.

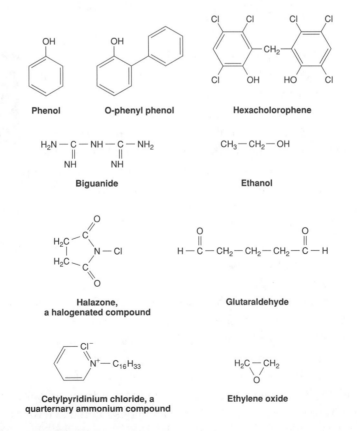

FIGURE 22.5. Chemicals used to control microbial growth.

The halogen chlorine is widely used in sanitization of restaurant and dairy equipment and in the disinfection of public water. In water, chlorine forms the germicidal compound *hypochlorous acid (HOCl)* by the following reaction:

$$Cl_2 + H_2O \rightarrow H^+ + Cl^- + HOCl$$

> **REMEMBER**
> Common household bleach contains the halogen chlorine, which forms the germicidal molecule hypochlorous acid when it is dissolved in water.

One disadvantage to the use of chlorine is that it binds to organic matter and becomes inactivated. Because organic matter is present in public water lines, chlorine is added in excess of the concentration needed for disinfection. Also, water lines are constructed so that they do not have any dead ends in which organic matter can accumulate. Soldiers in the U.S. military use tablets of sodium dichloroisocyanurate to disinfect water in the field. In emergencies, household bleach such as Clorox can be used to treat water and make it safe for drinking. If two to four drops are added to a liter of water, the water is considered safe for drinking after a period of 30 minutes.

Alcohols denature proteins and have a solvent action on lipids such as those in the plasma membrane. The most commonly used alcohols are ethanol (Figure 22.5) and iso-propanol (rubbing alcohol). Because the presence of some water improves the action of alcohols, they are most effective at concentrations of 70 to 95 percent. At these concentrations, they kill bacteria and fungi, but do not kill endospores and non-enveloped viruses. Common uses for alcohols include disinfection of thermometers and small instruments by soaking them for 15 minutes and the degerming of skin prior to injection.

Although **heavy metals** denature proteins and can kill microorganisms, most of them are too toxic for medical use. The organic mercury compounds, mercurochrome and merthiolate, have long been used as antiseptics. Silver can also be used as an antiseptic as a 1 percent silver nitrate solution. This solution used to be routinely dropped into the eyes of newborn infants to prevent infection by *Neisseria gonorrhoeae*. Although this practice has been replaced by the use of antibiotics, other medicinal uses of silver have recently been developed. Silver-impregnated bandages are used as dressings on wounds infected by antibiotic-resistant bacteria, and silver-coated catheters help prevent infection from indwelling catheters. Nonmedicinal use of heavy metals includes the use of copper in the form of copper sulfate ($CuSO_4$) as an algicide in swimming pools and the use of zinc as an antifungal agent in household paints.

Surface-active agents, or **surfactants**, such as soaps and detergents disrupt plasma membranes and are useful in the *emulsification*, or breaking up, of oily debris. Soap is very effective in the physical removal of microbes and debris from the surface of the skin. Acid-anionic surfactants are important in the sanitization of dairy equipment. The most widely used detergents are the cationic detergents. Because these detergents are made of positively charged ions, or cations, and hydrophobic groups, they are very effective at disrupting plasma membranes. Quaternary ammonium compounds, or quats (Figure 22.5), such as those in Zephiran and Cepacol contain the positively charged ammonium ion. Quats are popu-

> **REMEMBER**
> Soap helps break up oily debris and remove materials, including microbes, from the surface of the skin.

lar because they are effective against a wide range of microbes, easy to use, colorless, odorless, and nontoxic. However, the effectiveness of quats is limited in the presence of organic matter.

Aldehydes, such as glutaraldehyde and formaldehyde, are very effective antimicrobials. They denature proteins and can sterilize when used at high concentrations. Glutaraldehyde (Figure 22.5) is the active ingredient in Cidex, which is used to disinfect hospital instruments. Glutaraldehyde is also used to sterilize catheters and dialysis machines. Both formaldehyde and glutaraldehyde are used as preservatives during embalming. Although these chemicals are very effective antimicrobials, they also pose risks for humans. Formaldehyde is classified as a probable human carcinogen under conditions of prolonged exposure by the U.S. Environmental Protection Agency, and studies have shown that embalmers have higher incidence for some types of cancer.

Many heat-sensitive items in hospitals are sterilized using gaseous chemosterilizers such as **ethylene oxide** (Figure 22.5). Ethylene oxide disrupts proteins and can achieve complete sterilization in 4 to 18 hours. It has good penetration and can sterilize large items such as mattresses. One concern with these gases is that they are mutagenic and therefore suspected carcinogens. Hospitals often have large gas facilities containing gas chambers and areas for equipment to aerate after gassing. Workers in these facilities must wear badges that monitor their exposure to the gas. Gas sterilization is used for a wide variety of hospital equipment, especially those made of plastics that would be destroyed by heat.

Peroxygens are toxic forms of oxygen that oxidize and destroy cellular components. Ozone is sometimes used to control the microbial growth in swimming pools. Hydrogen peroxide is an effective disinfectant of nonliving surfaces. Although it is also frequently used as an antiseptic on wounds, this use is not recommended because it may actually slow wound healing. Benzoyl peroxide is widely used in acne medication and is also good for treating wounds infected with anaerobic pathogens. Other peroxygens are important for the disinfection of food-processing and medical equipment.

Resistance to Control Efforts

The ability of control agents to limit microbial growth effectively is strongly affected by the composition of the target population of microbes. Certain microbes have structures or abilities that make it very difficult to control them (Table 22.1). Because of their strong spore coat, endospores are very difficult to destroy. Compounds that can destroy endospores are labeled sporicidal.

The waxy walls of mycobacteria prevent many chemicals from reaching the membrane of these cells, effectively protecting them from destruction. Disinfectants that can eliminate mycobacteria are labeled tuberculocidal.

Likewise, the outer membrane of gram-negative bacteria can exclude many compounds and makes them more resistant to antimicrobials than are the gram-positive bacteria.

TABLE 22.1. Resistance to Microbial Control Methods

Level of Resistance	Microorganisms or Structure
Greatest resistance	Prions
	Bacterial endospores
	Mycobacteria
	Protozoan cysts
	Protozoa
	Gram-negative bacteria
	Fungi
	Unenveloped viruses
	Gram-positive bacteria
Least resistance	Enveloped viruses

Among the gram-negative bacteria, *Pseudomonas aeruginosa* is a particular problem in hospital environments. In addition to the protection it gets from its outer membrane, *P. aeruginosa* also has the ability to break down a wide variety of carbon-containing compounds. This ability enables it to live in a wide variety of habitats and to even break down some disinfectants!

Enveloped viruses, also called *lipophilic viruses*, are relatively easy to control by any substance that disrupts plasma membranes. Non-enveloped viruses, however, are more resistant and must be controlled by substances that can denature their proteins.

Recently, infectious proteins called prions have become of greater medical concern because of their spread in infected beef, causing mad cow disease in cattle and bovine spongiform encephalopathy in humans. Although they are proteins, prions are extremely resistant to inactivation and cannot be destroyed by cooking temperatures or even standard autoclaving. The World Health Organization recommends a special protocol of cleaning contaminated equipment with a combination of autoclaving and a sodium hydroxide solution.

REVIEW EXERCISES FOR CHAPTER 22

1. State the definitions for the following terms: antibiotic, sterilization, disinfection, degerming, sanitization, asepsis, bacteriocidal, bacteriostatic, virucidal, sporicidal, tuberculocidal, algicide, and fungicide.

2. State how the following factors affect the efficacy of disinfectants: temperature, concentration of disinfectant, presence of organic matter, numbers of microbes present, types of microbes present, duration of exposure.

3. Describe the different methods for using heat to control microorganisms and give examples of their uses.

4. Explain how cold temperatures are used to control microbial growth.

5. Explain how drying is used to control microbial growth.

6. Describe two different methods for control microbial growth on items that are destroyed by heat.

7. Describe how the disk diffusion test is used to test the ability of chemicals to control microbial growth.

8. Complete the following table:

Chemical	Effect on Cells	Uses
Phenols		
Biguanides		
Halogens		
Alcohols		
Heavy metals		
Surface active agents		
Aldehydes		
Ethylene oxide		
Peroxigens		

9. Describe the basis for resistance to the action of disinfectants and other control measures by unenveloped viruses, endospores of bacteria, mycobacteria, gram-negative bacteria, and prions.

SELF-TEST

1. Which of the following does **not** achieve sterilization?

   A. Autoclaving
   B. Pasteurization
   C. Filtration
   D. Dry heat
   E. Gas (ethylene oxide)

2. Which method would be appropriate for the sterilization of a plastic heart valve?

   A. Autoclaving
   B. Pasteurization
   C. Filtration
   D. Dry heat
   E. Gas (ethylene oxide)

3. True or False. Sterilization can be achieved with heat, but not with chemicals.

4. True or False. The presence of organic matter does not affect the ability of disinfectants to control microbial growth.

5. True or False. Surface-active agents (surfactants) kill microbes by disrupting their plasma membranes.

6. Which of the following is most resistant to the action of disinfectants?

   A. Endospores
   B. Enveloped viruses
   C. Unenveloped viruses
   D. Gram-negative bacteria
   E. Gram-positive bacteria

7. Heat-sensitive vitamins and antibiotics are sterilized by

   A. gas (ethylene oxide).
   B. formaldehyde.
   C. autoclaving.
   D. filtration.
   E. radiation.

8. True or False. Quaternary ammonium compounds (quats) are effective against a very limited range of microbes.

9. Which method would most likely be used to sterilize liquid media?

    A. Autoclaving
    B. Pasteurization
    C. Filtration
    D. Dry heat
    E. Gas (ethylene oxide)

10. The disk-diffusion method was used to determine the effectiveness of several disinfectants against *Staphylococcus aureus*. Based on the following data, which disinfectant was most effective?

Disinfectant	Zone of Inhibition
A	10 mm
B	2 mm
C	14 mm
D	9 mm
E	0 mm

Answers

Review Exercises

1. An antibiotic is a substance produced by a microbe that inhibits other microbes. Sterilization is the complete killing of all organisms, including endospores and viruses. Disinfection kills vegetative cells, but viruses and endospores may remain capable of reproduction. Degerming is the physical removal of organic matter and some microbes. Sanitization is the reduction of microbes to safe levels for human consumption. Asepsis is the practice of maintaining a sterile environment. Bacteriocidal compounds kill bacteria. Bacteriostatic compounds inhibit bacterial growth. Virucidal compounds destroy viruses. Sporicidal compounds destroy endospores. Tuberculocidal compounds are effective against mycobacteria. Algicides kill algae, and fungicides kill fungi.

2. In general, both increasing temperatures and chemical concentration improve the efficacy of disinfectants. If organic matter is present, it may interfere with the action of disinfectants and make them less effective. The greater the number of microbes present, the longer it takes for a disinfectant to be effective. If resistant organisms are present, the disinfectant may not be completely effective. The longer disinfectants are allowed to contact the surface to be cleaned, the more effective they are.

3. Autoclaving uses heat and pressure to sterilize items. It is typically used for liquid media, glassware, equipment, utensils, dressings, and linens. Dry heat can be used as a direct flame to sterilize inoculating loops, to incinerate contaminated medical waste, or to sterilize glassware and medical instruments in large ovens. Pasteurization is brief heating that is used to lower the numbers of microbes in foods such as milk, wine, and juice.

4. Cold temperatures decrease the rate of microbial metabolism, slowing or stopping growth. Refrigeration and freezing are used to preserve food.

5. Drying can also be used to preserve food. Because they require water for growth, microbes can be stopped from growing if water is removed.

6. Filter sterilization, gas, and radiation can all be used to sterilize items that are damaged by heat. Filter sterilization passes solutions through membrane filters that contain holes small enough to block the passage of bacteria and viruses. Gas sterilization utilizes the toxic qualities of chemicals such as ethylene oxide. Ionizing radiation causes water to ionize, forming highly reactive hydroxyl radicals that damage molecules and destroy microbial cells. Ultraviolet radiation damages DNA by inducing thymine dimers, but because it has poor penetrance, it is only useful for sterilizing surfaces.

7. A sample of bacteria is spread evenly over the surface of a petri plate. Sterile paper disks that have been treated with the chemical to be tested are placed on the surface of the plate. The plate is incubated so that the bacteria can grow. If the chemical in the disk inhibits the bacterium, it will not grow near the disk. This creates a clear zone of inhibition around the disk.

8.

Chemical	Effect on Cells	Uses
Phenols	Disrupt plasma membranes and denature proteins	Disinfectants like Lysol, antibacterial agent in soaps and plastics
Biguanides	Disrupt plasma membrane	Skin disinfectant
Halogens	Oxidize cellular components	Water sanitation (chlorine), disinfection of eating utensils and equipment, antiseptic (iodine)
Alcohols	Disrupt plasma membrane, denature proteins	Disinfection of small instruments, degerming
Heavy metals	Denature proteins	Bandages, catheters
Surface active agents	Disrupt plasma membranes	Degerming of skin (soap), antiseptics, disinfectant of instruments and utensils
Aldehydes	Denature proteins	Sterilization and disinfection of equipment, embalming
Ethylene oxide	Denature proteins	Sterilization of plastic medical equipment
Peroxigens	Oxidize cellular components	Water sanitation, cleaning of deep wounds to rid them of oxygen-sensitive organisms

9. Unenveloped viruses consist of nucleic acid surrounded by a protein coat. Thus, they are resistant to any control methods that rely on disruption of the plasma membrane. Only control methods that denature proteins are effective against them. Endospores are surrounded by a very tough protein coat. Again, only control methods that are capable of denaturing proteins are effective. Mycobacteria have a waxy wall that is capable of preventing many disinfectants from reaching the cell, thus protecting the cell from their action. Gram-negative bacteria have an outer membrane containing lipopolysaccharide, which restricts the passage of many disinfectants. In addition, the porin proteins in the outer membrane are capable of resisting the passage of some chemicals. Prions are infectious proteins that are extremely resistant to all control methods, even those that typically denature proteins. Apparently, there is something about the structure of prions that enables them to very strongly resist denaturation.

Self-Test
1. B   5. T   9. A
2. E   6. A   10. C
3. F   7. D
4. F   8. F

Antimicrobial Drugs

WHAT YOU WILL LEARN

This chapter introduces the various types of antimicrobial drugs and how they work. As you study this chapter, you will:

- learn how antibiotics were discovered;
- review how medical professionals balance the benefits and risks of a drug;
- explore how antibiotics kill bacteria;
- investigate the kinds of drugs that are useful against viruses;
- examine how scientists test the antimicrobial activity of a drug;
- consider why using antibiotics leads to antibiotic resistance.

SECTIONS IN THIS CHAPTER

- History of the Discovery of Antimicrobial Drugs

- Important Considerations for Targeting Microbial Infections

- Mechanism of Action of Antimicrobial Drugs

- Determining the Level of Antimicrobial Activity

- Resistance to Antimicrobial Drugs

We are very fortunate to live in a time when many infectious diseases are treatable with drug therapy. It was not so long ago that humans were fairly helpless in the face of diseases such as pneumonia, tuberculosis, and cholera. Today, these diseases can be controlled with drugs called **antimicrobial drugs** or **chemotherapeutic agents**. Antimicrobial drugs are separated into two general categories depending on their source and target organisms. **Antibiotics**, the most familiar antimicrobial drugs, are produced by microorganisms and inhibit the growth of bacteria. Drugs that do not have a microbial source are referred to as **synthetic agents**. Synthetic agents that block viral replication may also be referred to as **antiviral drugs**.

> **REMEMBER**
> Antibiotics are produced by microbes and kill bacteria. Antibiotics do not work against viruses.

History of the Discovery of Antimicrobial Drugs

The first antimicrobial drug was a synthetic agent discovered in 1910 by the German scientist Paul Ehrlich. When preparing cells for microscopy, Ehrlich noticed that some dyes would stain microbes but not animal cells. Because the dyes stuck to the components of microbial cells, he thought that they might inhibit microbial growth. And, because they did not stick to the components of animal cells, he thought they would not harm them. Thus, if a microbe infected an animal, he thought the dyes might act like *magic bullets* that would attack microbial cells without harming the cells of the host. This concept, called **selective toxicity**, is at the heart of all drug therapy. Ehrlich tested hundreds of different chemicals for selectivity, finally finding salvarsan, an arsenic-containing drug that inhibits the growth of the syphilis spirochete without harming the animal host.

The next breakthrough in antimicrobial chemotherapy came in the 1930s with the discovery of **sulfa drugs**, the first drugs that were found to be effective against a wide range of bacteria. Gerhard Domagk at the Bayer Chemical Company in Germany discovered sulfa drugs. While testing synthetic dyes for their ability to affect streptococcal infections in mice, he discovered that the breakdown product of one of the dyes could inhibit the bacteria without harming the mice. The active chemical, *sulfanilamide*, was then used as the pattern for the synthesis of a number of effective sulfa drugs.

The first use of antibiotics to fight infection occurred when penicillin was used during the Second World War in the early 1940s. The discovery of the antimicrobial properties of penicillin had actually occurred in 1928 when a Scottish physician named Alexander Fleming noticed an unusual growth pattern on some of the plates in his lab. A mold had apparently contaminated some of his plates that contained growing bacteria. What Fleming found interesting about this was the presence of a clear zone around the mold in which no bacteria would grow. He realized that the fungus was producing a chemical that inhibited the growth of the bacteria. Fleming characterized the properties of this chemical, which he named *penicillin* in honor of *Penicillium notatum*, the mold that produced it. In the late 1930s, as the threat of war in Europe grew, a group of British scientists led by Howard Florey tested penicillin for its ability to control infections in humans. When these tests were successful, they developed methods for the large-scale production of penicillin. Penicillin was used to treat soldiers during World War II and became available to civilians after that.

The success of penicillin in treating certain types of infection encouraged scientists to look for more antibiotics. The microbiologist Selman Waksman screened 10,000 strains of soil bacteria looking for antibiotics. He and his students isolated several antibiotics, including the important drug *streptomycin* in 1943. For his discovery of streptomycin, which is produced by the actinomycete *Streptomyces griseus*, Waksman was awarded the Nobel Prize in 1952. Over the next 10 years, further screening of soil bacteria led to the discovery of many more antibiotics, including *chloramphenicol, tetramycin, neomycin,* and *tetracycline*.

Following the initial boom in the discovery of antibiotics, research into new antibiotics tapered off. The antibiotics that had been discovered in the 1940s and 1950s proved to be valuable assets to the medical community, and many dangerous infections were brought under control. Recently, however, there has been a shift in the battle of humans versus bacterial pathogens. Strains of bacteria have emerged with **antibiotic resistance** to our commonly used antibiotics. In fact, some strains exist that cannot be controlled with antibiotics at all. This has renewed interest in the discovery of antibiotics, and many approaches are being taken to try to develop new drugs that are effective against antibiotic-resistant strains.

Important Considerations for Targeting Microbial Infections

Although there are many chemicals in the world that would stop bacterial growth, most of these cannot be used to treat infections because they would also block the functioning of human cells. In order to use a drug against a microbial infection, it must act like one of Ehrlich's magic bullets. In other words, the drug must have **selective toxicity**, the ability to poison the microbe, without harming the host. Bacterial cells have many differences from human cells, such as the presence of a peptidoglycan wall, which can be used as targets for drugs. Because human cells do not have a peptidoglycan wall, any drug that affects that structure will have no effect on the human cell. Such a drug would have a high degree of selective toxicity.

> **REMEMBER**
> In order for an antimicrobial drug to be useful, it must have selective toxicity, the ability to target the microbe without harming the host.

Because of the issue of selective toxicity, infectious diseases caused by eukaryotes and viruses are difficult to treat with drug therapy. Eukaryotic pathogens, such as fungi and parasitic worms, have cell structures that are much more similar to those of human cells. It is therefore much harder to find drugs that are selectively toxic against these pathogens. It is even harder to find selective drugs for the treatment of viral infections. Viruses use human cells to reproduce and have very few characteristics of their own. To find an effective and safe drug for the treatment of viral infections, it takes a great deal of research. For some highly significant viruses, like HIV, herpes, and influenza, research has led to the discovery of helpful drugs. However, for most viral infections, we must rely solely upon our immune systems.

The degree of selective toxicity of a drug can be determined by its **therapeutic index**. The therapeutic index is the ratio of the toxic dose to the therapeutic dose (toxic dose : therapeutic dose). The toxic dose is the dose that would cause harm to the host cell. The therapeutic dose is the minimum dose that would benefit the host by inhibiting the microbe. For example, a drug that had a therapeutic index of 100 milligram per cubic centimeter to 1 milligram per cubic centimeter would be a safe and effective drug because it would have to be taken in a dose of 100 milligrams per cubic centimeter before it would harm your cells, yet a dose as small as 1 milligram per cubic centimeter would help fight the infection.

In addition to selective toxicity, there are many other considerations to be made when choosing a drug to fight an infection. One of these is the spectrum of antimicrobial activity of the drug; in other words, what types of microbes does the drug inhibit. Obviously, it is important to choose a drug that can inhibit the microbe causing the infection. Some antibiotics have a **broad spectrum** of activity and are effective against a wide variety of bacteria. For example, the broad-spectrum antibiotic tetracycline inhibits both gram-positive bacteria and many gram-negative bacteria. Other antibiotics have a **narrow spectrum** of activity and only inhibit a few types of bacteria (Table 23.1). For example, natural penicillins are only effective against gram-positive bacteria. Another consideration before administering an antimicrobial drug therapy is whether the drug will produce an allergic reaction. Finally, the fate of the drug in the body is considered, including how long the drug persists, how well it penetrates tissues, and whether it will be destroyed by the gastric juice if it is taken orally. All of these characteristics must be considered when choosing a drug to fight a particular microorganism.

TABLE 23.1. Spectrum of Activity of Selected Antibiotics

Antibiotic	Spectrum of Activity
Erythromycin	Narrow (gram-positive, mycoplasmas)
Tetracycline	Broad (gram-positive, gram-negative, rickettsias, chlamydias)
Streptomycin	Broad (gram-positive, gram-negative, mycobacteria)
Novobiocin	Narrow (gram-positive)
Neomycin	Broad (gram-positive, gram-negative, mycobacteria)
Ampicillin	Broad (gram-positive, some gram-negative)
Kanamycin	Broad (gram-positive, gram-negative, mycobacteria)
Penicillin	Narrow (gram-positive)

Mechanism of Action of Antimicrobial Drugs

To fight an infecting microbe effectively, a drug must stop a microbe from growing and then trigger its destruction. To stop a microbe from growing, a drug must stop its cells from functioning. To maintain themselves and multiply, the cells of microbes must be able to make proteins, copy their DNA, perform metabolic reactions, and control what enters and exits the cell. So, to stop a microbe, a drug needs to target one of these essential cell functions (Figure 23.1). Some of the commonly used antibiotics are summarized in Table 23.2.

> **REMEMBER**
> Antibiotics work by stopping bacteria from performing essential cell functions like cell wall synthesis and protein synthesis.

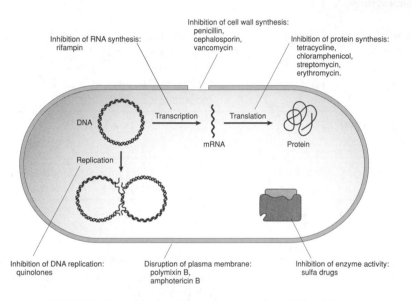

FIGURE 23.1. Mechanisms of action of antibacterial drugs.

INHIBITION OF ENZYME ACTIVITY

The **sulfa drugs**, or **sulfonamides**, were the first synthetic agents to be widely used against bacterial infections. They inhibit bacterial growth by interfering with an important metabolic pathway that synthesizes folic acid. For example, *sulfanilamide* is very similar in structure to para-aminobenzoic acid (PABA), a precursor to folic acid (Figure 23.2). If sulfanilamide is present in a bacterial cell, it inhibits the activity of an enzyme in the pathway that synthesizes folic acid. It does this by mimicking PABA and slipping into the active site of the enzyme, blocking PABA from entering. The bacterium cannot synthesize enough folic acid, which it needs for the synthesis of DNA. The host is not affected because humans get their folic acid from their food and do not need to make their own.

TABLE 23.2. Some Commonly Used Antibiotics
Organized by Chemical Structure

Antibiotic Classification	Example	Mode of Action and Use
Carbohydrate-containing compounds		
Aminoglycosides	Streptomycin	Inhibits protein synthesis; broad spectrum, including mycobacteria
C-Glycosides	Vancomycin	Inhibits cell wall synthesis; used to treat methicillin-resistant *Staphylococcus aureus*
Macrocyclic lactones		
Macrolides	Erythromycin	Inhibits protein synthesis; used as alternative to penicillin
Ansamycins	Rifampin	Inhibits mRNA synthesis; used to treat tuberculosis
Quinones and related compounds		
Tetracyclines	Tetracycline	Inhibits protein synthesis; broad spectrum, including chlamydias and rickettsias
Amino acid and peptide analogs		
Beta-lactam antibiotics	Penicillin	Inhibits cell wall synthesis; natural penicillins are narrow spectrum, some semisynthetic penicillins are broad spectrum
Peptide antibiotics	Bacitracin	Inhibits cell wall synthesis; used topically against gram-positive bacteria
Aromatic compounds		
Benzene derivatives	Chloramphenicol	Inhibits protein synthesis; broad spectrum, high toxicity
Quinolone compounds		
Fluoro-4-quinolones	Ciprofloxacin	Inhibits DNA synthesis; broad spectrum

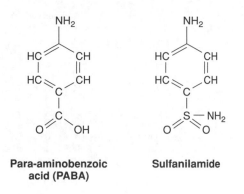

FIGURE 23.2. Comparison of PABA and sulfanilamide.

The enzyme inhibitor sulfanilamide is so similar in structure to PABA that it can fit into the active site of the enzyme that uses PABA as a substrate in folic acid synthesis.

INHIBITION OF CELL WALL SYNTHESIS

Because human cells do not make peptidoglycan, drugs that inhibit its formation are highly selective. This includes the **penicillins**, which block formation of the peptide cross-links that stabilize peptidoglycan structure (Chapter 4). The penicillins are a large group of antibiotics that all have a common structure called the **beta-lactam ring** (Figure 23.3), which is essential to the function of the penicillins. There are two main categories of penicillins: **natural penicillins** that are produced directly by fungi and **semisynthetic penicillins** that are chemically modified from natural penicillins (Figure

> **REMEMBER**
> Because peptidoglycan isn't found in human cells, drugs that target bacterial cell wall synthesis are highly selective.

23.3). Whereas natural penicillins like *penicillin G* are effective against only a narrow spectrum of gram-positive bacteria, semisynthetic penicillins like *amoxicillin* and *ampicillin* are broader spectrum and also affect some gram-negative bacteria.

Several other antibiotics also target the formation of peptidoglycan. **Cephalosporins** are similar to the penicillins in that they have a beta-lactam ring and block formation of peptide cross-links just like the penicillins. Cephalosporins tend to be broader spectrum and more effective against some antibiotic-resistant bacteria than the penicillins. A new class of antibiotics, the **monobactams**, has a single ring instead of the typical double-ring of the penicillins or cephalosporins. They block cell wall synthesis and are effective against certain gram-negative bacteria.

Two **polypeptide antibiotics**, *bacitracin* and *vancomycin*, are also effective against the peptidoglycan cell wall. Bacitracin blocks synthesis of peptidoglycan and is used topically to treat superficial infections. Vancomycin is an extremely important antibiotic for the treatment of antibiotic-resistant *Staphylococcus aureus*. In particular, it is the last line of defense against *MRSA*, or *methicillin-resistant Staphylococcus aureus*. It prevents peptidoglycan synthesis by preventing cross-linking of the glycan layers. Recently, strains of *Staphylococcus aureus* have emerged that are resistant to vancomycin, leaving us with no effective drug treatment for infections by these strains.

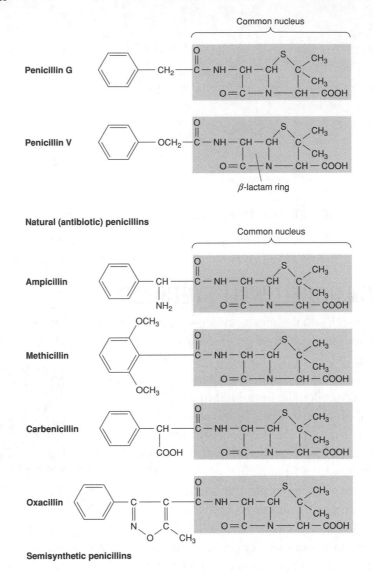

FIGURE 23.3. The structure of penicillins.

All penicillins have the common nucleus as part of their structure. The common nucleus contains the beta-lactam ring, which is essential for function of penicillin. Semisynthetic penicillins are chemically modified from natural penicillins to change the chemical group attached to the common nucleus.

The cell walls of mycobacteria, such as *Mycobacterium tuberculosis* and *Mycobacterium leprae*, are different from other bacteria because they have an additional waxy outer layer made primarily of mycolic acid. *Isoniazid (INH)*, which blocks the synthesis of mycolic acid, is used as part of the drug therapy given to tuberculosis patients. Another drug, *ethambutol*, blocks incorporation of mycolic acid into the cell and is also used against mycobacteria.

INHIBITION OF PROTEIN SYNTHESIS

No cell can survive for long without being able to synthesize proteins (Chapter 7), which makes the ribosome an attractive target for drug therapy. Although both eukaryotic and prokaryotic cells have ribosomes, eukaryotic and prokaryotic ribosomes differ slightly in their structure. Prokaryotic ribosomes are smaller and have slightly different protein and rRNA components than those of eukaryotic ribosomes (Chapter 4). The prokaryotic ribosome is referred to as a 70S ribosome, which is an indication of its size based on centrifugation (Chapter 4). When the subunits of the ribosome are measured separately, the large subunit is 50S, and the small subunit is 30S. Several antimicrobial drugs target the bacterial ribosome (Figure 23.4). Because the mitochondria inside eukaryotic cells have 70S ribosomes, there can be side effects of these types of drugs on the host.

> **REMEMBER**
> Prokaryotic ribosomes are slightly different in structure from eukaryotic ribosomes, so drugs that target the prokaryotic ribosome will harm bacterial cells faster than they harm human cells.

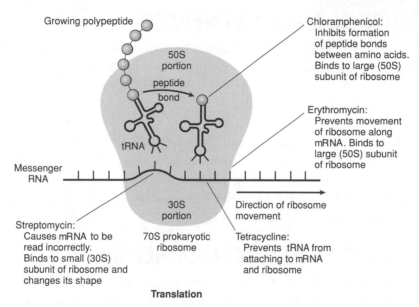

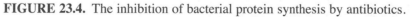

FIGURE 23.4. The inhibition of bacterial protein synthesis by antibiotics.

Drugs that target the bacterial ribosome include *chloramphenicol* (Figure 23.4), an inexpensive broad-spectrum antibiotic. Chloramphenicol binds to the large subunit (50S) of the ribosome and prevents formation of the peptide bond between amino acids. Because it has potentially serious side effects, chloramphenicol is only recommended for situations in which other alternatives are not available. The **macrolides** include *erythromycin* (Figure 23.4), which also binds to the large subunit of the ribosome and prevents the ribosome from moving along the mRNA. Because erythromycin cannot penetrate the walls of gram-negative bacteria, it is a narrow-spectrum antibiotic. Other macrolides, like *azithromycin* and *clarithromycin*, are broader spectrum. Two new groups of antibiotics, the **streptogramins** and the **oxazolidinones**, also block protein synthesis by binding to the 50S ribosomal subunit. These groups of antibiotics are effective against gram-positive bacteria, such as MRSA, that are resistant to other antibiotics.

Bacterial protein synthesis is also inhibited by the **aminoglycosides**, which include *streptomycin* (Figure 23.4). Streptomycin binds to the small subunit (30S) of the ribosome, changing its shape and causing errors in translation. *Neomycin* is an aminoglycoside that is used in topical preparations. *Gentamicin* is a valuable antibiotic for the control of *Pseudomonas*. Aminoglycosides are broad-spectrum but have been reported to cause kidney damage and hearing loss.

Tetracyclines are broad-spectrum antibiotics that penetrate tissues well. Because of this, they are useful in the treatment of the intracellular rickettsias and chlamydias, as well as for various other infections. *Tetracycline* (Figure 23.4) inhibits protein synthesis by preventing the attachment of tRNAs to the mRNA in the ribosome.

DISRUPTION OF PLASMA MEMBRANE

The basic structure of the plasma membrane of bacterial and eukaryotic cells is very much alike. Thus, the selective toxicity of most drugs that damage membranes is low, and they would not be useful in treating infections. There are some differences in membranes, however, particularly in the sterol components, which can differ among eukaryotes and are largely absent in bacterial membranes. One compound, *polymyxin B*, disrupts bacterial membranes by binding to phospholipids. It is used primarily in topical preparations such as triple-antibiotic ointment, which also contains bacitracin and neomycin. Several drugs, including *amphotericin B,* bind to the sterols that are found in fungal membranes, disrupting the membranes and lysing the cells. These drugs are used to treat a wide variety of fungal infections.

INHIBITION OF NUCLEIC ACID SYNTHESIS

Just as it is difficult to find selective drugs that target the plasma membrane, the similarity in nucleic acid synthesis of bacterial and eukaryotic cells makes it difficult to find selective drugs that affect the processes of DNA replication and transcription (Chapter 7). All cells have DNA and all cells copy their DNA using a DNA polymerase. Likewise, RNA is synthesized using an RNA polymerase. However, there are some small differences in the bacterial and eukaryotic versions of these enzymes, as well as in other enzymes associated with nucleic acid replication.

The **quinolones** inhibit an enzyme, DNA gyrase, which is required during DNA replication to release tension in the bacterial DNA molecule as the strands are separated but that is not found in eukaryotic cells. Derivatives of

> **REMEMBER**
> Although DNA replication is generally very similar between prokaryotic and eukaryotic cells, quinolones are selective because they target an enzyme involved in this process, called DNA gyrase, which is unique to bacterial cells.

quinolones called **fluoroquinolones** include the antibiotic *Ciprofloxacin*, which is used to treat infections by *Bacillus anthracis*, the causative agent of anthrax. Because both gram-positive and gram-negative bacteria have DNA gyrase, quinolones and fluoro-quinolones are broad-spectrum.

The **rifamycins**, which include *rifampin*, preferentially bind to bacterial RNA poly-merases, inhibiting transcription in bacterial cells and mitochondria. Because rifampin readily penetrates tissues, it is an important antibiotic for the treatment of mycobacter-ial infections such as tuberculosis and leprosy. This is because mycobacteria may repli-cate inside of tissues or macrophages where it is difficult for many drugs to penetrate.

TRIGGER CELL SUICIDE

One question that has remained about the activity of many antimicrobial drugs is why do they *kill* bacteria? For example, the penicillins block peptidoglycan synthesis, which will obviously stop bacterial cells from growing, but why don't existing cells survive and wait until the penicillin is gone? For some antibiotics, the answer seems to be that the activity of antibiotics triggers a normal cell suicide pathway. As cells age, or if they are damaged, programmed cell death, or **apoptosis**, occurs. For example, when a human fetus is developing in the uterus, it initially has webbing between the fingers. Apoptosis occurs between the fingers to remove the webbing. It seems that some antibiotics trigger apoptosis in bacterial cells by activating proteins called **autolysins** that dissolve the cell wall during apoptosis. Mutant pneumococci that can-not trigger the release of autolysins are resistant to killing by penicillin. In these mutant bacteria, penicillin has a bacteriostatic effect, whereas in nonmutated pneumo-cocci, it has a bacteriocidal effect. This evidence supports the idea that at least some antibiotics trigger cell death.

ANTIVIRAL DRUGS

Because viruses have very few characteristics of their own and use the host cell to replicate, it is very diffi-cult to find drugs that are selectively toxic for viruses. Most drugs that would affect the virus would also affect the host cell. After extensive research, a few

> **REMEMBER**
> It is difficult to find drugs that would be selective against viruses because they have very few proteins of their own, relying instead upon their host cells to provide what they need for replication.

drugs that are selectively toxic against certain viral infections have been identified. In some cases, unique viral enzymes that serve as drug targets have been identified. In others, drugs that affect viral enzymes more strongly than they do host enzymes have been found. A third way that selective toxicity was achieved was the discovery of a drug that was only activated in the presence of viral enzymes.

Several antiviral drugs block replication of viral nucleic acid by acting as **nucleotide analogs**, or mimics of the building blocks of DNA. These molecular mimics enter the active site of the DNA polymerase and even get incorporated into the growing chain of viral DNA. However, because their chemical structure is not actually that of a nucleotide, no more nucleotides can be added to the chain, and replication is stopped.

For example, the drug *zidovudine* (AZT, Figure 23.5), which is used to treat HIV infection, mimics the nucleotide thymine. The viral enzyme **reverse transcriptase** that makes DNA from RNA during HIV infection (Chapter 13) inserts the AZT into the growing DNA strand. However, AZT lacks a hydroxyl group at its 3' end, so no further nucleotides can be added (Chapter 7). AZT is selectively toxic because the viral reverse transcriptase has a higher affinity for the AZT than does human DNA polymerase.

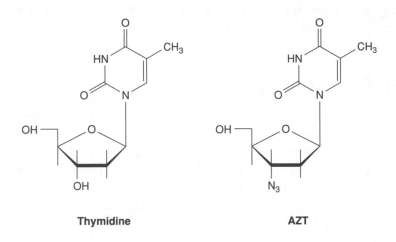

Thymidine   **AZT**

FIGURE 23.5. A comparison between AZT and thymidine.

AZT is used to treat HIV infection. It is so similar to thymidine that the viral DNA polymerase inserts AZT into newly replicating DNA. After AZT has been added to the DNA chain, further elongation of the chain is blocked.

Another example of a nucleotide analog is the drug *acycloguanosine (*acyclovir, Figure 23.6), which is used to treat herpes infections. Acyclovir actually mimics a precursor to the nucleotide guanine. In a cell infected with herpesvirus, a viral enzyme called thymidine kinase helps convert acyclovir into a mimic of guanine that is ready to be inserted into DNA (Figure 23.6). As the herpesvirus DNA polymerase replicates its DNA, the false nucleotide is added to the growing chain, preventing further elongation. Acyclovir is selectively toxic because it is only activated in cells that are infected by herpesvirus. In uninfected cells, the acyclovir remains in its prenucleotide form and causes no harm to the host cell.

Several drugs that target proteins unique to viruses have been discovered. For example, the antiflu drugs *zanamivir* (Relenza) and *oseltamivir* (Tamiflu) act as **neuraminidase inhibitors**. Neuraminidase is a viral enzyme on the surface of the flu virus that is essential in the release of virus from infected cells (Chapter 13). Likewise, the drugs *amantadine* and *rimantadine* inhibit the influenza protein M2 that is necessary for uncoating the influenza virus. Because they block aspects of the viral replication cycle of influenza, these drugs are supposed to shorten flu symptoms if taken very soon after exposure to the flu. Other examples of drugs that inhibit viral enzymes are the **protease inhibitors** used to treat HIV. Protease is a viral enzyme that is needed to cut viral proteins during maturation of the virus. Thus, protease inhibitors such as *indinavir* and *saquinavir* prevent completion of the viral replication cycle.

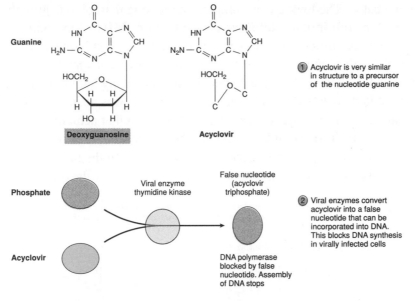

FIGURE 23.6. Structure and function of acyclovir.

Acyclovir is used to treat viral infections like herpes.

Interferons are cytokines that are made by virally infected cells and signal neighboring cells to protect themselves against viral attack (Chapter 18). Interferons have been produced as a pharmaceutical agent using recombinant DNA technology. The human gene sequence for interferon was inserted into bacteria, which then transcribed and translated the gene to produce high levels of the human protein. Interferons have been used to treat hepatitis, influenza, and herpes infections.

Determining the Level of Antimicrobial Activity

To determine the spectrum of activity of an antimicrobial drug, it must be tested against different microbes. A common method for doing this is to use the **disk-diffusion method** (Chapter 22), also called the *Kirby-Bauer test*. In this test, bacteria are first swabbed over the surface of a petri plate to create a lawn. Then, paper disks that contain known concentrations of antibiotics are placed on the plates. The plates are incubated and examined for clear zones, or *zones of inhibition*, around the disks. A clear zone around the disk indicates that the antibiotic does inhibit the growth of the bacterium (Figure 22.5). The size of the zone is compared to set standards for the drug and the bacterium is determined to be *susceptible, intermediate,* or *resistant*. Sensitivity tests like this one are particularly important to perform in cases of infection with microbes that might be resistant to antibiotics.

To determine the proper dose for an antimicrobial drug, **dilution susceptibility tests** are run. These tests require a set of plates or broths that contain different dilutions of one antibiotic. Each sample of media is inoculated with a standard number of bacteria and then incubated. The lowest concentration of the drug that inhibits growth (Figure 23.7) is called the **minimum inhibitory concentration (MIC)**. Any sample that did not show growth is further tested to see if the microbe was just inhibited or was actually killed. To do this, an inoculum is taken from each of the samples that did not show growth and is placed in fresh media that does not contain any antibiotic. If the microbe was only inhibited, it should grow in the new media. If it was actually killed, it will not grow (Figure 23.7). The lowest concentration of the drug that actually killed the microbe is called the **minimum bactericidal concentration (MBC)**. Sometimes, in cases of a life-threatening infection, the level of antibiotic in the blood of a patient must be monitored to ensure that it remains above the MIC for the drug.

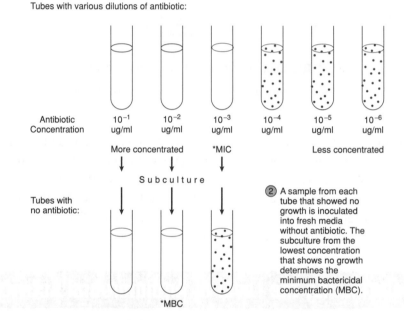

FIGURE 23.7. The broth dilution test.

Resistance to Antimicrobial Drugs

In recent years, there has been growing concern among medical personnel over the rise of antibiotic-resistant bacteria. Diseases that were once firmly under control like tuberculosis are becoming more difficult to manage owing to strains that are resistant to many of the standard drugs. Resistance is becoming especially common in hospital-acquired strains of staphylococci and enterococci. Some of these strains are resistant to all available antibiotics, including vancomycin. The hospital environment contributes to the rise and spread of antibiotic-resistant bacteria primarily because of the high use of antibiotics and contact between patients and staff.

TYPES OF RESISTANCE

There are two types of resistance in microbes: **inherent resistance** and **acquired resistance**. Some microbes have inherent resistance to antibiotics owing to the fundamental structure or chemistry of the cell. For example, natural penicillin is not effective against gram-negative bacteria because it cannot cross the outer membrane. However, the recent rise in antibiotic-resistant bacteria is caused by **acquired resistance**, where normally sensitive organisms become resistant. The engine that drives acquired resistance is *genetic change*. Bacteria are continuously mutating, either from spontaneous mutations during DNA replication or by the acquisition of genes from another bacterium (Chapter 7). This genetic change is random and generates many different strains of each species.

> **REMEMBER**
> Because of their outer membrane, gram-negative bacteria typically have greater inherent resistance to antimicrobial drugs than do gram-positive bacteria.

CAUSES OF ANTIBIOTIC RESISTANCE

The reason that the antibiotic-resistant strains are becoming so common is because of human behavior. When we use antibiotics, the strains that die fastest are those that are most sensitive. This leaves more resistant strains behind to multiply and take over. This process of *survival of the fittest* is called **selection**. A common confusion about antibiotic resistance is to think that somehow the antibiotic causes the bacterium to mutate against it. This is not so. A bacterium cannot control its genetic change. They change randomly, and those that can survive the antibiotic reproduce to make more antibiotic-resistant bacteria. Thus, *mutation plus selection* is causing the rise of antibiotic-resistant bacteria.

> **REMEMBER**
> Bacteria don't mutate in order to try to survive an antibiotic. Mutations are random—they happen all the time.

Human behavior has created the conditions that selected for antibiotic strains. Even normal use of antibiotics selects for resistance, but humans do not always use antibiotics responsibly. One misuse of antibiotics occurs when they are incorrectly prescribed for viral infections, which are not susceptible to antibiotics. Sometimes, people stop taking their antibiotics as soon as they feel better, rather than finishing the fully prescribed course. This only makes it easier for some resistant bacteria to escape destruction. Continuous low-dose antibiotics are sometimes prescribed for acne, ear infections, or immunosuppressed patients. Even though some of these may be legitimate, medically necessary uses of antibiotics, they do contribute to the overall problem. Antibiotics are often used for nonmedically necessary goals, such as the use of antibiotics in animal feed to increase animal yield. This encourages the development of resistant strains of bacteria that infect animals, many of which can also infect humans or which might share genes with human pathogens. In some

> **REMEMBER**
> Resistance in populations develops when bacteria are lucky enough to have a mutation that helps them survive a drug. These bacteria are more likely to reproduce, passing their resistance on to the next generation.

parts of the world, antibiotics can be obtained without a prescription, which encourages improper use. And if antibiotic-resistant strains arise in another part of the world, it is only a matter of a very short time before they will be spread around the world by global travel.

Bacteria can become resistant to antibiotics in an amazingly short period of time. In 1936, *Neisseria gonorrhoeae* was treatable with sulfa drugs. By 1942, most strains were resistant to sulfa drugs, and penicillin became the drug of choice. By 1956, penicillin-resistant strains emerged. Likewise, if you had a staph infection, caused by *Staphylococcus aureus*, in 1946, you would have been treated with penicillin. Today, most hospital strains of *S. aureus* are resistant to the natural penicillins, and some are resistant to the semisynthetic penicillins. Some strains of *MRSA* can only be treated with vancomycin. Even more dangerous strains are emerging: In 1996, the first case of *S. aureus* that showed intermediate resistance to vancomycin was reported, followed shortly by the first strain showing complete resistance in 2002. These strains are called *VISA* and *VRSA*, for *vancomycin intermediate-resistance Staphylococcus aureus* and *vancomycin-resistant Staphylococcus aureus*, respectively.

MECHANISMS OF ANTIBIOTIC RESISTANCE

Over time, bacteria have acquired many ways to evade the action of antibiotics. One method is to simply destroy the antibiotic. For example, many penicillin-resistant bacteria have the ability to make enzymes called **penicillinases** or **beta-lactamases** that destroy the beta-lactam ring of the penicillins and inactivate them (Figure 23.8). Another mechanism of resistance is to block the drug from reaching its target. This may be caused by inherent structures such as the outer membrane of gram-negative bacteria or waxy wall of mycobacteria that prevent certain drugs from entering the cell. Other bacteria have developed proteins in their plasma membranes that pump out a variety of antibiotics as quickly as they enter the cell. Bacteria that have these pumps often show **multidrug resistance** and include strains of *Mycobacterium tuberculosis, E. coli, Pseudomonas aeruginosa,* and *Staphylococcus aureus.* Another way that bacteria become resistant is if a mutation alters the target site so that the drug is no longer effective. For example, a single amino acid change in proteins of the ribosome could prevent erythromycin from binding. Resistance to drugs that block a metabolic pathway can arise if bacteria develop the ability to use another metabolic pathway. For example, bacteria that gain the ability to obtain folic acid from their food develop resistance to sulfa drugs. These methods all seem so ingenious that it is hard to believe the bacteria did not deliberately think them up. However, these antibiotic-resistant strains all resulted from random genetic change.

> **REMEMBER**
> Bacteria can evade antibiotics by destroying the antibiotic, blocking the drug from reaching its target, altering the target site, or switching to a different metabolic pathway.

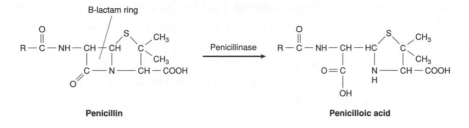

FIGURE 23.8. Inactivation of penicillin.

Bacteria that have enzymes called penicillinases or beta-lactamases can break the beta-lactam ring in penicillin, making it inactive.

OVERCOMING ANTIBIOTIC RESISTANCE

The rise of antibiotic-resistant bacteria is a serious medical issue, but it is not a hopeless one. In recent years in the United States, there has been a great deal of effort spent on educating the medical personnel and the public about the proper use of antibiotics. If antibiotics are used only when medically necessary, we might slow down the increase in antibiotic-resistant bacteria. In addition, many existing drugs are being chemically modified to generate new forms to which bacteria may yet be resistant. For example, the semisynthetic penicillin methicillin was originally resistant to the action of penicillinases. Semisynthetic macrolides called *ketolides* are being developed to cope with resistance in that group of drugs.

Another strategy for fighting antibiotic resistance is to use *drug combinations*. Sometimes two drugs in combination will have a *synergistic effect*; that is, the two drugs used together will have a better effect than expected based on the effect of both drugs when used alone. One example of this is the drug combination Augmentin. Augmentin consists of amoxicillin and a penicillinase-inhibitor, potassium clavulanate. When used against bacteria that make penicillinases, the potassium clavulanate inhibits the activity of the penicillinase. This allows the amoxicillin to survive and attack the peptidoglycan of the bacterial cell.

Finally, the recent increase in antibiotic-resistant bacteria has been a wake-up call for medical personnel and the pharmaceutical industry. Many efforts are being made to discover new antibiotics. Small, hollow proteins that insert preferentially into the membranes of bacterial cells and cause bacterial cell lysis are being examined. In some countries, bacteriophage are being tested for their ability to control bacterial infections. Other researchers are returning to the soil bacteria, searching among them for new compounds. To examine strains that will not grow in the lab, some groups are extracting DNA from environmental samples and then introducing that DNA into commonly used lab bacteria like *E. coli*. The recombinant bacteria are then examined for their ability to produce antimicrobial compounds. Other researchers are looking for ways to interfere with the interaction between bacteria and host cells that would block the disease process without promoting the proliferation of resistant bacteria. For example, many bacteria secrete harmful molecules that get transported into host cells. If the process for the secretion of these molecules could be blocked with drugs, then host cells would not be harmed. If the drug did not actually kill the bacteria, then there would be no selection for drug-resistant strains. These are just a few examples of some of the novel approaches that are being taken, some of which may yield new tools in the human struggle to control infection.

REVIEW EXERCISES FOR CHAPTER 23

1. Explain the concept of selective toxicity.

2. Use the concept of selective toxicity to explain why it is easier to discover drugs that are useful against bacteria than it is to discover drugs that are useful against eukaryotic or viral pathogens.

3. Distinguish between broad-spectrum and narrow-spectrum antibiotics.

4. State which cellular process the following drugs block: quinolones, rifampin, chloramphenicol, streptomycin, tetracycline, erythromycin, penicillin and its derivatives (-cillins), sulfur drugs, amphotericin B, polymyxin B.

5. Describe how semisynthetic penicillins differ from natural penicillins.

6. Distinguish between inherent and acquired resistance.

7. Describe how acquired resistance arises in individuals and in populations.

8. Describe the human behaviors that have contributed to the rise of populations of antibiotic resistant bacteria.

9. Describe the strategies that microbes use to avoid the action of antibiotics.

10. Describe the strategies people are employing to overcome antibiotic resistance in bacteria.

SELF-TEST

1. Which of the following drugs affects protein synthesis?

   A. Penicillin
   B. Vancomycin
   C. Sulfanilamide
   D. Chloramphenicol
   E. Ciprofloxacin

2. True or False. Chemicals that affect DNA replication are generally selectively toxic.

3. Drugs that affect the bacterial (70S) ribosome usually have side effects because

   A. eukaryotic cells have 70S ribosomes in their cytoplasm.
   B. eukaryotic cells have 70S ribosomes in their mitochondria.
   C. these drugs also affect nucleic acid synthesis.
   D. these drugs also affect the plasma membrane.
   E. None of the above is correct.

4. Which of the following statements best reflects the cause of antibiotic resistance?

   A. Antibiotics cause mutations that lead to resistance.
   B. Bacteria respond to antibiotics by changing their DNA so they can fight the drug.
   C. Antibiotics kill susceptible cells, leaving resistant cells to form new populations.
   D. Both A and B are correct.
   E. None of the above is correct.

5. Antiviral drugs target all of the following **except**:

   A. Uncoating
   B. DNA replication
   C. 70S ribosomes
   D. RNA synthesis
   E. Viral proteins

6. True or False. Antibiotic resistance can be overcome by using two drugs in combination.

7. True or False. The greater the therapeutic index, the more selective the drug.

8. True or False. Selective toxicity of antiviral drugs is increased if they require activation by viral enzymes.

9. True or False. Natural penicillins have a broader spectrum of activity than semi-synthetic penicillins.

10. Which of the following represents inherent resistance to natural penicillin?

    A. Production of penicillinase
    B. Presence of an outer membrane that excludes penicillin
    C. Presence of a gene for penicillinase
    D. Both A and C are correct.
    E. A, B, and C are correct.

11. Given the following data from a broth dilution test, what is the minimum inhibitory concentration and minimum bactericidal concentration of the drug?

Antibiotic Concentration (micrograms per milliliter)	Growth in Antibiotic	Growth in Subculture That Lacks Antibiotic
150	No	No
100	No	No
50	No	Yes
25	Yes	Yes
10	Yes	Yes

12. A 3-year-old boy is given amoxicillin for an ear infection. His parent is instructed to give the amoxicillin three times a day. The parent is very busy and regularly forgets to give the midday dose and stops giving the drug early. A few days after the drug has been stopped, the boy's symptoms return. Explain the probable cause of the relapse.

Answers

Review Exercises

1. Drugs that are selectively toxic target the pathogen without harming the host. For this to be possible, the drug must target features of the pathogen that are not present in the host cell.

2. Bacterial cells have several features that are different from those of eukaryotic cells. These features can be safely targeted by drugs; they include peptidoglycan cell walls, 70S ribosomes, and unique enzymes. Because these features are not present in eukaryotic cells, the drugs will not harm the human host. It is much harder to target eukaryotic pathogens because their cells are more similar to those of humans. Viruses have very few proteins of their own and so are even harder to target. They use resources in the human cell to multiply themselves.

3. Broad-spectrum antibiotics affect a wide range of bacteria. Narrow-spectrum antibiotics only affect a particular group.

4. Quinolones block DNA replication. Rifampin targets transcription. Chloramphenicol, streptomycin, tetracycline, and erythromycin all target protein synthesis at the 70S ribosome. Penicillin and its derivatives target peptidoglycan synthesis. Sulfur drugs target enzymes in the folic acid synthesis pathway. Amphotericin B targets fungal cell membranes, and polymixin B targets bacterial cell membranes.

5. Both semisynthetic penicillins and natural penicillins contain beta-lactam rings. The semisynthetic penicillins are created by chemically modifying the chemical groups that are attached to the beta-lactam ring. This tends to make them more resistant to penicillinases and therefore broader spectrum.

6. Inherent resistance is resistance attributable to cell structure or physiology that is common to that type of bacterium. Acquired resistance develops during the life of the cell through mutation or acquisition of genes by particular strains.

7. Acquired resistance arises in individuals through mutation. Mistakes during DNA replication or acquisition of genes from other cells can change bacteria to make them resistant to antibiotics. When antibiotics are used, susceptible cells die first, leaving resistant cells to multiply. Thus, when antibiotics are used, populations become resistant to them.

8. The simple use of antibiotics leads to antibiotic resistance because it kills susceptible bacteria and leaves resistant bacteria to multiply. Humans have made this situation worse by overusing and misusing antibiotics. For example, humans do not always finish prescriptions, or they take antibiotics for viral infections. Antibiotics are used in animal feed and as long-term measures to control acne and ear infections. The more antibiotics are used, the worse the problem gets.

9. Microbes may acquire the ability to make enzymes that deactivate antibiotics. Mutation can lead to changes in the target of antibiotics, rendering them ineffective. Some microbes produce pumps that are capable of pumping antibiotics out of the cell before reaching their target. If an antibiotic blocks a metabolic process, microbes may develop that have the ability to bypass that pathway.

10. One of the major initiatives in recent years is to control the use of antibiotics, preventing their misuse and overuse. Also, people are trying to discover and develop new antibiotics. When antibiotics are used in combination, they are sometimes more effective than either individual antibiotic alone.

Self-Test

1. D	5. C	9. F
2. F	6. T	10. B
3. B	7. T	
4. C	8. T	

11. MIC is 50 micrograms per milliliter. MBC is 100 micrograms per milliliter.

12. Because the antibiotic was not used properly, levels in the boy's system were probably not sufficient to kill all the bacteria. Those bacteria that were most resistant to the antibiotic may have been inhibited, but when the drug was stopped, these microbes were able to grow again.

Infectious Diseases of the Skin and Eyes

WHAT YOU WILL LEARN

This chapter presents some of the most important diseases of the skin and eyes. As you study this chapter, you will:

- review the structure of the skin and mucous membranes;
- learn the different types of skin lesions;
- compare the skin infections caused by *Staphylococcus* and *Streptococcus*;
- examine the causes of acne;
- explore the viral diseases of the skin;
- consider the effects of microbial infections of the eye.

SECTIONS IN THIS CHAPTER

- The Skin
- Mucous Membranes
- Skin Lesions
- Bacterial Diseases of the Skin
- Viral Diseases of the Skin
- Microbial Diseases of the Eye

Occasionally, our nonspecific and specific defenses fail, and microorganisms colonize the body and cause disease. Depending on their abilities and the way they are introduced into the body, various pathogenic microorganisms affect different areas of the body and cause different signs and symptoms. In the remaining chapters of this book (Chapters 24 through 29), we will consider the various major systems of the body and some of the most important and common diseases that occur as the result of infection by pathogenic microorganisms. When studying infectious disease, it is important to know the causative agent, the mode of transmission, the effect of the pathogen on the body, and the possible treatment for the disease. Whenever possible, this information is provided for the diseases presented in the following chapters.

Your skin is not only the largest organ in your body but also your best defense against invasion by pathogenic microorganisms. Most areas of the skin are dry and do not support microbial growth well. Antimicrobial substances in perspiration also help discourage microbial growth. The closely packed layers of cells create a physical barrier to microbial entry into the body. However, even small breaks in the skin can allow microbes to penetrate into the body.

The Skin

The skin (Figure 18.1) consists of two main layers, the **epidermis** and the **dermis**. The epidermis is the thin outer portion of the skin and is made up of several layers of epithelial cells. The outermost layer of the epidermis is the **stratum corneum**, which is made up of dead cells that contain the protein keratin. Keratin, the same protein that makes up your fingernails and hair, is a very tough protein. The presence of keratin waterproofs and protects the outermost layer of the skin.

Underneath the epidermis is the **dermis** (Figure 18.1), a thick layer that contains large amounts of connective tissue. Hair follicles, sweat glands, and oil gland ducts are found in the dermis, and these may sometimes allow microbes to access the body. The sweat glands produce *perspiration*, which inhibits microbial growth because it contains salt (Chapter 6), and the antimicrobial compounds *lysozyme* (Chapter 18) and *dermicidin*. The oil glands produce *sebum*, which is a mixture of lipids, protein, and salt. When the lipid component of sebum breaks down, it releases fatty acids that inhibit microbial growth by lowering the pH of the skin. However, both perspiration and sebum can provide nutrients that support microbial growth.

NORMAL MICROBIOTA OF THE SKIN

The microbes that can normally be found living on the skin are resistant to dry conditions and to relatively high concentrations of salt. Many of these microbes are gram-positive staphylococci and micrococci. The anaerobic rod-shaped bacterium, *Propionibacterium acnes,* may inhabit hair follicles and contribute to the development of acne. The aerobic, gram-positive rod, *Corynebacterium xerosis*, may be found on the skin surface. A yeast, *Pityrosporum ovale*, grows on oily skin secretions and may be responsible for dandruff.

Mucous Membranes

The body cavities of the gastrointestinal, genital, urinary, and respiratory tracts are lined with **mucous membranes**. Mucous membranes consist of layers of tightly packed epithelial cells that are attached to a layer of extracellular material called the *basement membrane*. However, unlike the skin, the epithelia of the mucous membranes are moist and contain cells that secrete thick mucus, which traps microorganisms. In the respiratory tract, trapped microbes can then be swept toward the throat by ciliated cells of the *ciliary escalator*. This makes it possible to eliminate the trapped microbes from the body by coughing. Other mucous membranes, like those that cover the eye, are protected from microbes as secretions wash over their surfaces.

SKIN LESIONS

> **REMEMBER**
> Skin lesions may result from an infection of the skin or an underlying systemic infection.

The appearance of lesions on the skin may indicate a skin infection or may be a sign of an underlying systemic infection. The type of lesion (Figure 24.1) that is present can provide valuable clues to the identity of the pathogen. Small, fluid-filled lesions are called **vesicles**. Vesicles that are larger than 1 centimeter are called **bullae**. **Macules** are flat, reddened lesions. **Papules** are raised lesions; if they contain pus, they are called **pustules**. A rash on the skin is called an **exanthem**. If a rash occurs on the mucous membranes, it is called an **enamthem**.

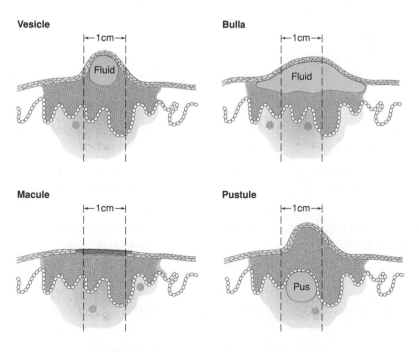

FIGURE 24.1. Types of skin lesions.

Vesicles are small, fluid-filled lesions. Bullae are larger, fluid-filled lesions. Macules are flat, reddened lesions. Pustules are raised lesions (papules) that contain pus.

Bacterial Diseases of the Skin

The two most significant genera of bacteria that cause skin disease are *Streptococcus* and *Staphylococcus*. Members of these genera are frequently found on the skin, and some produce enzymes and toxins that lead to the development of skin disease. Because of their importance, these two genera will be considered in detail here. In addition, we will consider acne because it is the most common skin disease. Other bacterial skin diseases are summarized in Table 24.1.

TABLE 24.1. Selected Bacterial Diseases of the Skin

Disease	Causative Agent	Most Common Mode of Transmission	Clinical Symptoms	Treatment
Scarlet fever	*Streptococcus pyogenes*	Droplet	Fever and rash that begins on chest and abdomen that spreads over body, peeling of skin	Penicillin
Impetigo	*Staphylococcus aureus* or *Streptococcus pyogenes*	Contact	Superficial skin infection, isolated pustules	Penicillin
Acne	*Propionibacterium acnes*	N/A (not thought to be a communicable disease)	Inflammatory lesions	Benzoyl peroxide, isotretinoin, azelaic acid
Erysipelas	*Streptococcus pyogenes*	Respiratory, contact	Fever and reddish patches on skin	Penicillin
Necrotizing fasciitis	*Streptococcus pyogenes*	Parenteral route	Rapid and extensive tissue destruction	Surgical removal of tissue, penicillin
Scalded skin syndrome	*Staphylococcus aureus*	Contact	Blisters, fever, large areas of peeling skin	Fluid rehydration, IV antibiotics (Usually Nafcillin, Oxacillin, or Vancothycin)

STREPTOCOCCUS

Members of the genus *Streptococcus* are gram-positive, oval cells that form chains. The species that cause human disease often produce **hemolysins**, enzymes that lyse red blood cells, and other cells. This characteristic, along with the presence of specific molecules, is used to categorize the streptococci. The ability to lyse red blood cells completely is called **beta hemolysis**, partial lysis is **alpha hemolysis**, and no lysis is **gamma hemolysis**. Serological testing for the presence of specific carbohydrates of the cell wall is also used to divide the streptococci into serological groups that are lettered from A to T.

> **REMEMBER**
> Group A streptococci (GAS) can cause a wide variety of skin diseases, including impetigo, erysipelas, and necrotizing fasciitis.

The most significant human pathogens among the streptococci are those that both are beta-hemolytic and fall into the serological category A. These bacteria are referred to as **Group A streptococci (GAS)** or by the species name *Streptococcus pyogenes*. Members of this group can be distinguished by the properties of a specific antigen, called the **M protein**, which is located on a layer of loose fibrils that surrounds the cell wall. The presence of the M protein helps Group A streptococci attach to mucous membranes and resist destruction by phagocytosis. Different strains of Group A streptococci cause a wide variety of human diseases, from strep throat to flesh-eating disease (necrotizing fasciitis). The disease they cause depends on the route of entry into the body and on the bacterial strain.

STAPHYLOCOCCUS

Members of the genus *Staphylococcus* are gram-positive, spherical cells that form clusters. The pathogenic and nonpathogenic strains are distinguished by the presence of an enzyme, *coagulase*, which triggers the formation of blood clots. Species that lack coagulase, such as *Streptococcus epidermidis*, are part of the normal microbiota of the skin and do not usually cause disease. Species that produce coagulase, the most important of which is *Staphylococcus aureus*, are significant opportunistic pathogens.

Most human diseases caused by staphylococci are caused by strains of *Staphylococcus aureus*. The severity of the disease that results from an infection by *S. aureus* depends on the route of entry into the body and on the enzymes and toxins produced by the bacteria. In addition to producing destructive enzymes and toxins, many strains of *S. aureus* are resistant to antibiotics, which makes treatment of these infections difficult. Most strains of *S. aureus* are resistant to penicillin, and some strains have also become resistant to methicillin. These bacteria, called methicillin-resistant *Staphylococcus aureus*, can only be treated with the antibiotic vancomycin. Recently, strains of *Staphylococcus aureus* that show intermediate or complete resistance to vancomycin have also emerged. We currently have no effective antibiotics for treating infections by these latter strains.

> **REMEMBER**
> Some staphylococci are normal microbiota of the skin, whereas others cause skin diseases such as impetigo and scalded skin syndrome.

ACNE

For many people, acne is a relatively benign disease, but for some it can lead to permanent scarring. The long-term effects of acne depend on the type of acne that a person has. **Comedonal acne** occurs when follicles become clogged with a mixture of sebum and dead skin cells. Because the sebum can no longer exit to the surface of the skin, it builds up, forming whiteheads, or **comedos**. If the sebum protrudes out of the skin, a blackhead, or **comedone** or *open comedo*, is formed. The black color is due to the oxidation of the sebum. This type of acne is not usually severe and can be treated successfully with topical agents.

If bacteria present on the skin begin to metabolize the sebum, **inflammatory acne** can result. The bacterium that is most commonly associated with this type of acne is *Propionibacterium acnes*. This bacterium metabolizes the sebum, producing fatty acids. This triggers an inflammatory response and leads to the formation of **pustules** and **papules**. This type of acne may be treated with drugs to block sebum formation or with antibiotics to discourage growth of *P. acnes*. Drugs that block sebum formation, such as Accutane, can cause birth defects and are therefore not advised for women who might become pregnant.

In some cases, acne develops into the most serious form, **nodular cystic acne**. In this case, inflamed pockets of pus called nodules or cysts, develop deep within the skin. This type of acne can lead to severe, permanent scarring and is usually treated with drugs that block sebum formation.

Viral Diseases of the Skin

Distinctive rashes or lesions accompany several systemic viral diseases, such as measles and chicken pox, on the skin. Several viral diseases of the skin are summarized in Table 24.2.

TABLE 24.2. Selected Viral Diseases of the Skin

Disease	Causative Agent	Most Common Mode of Transmission	Clinical Symptoms	Treatment and Prevention
Herpes	HSV-1 and HSV-2	Contact	Cold sores around the mouth or blisters in the genital area	Acyclovir may alleviate symptoms
Rubella	Rubella virus	Airborne	Mild rash that disappears quickly quickly	No treatment; vaccine is available

TABLE 24.2. (continued)

Disease	Causative Agent	Most Common Mode of Transmission	Clinical Symptoms	Treatment and Prevention
Measles	Measles virus	Airborne	Reddish macules that appear on face and then spread to trunk and extremities	No treatment; vaccine is available
Chicken pox	Human herpes virus 3 (varicellazoster)	Airborne	Vesicles on face, throat, and back	Acyclovir for immunocom-promised patients; vaccine is available
Shingles	Human herpes virus 3 (varicellazoster)	Airborne	Vesicles on face, scalp, chest, or in bands on waist	Acyclovir for immunocom-promised patients; vaccine is available
Warts	*Papillomavirus*	Contact	Rough, hard projections of skin	Physical removal by various methods, including freezing and treatment with acid

Microbial Diseases of the Eye

Most infections of the eye occur in the *conjunctiva*, the mucous membrane that lines the eyelids and covers the surface of the eyeball. Inflammation of the conjunctiva is called **conjunctivitis**, commonly known as pink eye or red eye. Bacteria, viruses, and protozoans can all cause conjunctivitis. The most common bacterial cause is infection by *Haemophilus influenzae*. The most common viral cause is infection by *Adenovirus*. A few more serious infections of the eye may result from infection by the bacteria *Neisseria gonor-rhoeae* and *Chlamydia trachomatis*. If eye infection by *Neisseria gonorrhoeae* occurs in newborns, it can result in blindness. For this reason, the eyes of newborns in the United States are treated with antibiotic eye drops. Eye infection by *Chlamydia trachomatis*, called **trachoma**, can lead to scarring of the conjunctiva. The rough scars on the conjunctiva, as well as under turned eyelashes, can cause abrasions on the cornea, damaging it and leading to blindness. Trachoma, which is spread by contact with contaminated fingers or fomites such as towels, is the most common infectious cause of blindness in the world today.

> **REMEMBER**
> Trachoma, caused by *Chlamydia trachomatis*, is the most common infectious cause of blindness in the world.

REVIEW EXERCISES FOR CHAPTER 24

1. What features of the skin protect against invasion by microbes? What features support microbial growth?

2. How are mucous membranes similar to the skin? How are they different?

3. Name and describe the different types of skin lesions.

4. What are Group A streptococci? What skin diseases do they cause?

5. What characterizes *Staphylococcus aureus?* What diseases does it cause?

6. What is the relationship between chicken pox and shingles?

7. What is conjunctivitis? What causes it?

SELF-TEST

1. Which disease results in extensive tissue destruction?

   A. Impetigo
   B. Acne
   C. Herpes simplex
   D. Chicken pox
   E. Necrotizing fasciitis

2. Which disease results in the formation of isolated pustules?

   A. Impetigo
   B. Erysipelas
   C. Chicken pox
   D. Herpes
   E. Necrotizing fasciitis

3. Which disease is caused by *Streptococcus pyogenes*?

   A. Toxic shock syndrome
   B. Acne
   C. Erysipelas
   D. Herpes
   E. Chicken pox

4. Which disease results in the formation of vesicles?

   A. Erysipelas
   B. Impetigo
   C. Necrotizing fasciitis
   D. Chicken pox
   E. Herpes

5. Which disease is caused by *Staphylococcus aureus*?

   A. Herpes
   B. Shingles
   C. Folliculitis
   D. Erysipelas
   E. Acne

6. Which disease is caused by a virus?

   A. Acne
   B. Impetigo
   C. Shingles
   D. Erysipelas
   E. Necrotizing fasciitis

7. Which organism can cause blindness?

   A. *Haemophilus influenzae*
   B. *Neisseria gonorrhoeae*
   C. *Chlamydia trachomatis*
   D. Both A and C are correct.
   E. Both B and C are correct.

8. Small fluid-filled lesions are called

   A. bullae.
   B. vesicles.
   C. macules.
   D. papules.
   E. pustules.

9. Acne that can result in severe permanent scarring is called

   A. inflammatory acne.
   B. comedonal acne.
   C. nodular cystic acne.
   D. scarring acne.
   E. None of the above are correct.

Answers

Review Exercises

1. The skin consists of many, tightly packed layers of cells, the uppermost layer of which contains a tough protein called keratin. These features, combined with the dryness, slight acidity, and normal microbiota of the skin, all protect against microbial invasion. However, natural openings in the skin such as hair follicles may provide an entry for microbes. In addition, some microbes can utilize the nutrition that is available in secretions such as sebum.

2. Like the skin, the mucous membranes consist of multiple layers of tightly packed cells. However, unlike the skin, the mucous membranes are moist and are not keratinized, which makes them easier to penetrate. Mucous membranes are protected by the presence of thick mucus, which traps microbes.

3. Small, fluid-filled lesions are called vesicles. Vesicles that are larger than 1 centimeter are called bullae. Macules are flat, reddened lesions. Papules are raised lesions; if they contain pus, they are called pustules.

4. Group A streptococci are a group of bacteria within the genus *Streptococcus*. They are gram-positive, oval cells that form chains. They are characterized by the presence of the A antigen on their cell walls. They produce hemolysins and other enzymes that help them invade the body. Many produce toxins that cause disease symptoms. They cause many skin diseases, erysipelas, necrotizing fasciitis, and impetigo.

5. *Staphylococcus aureus* is a species of *Staphylococcus* that produces the enzyme coagulase. It also may produce other enzymes and toxins that help it invade the body and cause disease. It causes a number of skin diseases, including impetigo, folliculitis, and scalded skin syndrome.

6. Chicken pox and shingles are both caused by the same virus, human herpes virus 3. During the initial infection by the virus, some viral particles become dormant in nerve cells. Later in life, or when the immune system is compromised, these viral particles may become active, causing vesicles to form along the bands of superficial nerves in the back.

7. Conjunctivitis is inflammation of the conjunctiva, the mucous membrane that lines the eyelids and covers the eye. It can be caused by a variety of bacterial, viral, and protozoan pathogens, including *Haemophilus influenzae*, Adenovirus, *Neisseria gonorrhoeae*, and *Chlamydia trachomatis*.

Self-Test

1. E	4. D	7. E
2. A	5. C	8. B
3. C	6. C	9. C

Infectious Diseases of the Respiratory System

WHAT YOU WILL LEARN

This chapter introduces the infectious diseases of the respiratory system. As you study this chapter, you will:

- review the structure of the respiratory system;
- explore the types of diseases caused by bacterial infection of the respiratory system;
- be introduced to some viral diseases of the respiratory system.

SECTIONS IN THIS CHAPTER

- The Respiratory System
- Bacterial Diseases of the Respiratory System
- Viral Diseases of the Respiratory System

If you were to list the illnesses you have had in the past year, the common cold probably tops the list as the most frequent offender. Thus, it is probably no surprise that respiratory infections are the most common type of infections in humans. Every time we take a breath, we draw microbes into our respiratory tract. Our defenses prevent most of these microbes from becoming established, but a few of them are able to colonize either the upper or lower respiratory tract. Diseases of the upper respiratory tract include sore throats, ear infections, and the common cold. Diseases of the lower respiratory tract include whooping cough, tuberculosis, and pneumonia. In addition to these diseases that are specific to the respiratory system, several systemic diseases also gain entry to the body through the respiratory tract. These diseases include chicken pox (Chapter 24), measles (Chapter 24), and mumps (Chapter 26) and will be discussed in other chapters.

The Respiratory System

Although the respiratory system is one connected system, the upper portion, or **upper respiratory system**, is typically exposed to a great many more microbes than is the lower section, or **lower respiratory system**. The upper respiratory system consists of the nose and throat (**pharynx**), as well as the associated ducts and tubes from the eyes, sinuses, and ears (Figure 25.1). The ducts from the sinuses and tear-forming (**lacrimal**) apparatus both empty into the nasal cavity. The auditory (**eustachian**) tubes from the middle ear empty into the upper part of the throat. Because the nose, throat, eyes, sinuses, and ears are all connected, infection can sometimes spread from one area of the upper respiratory system to another.

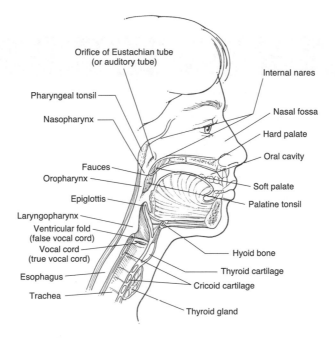

FIGURE 25.1. The upper respiratory system.

The lower respiratory system consists of the wind pipe (**trachea**), voice box (**larynx**), and the lungs (Figure 25.2). The lungs are lined with a double layer of membrane called the **pleura** and contain bronchial tubes and air sacs (**alveoli**). Blood is brought to the lungs and passes by the alveoli through capillary beds (Figure 25.2). As we breathe and fill our alveoli with air, oxygen from the air is exchanged for the carbon dioxide in the blood, supplying our body with oxygen and removing the carbon dioxide waste.

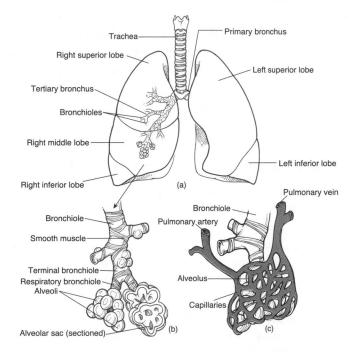

FIGURE 25.2. The lower respiratory system.

Both the upper and lower respiratory systems have defenses against infection. Both are lined with mucous membranes and are therefore protected by mucus, epithelial layers, and IgA antibodies (Chapter 18). In the upper respiratory tract, hairs in the nose help filter out inhaled dust particles, and the normal microbiota suppresses the growth of pathogens. The upper portion of the throat contains the tonsils, which are masses of lymphatic tissue that can help fight some infections. The lower respiratory tract is protected from invasion by the **ciliary escalator**, ciliated cells that move microbe-containing mucus toward the upper portion of the throat where it can be eliminated from the body by coughing. In addition, **alveolar macrophages** can destroy invading microbes that reach the lungs. The defenses of the respiratory system are effective barriers that usually prevent any microbes from even reaching the lower respiratory tract.

> **REMEMBER**
> The lower respiratory system is protected from infection by the ciliary escalator and shouldn't normally contain any microbes.

Bacterial Diseases of the Respiratory System

Today, the bacteria that most commonly infect the respiratory system are *Streptococcus pyogenes, Streptococcus pneumoniae,* and *Haemophilus influenzae* (Table 25.1). These bacteria typically cause inflammation of the tissue they infect and can cause a number of common ailments. These include sore throats, or **pharyngitis**, that result from inflammation of the mucous membranes of the pharynx, and **tonsillitis**, inflammation of the tonsils. Inflammation of the larynx, or **laryngitis**, can result in loss of speech, and inflammation of the sinuses, called **sinusitis**, can lead to headaches and heavy discharge of nasal mucus. When bacterial infection spreads to the middle ear, it results in **otitis media**, or earache. The pain in the ear is caused by the pressure caused by the build up of pus. If the bacteria penetrate the lower respiratory tract, they may cause inflammation of the bronchi, resulting in **bronchitis** or **bronchiolitis**.

TABLE 25.1. Selected Bacterial Diseases of the Respiratory System

Disease	Causative Agent	Most Common Mode of Transmission	Clinical Symptoms	Treatment
Strep throat (streptococcal pharyngitis)	*Streptococcus pyogenes* causes strep throat	Droplet	Inflammation of the throat, fever	Penicillin
Earache (otitis media)	*Streptococcus pneumoniae, Haemophilus influenzae, Moraxella catarrhalis, Streptococcus pyogenes*	Droplet (transmitted to Eustachian tubes following a respiratory infection)	Pain and pressure in the ear, fever	Amoxicillin; vaccines are available for *Streptococcus pneumoniae* and *Haemophilus influenzae* type b
Diphtheria	*Corynebacterium diphtheriae*	Droplet	Sore throat and fever followed by malaise and swelling of the neck, formation of membrane in the throat	Penicillin and erythromycin in conjunction with antitoxin; vaccine is available
Bacterial pneumonia	*Streptococcus pneumoniae*	Droplet	High fever, difficulty breathing, chest pain	Penicillin; vaccine is available
Haemophilus influenzae pneumonia	*Haemophilus influenzae*	Droplet	High fever, difficulty breathing, chest pain	Cephalosporin or erythromycin plus penicillin; vaccine is available

TABLE 25.1. (continued)

Disease	Causative Agent	Most Common Mode of Transmission	Clinical Symptoms	Treatment
Tuberculosis	*Mycobacterium tuberculosis*	Droplet	Weight loss, coughing (may contain blood), fatigue	Multidrug therapy including isoniazid, rifampin, and pyrazinamide. Multidrug resistant TB (MDRTB) requires second-line antibiotics and is harder to treat
Whooping cough	*Bordetella pertussis*	Droplet	General cold symptoms followed by violent coughing spasms	Erythromycin; vaccine is available

   Infection of the respiratory system by bacteria can lead to more serious conditions such as pneumonia and tuberculosis. Inflammation of the alveoli leads to **pneumonia**, which can be caused by several bacteria, as well as by viruses. Tuberculosis is a chronic, potentially fatal disease that begins in the lungs but can spread through the body via the blood or the lymph. Several other potentially serious infections of the respiratory system are now much more rare as a result of vaccination. These include diphtheria, whooping cough (**pertussis**), and **epiglottitis**, a life-threatening inflammation of the epiglottis. Several respiratory infections caused by bacteria are summarized in Table 25.1. *Streptococcus pyogenes*, which causes strep throat and many other diseases, is discussed in Chapter 24.

> **REMEMBER**
> Pneumonia refers to inflammation of the alveoli, which can be caused by a variety of bacterial and viral pathogens.

Viral Diseases of the Respiratory System

Viruses are the most common cause of respiratory infections in humans. Over 200 different viruses, only some of which have been identified, cause the common cold. The viruses that are most often identified as causing the common cold are rhinoviruses and coronaviruses. Influenza is also a very common viral respiratory infection, and can sometimes cause serious, life-threatening disease. The ability of influenza to mutate rapidly makes it especially dangerous (see Chapter 17). In infants, respiratory syncytial virus (RSV) is probably the most frequent cause of viral respiratory infection. It can range in severity from producing symptoms similar to the common cold or causing ear infections (**otitis media**) to more severe effects like **pneumonia** and even death. These viral diseases are summarized in Table 25.2.

TABLE 25.2. Selected Viral Diseases of the Respiratory System

Disease	Causative Agent	Most Common Mode of Transmission	Clinical Symptoms	Treatment
Common cold	Coronaviruses, rhinoviruses	Droplet	Coughing, sneezing, runny nose	None
Respiratory syncytial virus (RSV)	Respiratory syncytial virus	Direct contact	Stuffy nose, cough, wheezing, sometimes followed by fever, severe coughing, and difficulty breathing	Ribavirin reduces severity of symptoms, vaccine is in development
Influenza	Influenzavirus	Droplet	Fever, chills, headache, muscular aches, vomiting in young children	Antiviral drugs oseltamivir or zanamivir may alleviate symptoms; vaccine is available

REVIEW EXERCISES FOR CHAPTER 25

1. What features of the upper respiratory system protect the body against infection?

2. What features of the lower respiratory system protect the body against infection?

3. How is otitis media connected to respiratory diseases?

SELF-TEST

1. The most common mode of transmission for respiratory diseases is

   A. vehicle (airborne).
   B. direct (droplet).
   C. contact.
   D. vehicle (foodborne).
   E. vehicle (waterborne).

2. Which of the following is the causative agent of whooping cough?

   A. *Streptococcus pyogenes*
   B. *Bordetella pertussis*
   C. *Haemophilus influenzae*
   D. *Corynebacterium diphtheriae*
   E. *Mycobacterium tuberculosis*

3. Which of the following is the causative agent of acute pharyngitis?

   A. *Streptococcus pyogenes*
   B. *Bordetella pertussis*
   C. *Haemophilus influenzae*
   D. *Corynebacterium diphtheriae*
   E. *Mycobacterium tuberculosis*

4. True or False. Influenza is not caused by *Haemophilus influenzae*.

5. All of the following may cause symptoms similar to the common cold **except**:

   A. Rhinoviruses
   B. Coronaviruses
   C. *Bordetella pertussis*
   D. *Haemophilus influenzae*
   E. All of the above are correct.

6. True or False. The ciliary escalator protects the lower respiratory tract from infection.

Answers

Review Exercises

1. Hairs in the nose screen out large dust particles that may contain microbes. The mucous membranes have multiple layers of cells, mucus that traps microbes, and IgA antibodies that defend against microbes. The tonsils are lymphatic tissue that help screen out invading microbes. The normal microbiota of the upper respiratory tract also help prevent pathogens from establishing themselves.

2. The lower respiratory tract is also lined with mucous membranes (see answer to question 1). In addition, the ciliary escalator moves mucus that contains trapped microbes upward so that it can be eliminated from the throat. Alveolar macrophages in the lungs destroy microbes by phagocytosis.

3. Because the eustachian tubes drain into the upper part of the throat, it is possible for pathogens in the respiratory tract to migrate into the ear and cause an ear infection.

Self-Test

1. B   2. B   3. A   4. T   5. E   6. T

26

Infectious Diseases of the Digestive System

WHAT YOU WILL LEARN

This chapter presents the infectious diseases of the digestive system. As you study this chapter, you will:

- review the structure of the digestive system;
- explore the diseases caused by bacterial infection of the digestive system;
- learn about viruses that infect the digestive system;
- discover protozoans that can cause diseases of the digestive system.

SECTIONS IN THIS CHAPTER

- The Digestive System
- Bacterial Diseases of the Digestive System
- Viral Diseases of the Digestive System
- Protozoan Diseases of the Digestive System

Infection by intestinal pathogens is usually by ingestion of contaminated food or water. For most people in the United States, these infections result in unpleasant but short bouts of disease that may include nausea, vomiting, and diarrhea. However, for many people, these encounters are more serious and may result in hospitalization. In the United States, the CDC estimates that there are 76 million cases of foodborne disease each year. Of these, about 325,000 cases result in hospitalization, and 5000 result in death. Worldwide, the effects of intestinal pathogens are even more severe. Each year, diarrheal diseases account for the death of over two million people, most of whom are children under the age of five.

The high incidence of diarrheal diseases worldwide occurs primarily in countries that have poor sanitation. Transmission of intestinal pathogens occurs when fecal matter from an infected person or animal is introduced into the food or water supply. Drinking water can become contaminated if effective methods of sewage treatment are not available. Foodborne disease can occur by a variety of mechanisms, including the harvesting of shellfish from contaminated water, the spraying of crops with water or fertilizer that contain pathogens, or the handling of food by people who do not wash their hands. When pathogens are spread by the ingestion of food or water that has been contaminated with fecal matter, it is referred to as the **fecal–oral route of transmission**.

The Digestive System

The digestive system consists of the **gastrointestinal (GI) tract** and the **accessory structures** (Figure 26.1). The GI tract, which includes the **mouth**, throat (**pharynx**), food tube (**esophagus**), **stomach**, **small intestine**, and **large intestine**, is a muscular tube that is almost 30 feet in length! The accessory structures include the **teeth**, **tongue**, **salivary glands**, **liver**, **gallbladder**, and **pancreas**. The GI tract and accessory structures work together to break down food particles and enable the body to absorb water and nutrients from the food. Breakdown begins in the mouth and continues in the stomach. Absorption of nutrients primarily occurs in the small intestine. Water, vitamins, and some nutrients are absorbed in the large intestine. The remaining waste material is then released as feces.

The digestive system has many defenses against pathogenic organisms. The exterior of the digestive system is covered by mucous membranes that consist of multiple layers of cells, produce mucus, and are protected by phagocytes, IgA antibodies, and lysozyme (Chapter 18). The mouth and large intestines are protected by their normal microbiota, which inhibit colonization by other microbes. The low pH of the stomach and bile released from the gallbladder both prevent microbial growth. Any microbes that survive the stomach are then usually moved quickly through the small intestine, which prevents their establishment. For a microbe to cause a gastrointestinal disease, it must overcome these defenses.

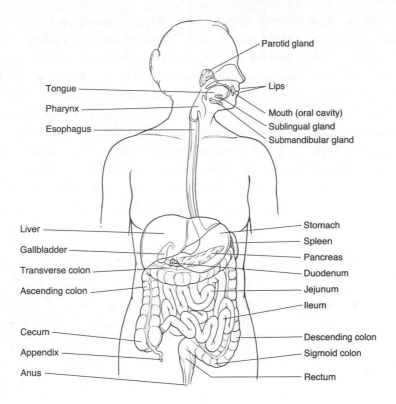

FIGURE 26.1. The digestive system.

Bacterial Diseases of the Digestive System

Diseases of the digestive system can be caused either by **infection**, which is actual colonization and multiplication of pathogenic bacteria, or by **intoxication**, which is ingestion of toxins that were produced by pathogenic bacteria. Disease resulting from infection usually occurs from 1 to 3 days after ingestion of contaminated food or water and may be accompanied by a fever, which is part of the body's response to the infecting bacterium. Disease caused by intoxication occurs more rapidly, often within a few hours after ingesting contaminated food or water, and is usually not accompanied by a fever. Both infection and intoxication can result in diarrhea, abdominal cramps, nausea, and vomiting. Severe diarrhea that contains blood or mucus is called **dysentery**. Any disease that causes inflammation of the stomach or intestinal mucosa is referred to as **gastroenteritis**.

> **REMEMBER**
> Illness due to ingestion of a toxin, called intoxication, is generally very rapid because the toxin is already present in the food, and growth of bacteria is not necessary to produce signs and symptoms.

Diseases of the digestive system range from mild gastroenteritis to life-threatening diseases such as cholera and typhoid. Some of the most important intestinal pathogens are gram-negative, rod-shaped bacteria that belong to the family Enterobacteriaceae. These include *Shigella,* which causes a gastroenteritis called shigellosis; *Salmonella,* which can cause gastroenteritis or typhoid fever; and *Escherichia coli*, which may cause mild gastroenteritis or severe, bloody diarrhea. *Staphylococcus aureus*, *Campylobacter*, and *Clostridium perfringens* may also cause gastroenteritis. Cholera is caused by a gram-negative, curved bacterium called *Vibrio cholerae*. Other diseases of the digestive system include cavities (dental caries) caused by *Streptococcus mutans*, periodontal disease caused by *Porphyromonas* sp., and peptic ulcers caused by *Helicobacter pylori*. These diseases are summarized in Table 26.1.

TABLE 26.1. Selected Bacterial Diseases of the Digestive System

Disease	Causative Agent	Most Common Mode of Transmission	Clinical Symptoms	Treatment and Prevention
Cavities (dental caries)	*Streptococcus mutans*	N/A	Softening of tooth enamel, holes in tooth enamel	Fluoride, physical removal of plaque
Gum disease (periodontal disease)	*Porphyromonas*	N/A	Accumulation of plaque, inflammation of the gum	Tetracycline, surgery
Staphylococcal food poisoning	*Staphylococcus aureus*	Vehicle (foodborne)	Rapid onset of nausea, vomiting, and diarrhea	Fluid and electrolyte replacement
Shigellosis	*Shigella* sp.	Fecal–oral	Diarrhea, fever, abdominal cramps, dysentery	Fluid and electrolyte replacement, fluoroquinolones
Salmonellosis	*Salmonella enterica*	Fecal–oral	Headache, chills, vomiting, and diarrhea	Fluid and electrolyte replacement
Typhoid fever	*Salmonella typhi*	Fecal–oral	High fever, headache, diarrhea	Fluid and electrolyte replacement, cephalosporin
Traveler's diarrhea	Enterotoxigenic *E. coli*	Fecal–oral	Watery diarrhea	Fluid and electrolyte replacement

TABLE 26.1. (continued)

Disease	Causative Agent	Most Common Mode of Transmission	Clinical Symptoms	Treatment and Prevention
Pediatric diarrhea, traveler's diarrhea	Enteroinvasive E. coli	Fecal–oral	Diarrhea, fever, dysentery	Fluid and electrolyte replacement
Pediatric diarrhea, traveler's diarrhea	Enteropathogenic E. coli	Fecal–oral	Diarrhea, dysentery	Fluid and electrolyte replacement
Bloody diarrhea (hemorrhagic colitis) and kidney failure (hemolytic uremic syndrome)	Enterohemor-rhagic E. coli	Fecal–oral	Bloody diarrhea, blood in urine, kidney failure	Fluid and electrolyte replacement, dialysis may be required
Cholera	Vibrio cholerae O:1 and O:139	Fecal–oral	Watery diarrhea	Fluid and electrolyte replacement
Campylobacter gastroenteritis	Campylobacter jejuni	Fecal–oral	Fever, abdominal cramps, diarrhea	Fluid and electrolyte replacement
Clostridium perfringens gastroenteritis	Clostridium perfringens	Vehicle (foodborne)	Diarrhea	Fluid and electrolyte replacement
Peptic ulcers	Helicobacter pylori	Fecal–oral	Stomach pain	Bismuth plus various antibiotics, including amoxicillin and tetracycline

Viral Diseases of the Digestive System

REMEMBER
Viruses that cause hepatitis can enter the body through the digestive system or through the parenteral route.

Several serious viral infections enter the body through the digestive system and affect the organs associated with this system. The most significant of these viral infections results in inflammation of the liver, which is called **hepatitis**. The common childhood disease, **mumps**, occurs when a virus causes swelling of the salivary glands. In addition, several viral species cause gastroenteritis. The characteristics of several viral diseases of the digestive system are summarized in Table 26.2.

TABLE 26.2. Selected Viral Diseases of the Digestive System

Disease	Causative Agent	Most Common Mode of Transmission	Clinical Symptoms	Treatment and Prevention
Mumps	Mumps virus	Droplet	Fever, headache, muscle ache, and swelling of the glands close to the jaw	Vaccine is available
Hepatitis	Hepatitis A virus	Fecal–oral	Usually subclinical, may cause fever, headache, malaise, jaundice	Immune globulin can be given to people at high risk for infection; vaccine is available for high-risk groups
Hepatitis	Hepatitis B virus	Parenteral	Often subclinical, may progress to severe liver damage and chronic liver disease	Alpha-interferon for chronic infections; vaccine is available for high-risk groups
Hepatitis	Hepatitis C virus	Parenteral	Often subclinical, likely to progress to severe liver damage and chronic liver disease	Alpha-interferon, ribavirin
Hepatitis	Hepatitis D virus	Parenteral	Severe liver damage, death	None
Hepatitis	Hepatitis E virus	Fecal–oral	Usually subclinical, may kill pregnant women	None
Viral gastroenteritis	*Rotavirus* and *Norovirus*	Fecal–oral	Vomiting, watery diarrhea, fever, abdominal pain	Fluid and electrolyte replacement, a vaccine is available for rotavirus

Protozoan Diseases of the Digestive System

Protozoa are eukaryotic, and their cells have many similarities to the cells of humans. Because of this, infection with protozoa can be very difficult to treat. Infection by protozoa commonly occurs by ingestion of **cysts**, special resistant cells formed by the protozoa that survive well in the environment. Several diarrheal diseases caused by protozoa are summarized in Table 26.3.

> **REMEMBER**
> Eukaryotic amoebae, like *Giardia*, that cause gastrointestinal disease are typically difficult to treat because their cells are so similar to ours.

TABLE 26.3. Selected Protozoan Diseases of the Digestive System

Disease	Causative Agent	Most Common Mode of Transmission	Clinical Symptoms
Giardiasis	*Giardia intestinalis*	Fecal–oral	Prolonged diarrhea, nausea, weakness, flatulence, weight loss, abdominal cramps
Cryptosporidiosis	*Cryptosporidium parvum*	Fecal–oral	Watery diarrhea, weight loss, fever, abdominal cramps, nausea, vomiting
Cyclosporidiosis	*Cyclospora cayetanensis*	Fecal–oral	Watery diarrhea, loss of appetite, weight loss, flatulence, stomach cramps, nausea, vomiting
Amebiasis	*Entamoeba histolytica*	Fecal–oral	Diarrhea, abdominal cramping
Amebic dysentery	*Entamoeba histolytica*	Fecal–oral	Bloody diarrhea, abdominal cramps, fever

REVIEW EXERCISES FOR CHAPTER 26

1. Compare and contrast the characteristics of gastrointestinal disease that result from infection versus those that result from intoxication.

2. Summarize the defenses of the digestive system that protect against disease.

3. Compare and contrast salmonellosis and typhoid fever.

4. Compare and contrast the following: giardiasis, cryptosporidiosis, cyclosporidiosis, and amebic dysentery.

SELF-TEST

1. True or False. All cases of hepatitis are caused by the same virus.

2. Which of the following is mostly likely to cause chronic liver disease?

   A. Hepatitis A virus
   B. Hepatitis B virus
   C. Hepatitis C virus
   D. Hepatitis D virus
   E. Hepatitis E virus

3. Which of the following is most commonly associated with periodontal disease (gum disease)?

   A. *Streptococcus mutans*
   B. *E. coli*
   C. *Porphyromonas* sp.
   D. *Campylobacter jejuni*
   E. *Helicobacter pylori*

4. Which of the following is **not** commonly transmitted by contaminated water?

   A. *Vibrio cholerae*
   B. *Giardia intestinalis*
   C. *Norovirus*
   D. *Streptococcus mutans*
   E. *Salmonella typhi*

5. True or False. Diarrheal disease can be caused by eating contaminated foods that have been cooked thoroughly.

6. Which of the following is caused by a virus?

   A. Shigellosis
   B. Salmonellosis
   C. Giardiasis
   D. Hepatitis
   E. Typhoid fever

7. Which of the following is **not** spread by the fecal–oral route of transmission?

   A. Shigellosis
   B. Giardiasis
   C. Typhoid fever
   D. Cryptosporidiosis
   E. Mumps

8. Which of the following may infect the liver?

   A. *Entamoeba histolytica*
   B. *Giardia intestinalis*
   C. *Cryptosporidium parvum*
   D. *Cyclospora cayetanensis*
   E. None of the above is correct.

Answers

Review Exercises

1. For an infection to occur, the organism itself must be present in the ingested food or water. Disease resulting from infection usually occurs from 1 to 3 days after ingestion of contaminated food or water and may be accompanied by a fever, which is part of the body's response to the infecting bacterium. For an intoxication to occur, a toxin must be present in the ingested food or water, but the organism that produced the toxin may be absent. Disease caused by intoxication occurs more rapidly, often within a few hours after ingesting contaminated food or water, and is not usually accompanied by a fever. Both infection and intoxication can result in diarrhea, abdominal cramps, nausea, and vomiting.

2. The exterior of the digestive system is covered by mucous membranes that consist of multiple layers of cells, produce mucus, and are protected by phagocytes, IgA antibodies, and lysozyme. The mouth and large intestines are protected by their normal microbiota, which inhibit colonization by other microbes. The low pH of the stomach and bile released from the gallbladder both prevent microbial growth. Any microbes that survive the stomach are usually moved too quickly through the small intestine to become established.

3. Both salmonellosis and typhoid fever are caused by species of *Salmonella*. Salmonellosis is a form of gastroenteritis and is much less severe than typhoid fever. Typhoid fever is caused by a particular species of *Salmonella, S. typhi*. *S. typhi* survives and multiplies inside of phagocytes, which carry it to organs such as the spleen and liver. The phagocytes are destroyed, and *S. typhi* is released into the blood. This results in a more severe disease that may involve high fever and may result in death.

4. All of these are diarrheal diseases caused by protozoans that are usually transmitted by the fecal–oral route. All of the protozoans produce special cell types that are capable of surviving well in the environment. *Giardia* and *Entamoeba* produce a special cell called a cyst. *Cryptosporidium* and *Cyclospora* produce a cell called an oocyst. All of these protozoans cause a watery diarrhea and are frequently transmitted by ingestion of contaminated water. *Entamoeba* can also infect the liver and may cause death.

Self-Test

1. F	4. D	7. E
2. C	5. T	8. A
3. C	6. D	

27

Infectious Diseases of the Urinary and Reproductive Systems

WHAT YOU WILL LEARN

This chapter introduces some infectious diseases of the urinary and reproductive systems. As you study this chapter, you will:

- review the structure of the urinary and reproductive systems;
- learn about the most common causes of urinary tract infections;
- compare bacterial, viral, and protozoan diseases of the reproductive system;
- consider the effects of diseases of the reproductive system.

SECTIONS IN THIS CHAPTER

- The Urinary System
- The Reproductive System
- Bacterial Diseases of the Urinary System
- Bacterial Diseases of the Reproductive System
- Viral Diseases of the Reproductive System
- Protozoan Diseases of the Reproductive System
- Fungal Diseases of the Reproductive System

The urinary and reproductive systems of humans are in close association with each other and share some common portals of entry for infectious organisms. Infections of the urinary system are usually caused by intestinal bacteria that enter the urinary tract from the outside. Urinary tract infections are estimated to be responsible for about 4 million visits to the doctor each year. In addition, invasive procedures such as catheterization make UTIs one of the leading types of hospital-acquired infections.

Infections of the reproductive system require intimate contact between an infected person and a new host. Most of these diseases are transmitted by sexual contact and are therefore called sexually transmitted diseases. There are estimated to be over 15 million new cases of STDs each year in the United States alone. Although many bacterial STDs can be successfully treated with antibiotics, there are several serious viral STDs for which there is no cure.

The Urinary System

The urinary system (Figure 27.1) functions to remove waste products from the blood. Blood circulates through the **kidneys**, where waste is removed. The fluid that contains the waste, called **urine**, passes from the kidneys to the **urinary bladder** through tubes called **ureters**. Urine is stored in the bladder until it is released from the body through the **urethra**. In women, the urethra is used only for the release of urine from the bladder. In men, the urethra is used both for the release of urine and for the release of seminal fluid from the reproductive system.

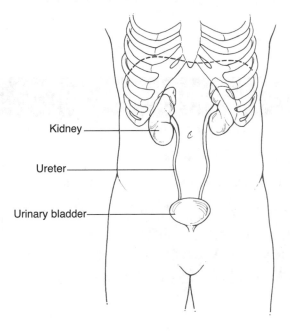

FIGURE 27.1. The urinary system.

The urinary system is protected from infection by several factors. As urine flows out through the urethra, it helps to wash away some potential pathogens. Urine is slightly acidic, which helps prevent the growth of many organisms. Valves prevent the backflow of urine into the kidneys, helping to protect the kidneys from infections that occur in the urethra. As a result of these defenses, normal urine is typically sterile, except near the opening to the outside of the body.

The Reproductive System

The function of the female reproductive system (Figure 27.2) is to produce eggs and support a developing embryo and fetus if fertilization occurs. Eggs are produced in the **ovaries**; they are released and travel through the **uterine (fallopian) tubes** to the **uterus**. If the egg is fertilized, it implants in the inner wall of the uterus where it is nourished as it grows into a fetus. The opening of the uterus, or **cervix,** leads to the **vagina**, a muscular canal that leads to the exterior of the body and the external genitalia. The vagina contains a complex microbiota that helps to protect the body from colonization by potential pathogens.

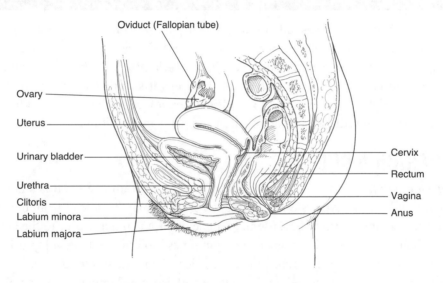

FIGURE 27.2. The female reproductive system.

The function of the male reproductive system (Figure 27.3) is to produce sperm cells for fertilization of the egg. Sperm are produced in the **testes** and pass through the **epididymis**, **ductus (vas) deferens, ejaculatory duct**, and **urethra** to exit the body. Except for the area right near the external opening, the male urethra is usually sterile.

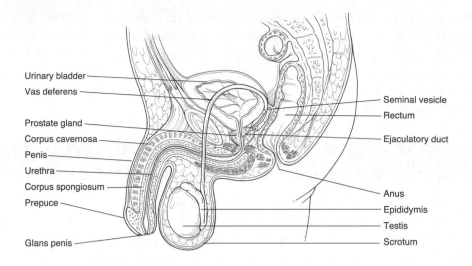

Urinary bladder

Vas deferens

Prostate gland

Corpus cavernosa

Penis

Urethra

Corpus spongiosum

Prepuce

Glans penis

Seminal vesicle

Rectum

Ejaculatory duct

Anus

Epididymis

Testis

Scrotum

FIGURE 27.3. The male reproductive system.

Bacterial Diseases of the Urinary System

Urinary tract infections are usually caused by *Escherichia coli*, but they may also be caused by other members of the Enterobacteriaceae. Urinary tract infections usually begin by causing inflammation of the urethra, or **urethritis**. If the bladder becomes infected, it causes inflammation of the bladder, or **cystitis**. If the ureters become infected, the inflammation is called **ureteritis**. If the infection moves to the kidneys, it is called **pyelonephritis**, which can result in serious complications.

ESCHERICHIA COLI

Escherichia coli is the most common cause of urinary tract infections. It is a normal inhabitant of the digestive system (see Chapter 26) and may be spread from the anus to the urinary opening. Perhaps because the female urethra is much closer to the anal opening than is that of males, women are eight times more likely to develop urinary tract infections than men. The female urethra is less than 2 inches long, so it is not uncommon for bacteria to travel up the urethra and cause inflammation of the bladder.

> **REMEMBER**
> Most urinary tract infections are caused by *Escherichia coli.*

Infections that are restricted to the lower urinary tract usually result in painful urination. Systemic infection may cause abdominal or back pain, fever, and sepsis. If pyelonephritis occurs, scar tissue may develop, resulting in decreased kidney function.

Bacterial Diseases of the Reproductive System

Bacteria cause many well-known sexually transmitted diseases, like gonorrhea, syphilis, and chlamydia. When these diseases are detected, they can usually be cured by antibiotics. However, several of these diseases cause little or no symptoms and may sometimes go undetected for long periods of time. In these cases, serious complications can occur. Selected bacterial diseases of the reproductive system are summarized in Table 27.1.

> **REMEMBER**
> Sexually transmitted diseases may go undetected, especially in women. Infertility and other serious consequences may develop as a result of these long-term infections.

TABLE 27.1. Selected Bacterial Diseases of the Reproductive System

Disease	Causative Agent	Most Common Mode of Transmission	Clinical Symptoms	Treatment
Toxic shock syndrome	*Staphylococcus aureus*	Indirect contact	Fever, chills, vomiting, diarrhea, muscle aches, rash, shock	Antitoxin, supportive care
Bacterial vaginosis	*Gardnerella vaginalis*	Unknown	Abnormal vaginal discharge, burning during urination, vaginal itch	Metronidazole
Gonorrhea	*Neisseria gonorrhoeae*	Direct contact during vaginal, anal, or oral sex	May not cause any signs or symptoms, or may cause abnormal discharge from penis or vagina and a burning sensation during urination	Fluorquinolone antibiotics
Syphilis	*Treponema pallidum*	Direct contact during vaginal, anal, or oral sex	During primary infection, a firm, round, painless sore called a chancre develops; during secondary infection, skin rashes and lesions of mucous membranes may develop	Benzathine penicillin
Chlamydia	*Chlamydia trachomatis*	Direct contact during vaginal, anal, or oral sex	Abnormal discharge, burning when urinating, pelvic pain	Azithromycin or other antibiotics

Viral Diseases of the Reproductive System

Because viruses cannot be treated with antibiotics, viral diseases of the reproductive system are usually incurable after they have been contracted. Although many infected people learn to live with these diseases, these diseases can cause serious complications and also increase the risk for infection by the human immunodeficiency virus, which causes acquired immunodeficiency syndrome. Although AIDS is a virus that is sexually transmitted, its main effect is on the immune system, so it will be considered in Chapter 28. Selected viral diseases that affect the reproductive system are summarized in Table 27.2.

TABLE 27.2. Selected Viral Diseases of the Reproductive System

Disease	Causative Agent	Most Common Mode of Transmission	Clinical Symptoms	Treatment and Prevention
Genital herpes	*Herpes simplex virus type 2*	Direct genital–genital contact or direct genital–oral contact	Recurrent painful lesions on the genitals	Acyclovir and related drugs may alleviate symptoms
Genital warts	*Human papilloma virus*	Direct genital contact	Warts on the genitals	Physical removal, imiquimod to stimulate production of interferon, a vaccine is available
Cytomegalovirus disease	*Cytomegalovirus*	Direct contact	Hearing loss, vision impairment, and mental retardation in infants infected as fetuses; Pneumonia, retinitis, and gastrointestinal disease in immuno-compromised people	Ganciclovir may alleviate symptoms

Protozoan Diseases of the Reproductive System

Inflammation of the vagina, or **vaginitis**, has multiple causes. In the case of **bacterial vaginosis**, the bacterium *Gardnerella vaginalis* overpopulates the vagina. The vagina can also be overpopulated by the protozoan, *Trichomonas vaginalis*.

TRICHOMONAS VAGINALIS

Trichomoniasis, which is caused by the protozoan, *Trichomonas vaginalis*, is a common sexually transmitted disease. The protozoan is transmitted by direct genital contact and typically infects the vagina in women and the urethra in men. Most men do not have signs or symptoms of *Trichomonas* infection; however, some experience a burning sensation during urination. In women, *Trichomonas* infection may not cause symptoms or it may cause a frothy, yellow-green vaginal discharge that has a strong odor. Discomfort during sexual intercourse and urination may also occur, as may itching of the vaginal area.

Women who are infected with *Trichomonas* are more susceptible to HIV infection and can more easily pass HIV infection to their partners. They are also more likely to have low birth weight babies.

Fungal Diseases of the Reproductive System

The normal balance of the vaginal microbiota can also be disrupted by overgrowth of yeast, which is a fungus. **Yeast infections** are very common, affecting nearly 75 percent of all adult women at one point in their lifetimes.

CANDIDA ALBICANS

The yeast *Candida albicans* causes genital candidiasis, also known as a **yeast infection**. In women, *Candida albicans* is normally present in low numbers in the vagina, but under certain circumstances it can multiply and outnumber the bacteria of the normal microbiota. These changes can result from hormonal changes, treatment with antibiotics, diabetes, and pregnancy. Women who have a yeast infection usually feel a burning or itching sensation during urination, and some may have an abnormal vaginal discharge that has the consistency of cottage cheese. Men can also get a yeast infection of the penis that causes an itchy rash. Yeast infections can be treated with antifungal drugs such as miconazole (Monistat-7) or clotrimazole (Gyne-Lotrimin).

> **REMEMBER**
> The yeast *Candida albicans* is an opportunistic pathogen that is normally present in low numbers in the vagina. If the bacterial population of the vagina is reduced for some reason—such as from the use of antibiotics—the yeast can multiply, resulting in a yeast infection.

REVIEW EXERCISES FOR CHAPTER 27

1. What defenses of the urinary system normally protect it from disease?

2. What is the most common cause of urinary tract infections? How are urinary tract infections transmitted?

3. What defenses of the reproductive system normally protect it from disease?

4. Name and briefly describe three causes of vaginitis.

5. What causes toxic shock syndrome?

6. Compare and contrast genital herpes and genital warts.

SELF-TEST

1. Which of the following is not usually transmitted by direct contact during sexual intercourse?

   A. Gonorrhea
   B. Syphilis
   C. Genital warts
   D. Urinary tract infection
   E. Chlamydia

2. True or False. Many people have sexually transmitted diseases but are unaware of them.

3. Which of the following is caused by a spirochete?

   A. Gonorrhea
   B. Chlamydia
   C. Urinary tract infection
   D. Toxic shock syndrome
   E. Syphilis

4. Which of the following is most commonly caused by *Candida albicans*?

   A. Yeast infection
   B. Gonorrhea
   C. Toxic shock syndrome
   D. Urinary tract infection
   E. Syphilis

5. True or False. Genital warts can be treated with antibiotics.

6. Cytomegalovirus is most likely to cause serious complications in which of the following?

    A. Sexually active women ages 20–24
    B. Sexually active men ages 35–39
    C. Immunocompromised people
    D. Both A and B are correct.
    E. All of the above are correct.

Answers

Review Exercises

1. The urinary tract is protected from many diseases by the flushing action of urine and by the slight acidity of urine, which prevents many microbes from growing. Valves that prevent the backflow of urine also help to prevent infectious agents from reaching the kidneys.

2. The most common cause of urinary tract infections is *E. coli*. *E. coli* can be transmitted to the urethra if fecal matter from the anus contacts the opening of the urinary system.

3. The reproductive system is lined with mucous membranes. These membranes produce mucus, which traps microbes, and by other defensive secretions such as IgA antibodies. The vagina is also protected by its normal microbiota and by the slightly acidic pH.

4. Bacterial vaginosis is usually associated with overgrowth of the bacterium *Gardnerella vaginalis*. Vaginitis can also be caused by the protozoan *Trichomonas vaginalis* and by the yeast *Candida albicans*.

5. Toxic shock syndrome is often caused by strains of the bacterium *Staphylococcus aureus*, which produces toxic shock syndrome toxin 1. This toxin causes overproduction of cytokines and leads to fever, rash, low blood pressure, tissue injury, and shock.

6. Viruses cause both genital herpes and genital warts. Genital herpes is usually caused by infection with the herpes simplex virus type 2. This virus may cause painful lesions on the genitals. Outbreaks of lesions may recur several times a year, with latent periods in between. Genital warts are caused by the human papilloma virus. These wart-like growths on the genitals may be large with many projections or smooth and flat. Women with HPV infections are at higher risk for cervical cancer. Once contracted, neither genital herpes nor genital warts can be cured.

Self-Test

1. D	3. E	5. F
2. T	4. A	6. C

28

Infectious Diseases of the Cardiovascular and Lymphatic Systems

WHAT YOU WILL LEARN

This chapter presents some infectious diseases of the cardiovascular and lymphatic systems. As you study this chapter, you will:

- review the structure of the cardiovascular and lymphatic systems;
- consider the causes and effects of sepsis;
- compare the bacterial, viral, and protozoan diseases of the cardiovascular and lymphatic systems.

SECTIONS IN THIS CHAPTER

- The Cardiovascular and Lymphatic Systems
- Bacterial Diseases of the Cardiovascular and Lymphatic Systems
- Viral Diseases of the Cardiovascular and Lymphatic Systems
- Protozoan Diseases of the Cardiovascular and Lymphatic Systems

Pathogens that gain access to the cardiovascular or lymphatic systems can be very dangerous because of their ability to spread throughout the body. They can colonize multiple systems of the body and may cause damage to internal organs. Bacteria that circulate in the blood may colonize the heart, ultimately leading to cardiac damage and death. Some pathogens have the ability to withstand or even destroy the very cells of the immune system that are supposed to defend us. For example, the pathogens that cause tularemia, brucellosis, and the plague can survive within macrophages and use them to circulate throughout the body. And, of course, the most devastating plague of our times, acquired immunodeficiency syndrome (AIDS), results from infection of helper T (T_H) cells by the HIV virus. T_H cells are crucial to the coordination of our immune response (Chapter 19), and their destruction by HIV makes the human body vulnerable to attack by opportunistic pathogens.

The Cardiovascular and Lymphatic Systems

The cardiovascular and lymphatic systems circulate fluids around the body. The cardiovascular system circulates **blood** through **blood vessels** and the **heart** (Figures 28.1 and 28.2). Blood consists of a fluid, called **plasma**, and a mixture of **red blood cells**, **white blood cells**, and **platelets**, which are called the **formed elements** of the blood. As blood circulates, the red blood cells carry oxygen to the cells of the body and remove carbon dioxide as waste. White blood cells help defend the body against disease and are also found in the lymphatic system.

The lymphatic system circulates **lymph** through **lymphatic vessels** and **lymph nodes** (see Figure 19.1). As blood moves around the body, blood plasma crosses out of blood capillaries into the tissues. The tissue fluids flow into lymphatic vessels and toward the heart to be returned to the blood. The lymphatic system screens the tissue fluids and filters out foreign microbes prior to returning the fluid to the heart. The white blood cells of the lymphatic system, such as phagocytes, T cells, and B cells, work together to destroy these microbes and protect the body from infection (see Chapter 19).

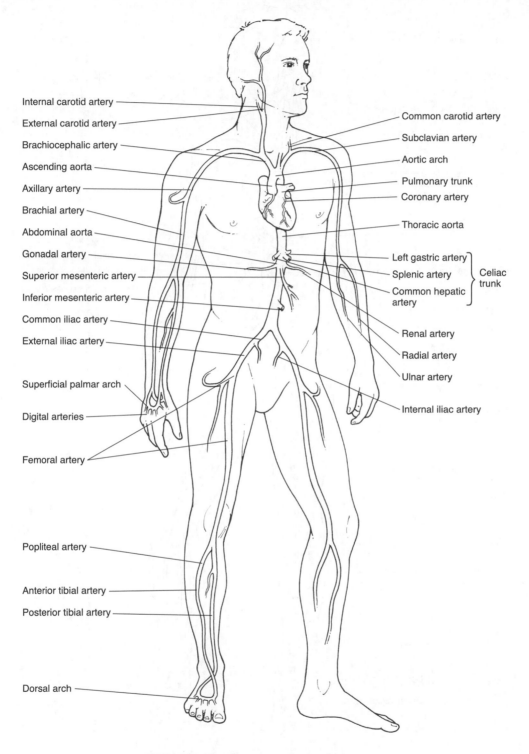

Internal carotid artery

External carotid artery

Brachiocephalic artery

Ascending aorta

Axillary artery

Brachial artery

Abdominal aorta

Gonadal artery

Superior mesenteric artery

Inferior mesenteric artery

Common iliac artery

External iliac artery

Superficial palmar arch

Digital arteries

Femoral artery

Popliteal artery

Anterior tibial artery

Posterior tibial artery

Dorsal arch

Common carotid artery

Subclavian artery

Aortic arch

Pulmonary trunk

Coronary artery

Thoracic aorta

Left gastric artery

Splenic artery

Common hepatic artery

Celiac trunk

Renal artery

Radial artery

Ulnar artery

Internal iliac artery

FIGURE 28.1. The human arterial system.

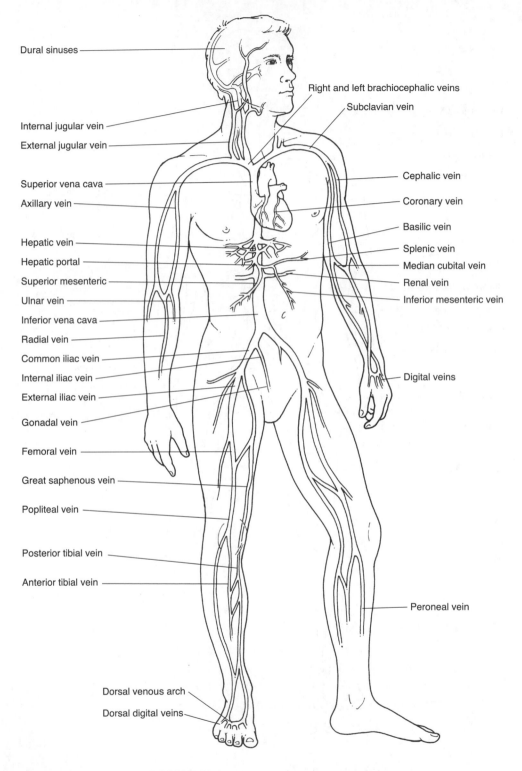

Dural sinuses

Internal jugular vein

External jugular vein

Superior vena cava

Axillary vein

Hepatic vein

Hepatic portal

Superior mesenteric

Ulnar vein

Inferior vena cava

Radial vein

Common iliac vein

Internal iliac vein

External iliac vein

Gonadal vein

Femoral vein

Great saphenous vein

Popliteal vein

Posterior tibial vein

Anterior tibial vein

Dorsal venous arch

Dorsal digital veins

Right and left brachiocephalic veins

Subclavian vein

Cephalic vein

Coronary vein

Basilic vein

Splenic vein

Median cubital vein

Renal vein

Inferior mesenteric vein

Digital veins

Peroneal vein

FIGURE 28.2. The human venous system.

Bacterial Diseases of the Cardiovascular and Lymphatic Systems

Because the cardiovascular and lymphatic systems circulate fluids around the body, pathogens that gain access to these systems can cause **systemic infections** that affect more than one area of the body. Normally, blood is sterile. If microbes enter the blood, they are typically destroyed by the white blood cells. However, if the defenses of the blood fail, microbes may begin to multiply in the blood. The multiplication of microbes in the blood is called **septicemia** or **sepsis**. Sepsis is the first step toward a dangerous systemic infection that may result in death. Frequently, when an infection becomes septic, red streaks leading away from the site of infection become visible under the skin. The appearance of these streaks, which are made by inflamed lymph vessels, is called **lymphangitis**.

> **REMEMBER**
> Pathogens that enter the blood or lymphatic systems can get dispersed around the body, causing systemic infections.

As the immune system attempts to control the spreading infection, cytokines, which trigger inflammation and fever, are released. A person with a systemic infection may display fever, chills, rapid breathing, and an accelerated heart rate. Vasodilation occurs, and blood vessel permeability increases, causing more fluid to leave the circulatory system and enter the tissues. Blood pressure drops and organs may begin to fail as **severe sepsis** occurs. If blood pressure cannot be controlled by the addition of fluids, **septic shock** results and is often followed by death.

Sepsis may be caused by several different bacterial species. As white blood cells destroy gram-negative bacteria, lipopolysaccharide, or endotoxin, is released from the cell wall. Because endotoxin stimulates white blood cells to release inflammatory cytokines, **septic shock** is most commonly caused by gram-negative bacteria. Treatment of **gram-negative sepsis** is problematic because antibiotics trigger destruction of the bacteria, releasing more endotoxin and worsening the symptoms. Sepsis may also be caused by gram-positive bacteria, including the following: *Staphylococcus aureus* (see Chapter 24), *Streptococcus pyogenes* (see Chapter 24), enterococci, and Group B streptococci. The mechanisms by which gram-positive bacteria induce septic shock are not fully known, although strains of *Staphylococcus aureus* produce toxins that trigger cytokine release.

In addition to causing shock, infection with *Staphylococcus aureus, Streptococcus pyogenes,* or enterococci can also lead to complications that damage the heart. Bacteria that are circulating in the blood may become trapped in the hearts of people who have abnormal valves. The bacteria multiply, causing inflammation of the inner layer of the heart, which is called the **endocardium**. This inflammation, referred to as **subacute bacterial endocarditis**, can result in weakness, fever, and a heart murmur. As the bacteria multiply in the heart, they become trapped in blood clots,

> **REMEMBER**
> Overproduction of cytokines in response to systemic infections can lead to a massive inflammatory response, causing a drop in blood pressure and resulting in septic shock.

which may eventually break free and enter the blood vessels, blocking blood flow and causing organ damage that can lead to death. *Staphylococcus aureus* can also cause a rapid destruction of heart valves, called **acute bacterial endocarditis**, which is usually fatal if not treated promptly.

Damage to the heart can also be caused by the response of the immune system to infections by *Streptococcus pyogenes*. Following a streptococcal sore throat, some people develop **rheumatic fever**. When the immune system targets the M protein on the bacterium, it may also mistakenly target the heart, causing inflammation and valve damage. Because the immune system mistakenly attacks part of the host, rheumatic fever is an example of an **autoimmune disorder**. In people who develop rheumatic fever, recurring infections with *S. pyogenes* can lead to further damage of the heart and may result in eventual death.

In addition to septic shock and heart damage, bacteria that spread through the blood or the lymph cause many other diseases. These diseases are summarized in Table 28.1.

TABLE 28.1. Selected Bacterial Diseases
of the Cardiovascular and Lymphatic Systems

Disease	Causative Agent	Most Common Mode of Transmission	Clinical Symptoms	Treatment and Prevention
Tularemia	*Francisella tularensis*	Vector or vehicle	Fever, chills, headache, diarrhea, aches, weakness, dry cough, may develop into pneumonia	Streptomycin
Brucellosis	*Brucella* sp.	Vehicle	Fever, chills, malaise, sweating, headache, backache	Tetracycline plus streptomycin; a vaccine is available
Anthrax	*Bacillus anthracis*	Parenteral, or vehicle	If parenteral, a blackened ulcer at the site of entry; if foodborne, severe abdominal pain, bloody diarrhea; if airborne, fever, chills, chest pain, systemic bleeding, death	Ciprofloxacin; a vaccine is available
Gas gangrene	*Clostridium perfringens*	Parenteral	Tissue death, swelling of tissue	Hyperbaric oxygen, tissue removal, penicillin
Plague or black death	*Yersinia pestis*	Vector (flea)	Fever, swelling of lymph nodes, thick urine, putrid breath	Streptomycin, tetracycline
Lyme disease	*Borrelia burgdorfii*	Vector (tick)	Bulls-eye rash, fever, headache, muscle pain	Various antibiotics, including amoxicillin and doxycycline, may be effective in the early stages; intravenous antibiotics may be effective in later stages
Epidemic typhus	*Rickettsia prowazekii*	Vector (louse)	Headache, chills, fever, prostration, confusion, photophobia, vomiting, rash that begins on trunk	Tetracycline, chloramphenicol
Rocky Mountain spotted fever	*Rickettsia rickettsii*	Vector (tick)	Headache, fever, abdominal pain, rash that begins on extremities	Tetracycline

Viral Diseases of the Cardiovascular and Lymphatic Systems

Viruses that infect the cardiovascular and lymphatic systems cause severe diseases that are endemic to tropical parts of the world and cause significant mortality, especially among children. Treatment for these diseases usually involves only treating the symptoms and is not always effective. Although vaccines are available for a few of these viral diseases, for some there is no prevention and no cure. Several of these diseases are summarized in Table 28.2.

TABLE 28.2. Selected Viral Diseases of the Cardiovascular and Lymphatic Systems

Disease	Causative Agent	Most Common Mode of Transmission	Clinical Symptoms	Treatment and Prevention
Infectious mononucleosis	Epstein–Barr virus	Direct contact	Fever, sore throat, swollen lymph glands	No specific treatment
Yellow fever	*Flavivirus*	Vector (mosquito)	Fever, headache, muscle ache, vomiting, diarrhea, jaundice, bleeding	No specific treatment; vaccine is available
Dengue fever	*Flavivirus*	Vector (mosquito)	Fever, severe muscle and joint pain	No specific treatment
Ebola hemorrhagic fever	Ebola virus	Contact, parenteral	Fever, weakness, muscle pain, sore throat, bleeding	No specific treatment
Hantavirus pulmonary syndrome	*Hantavirus*	Vehicle (airborne)	Fatigue, fever, muscle aches, possible nausea and vomiting, respiratory distress	No specific treatment
Acquired Immuno-deficiency Syndrome (AIDS)	Human Immuno-deficiency Virus (HIV)	Direct contact	Asymptomatic in early stages; increased susceptibility to infection in later stages	Highly active anti-retroviral therapy (HAART)

Protozoan Diseases of the Cardiovascular and Lymphatic Systems

Protozoan diseases of the cardiovascular and lymphatic systems, including malaria and Chagas' disease, have significant impacts worldwide. Several of these diseases are summarized in Table 28.3.

TABLE 28.3. Selected Protozoan Diseases of the Cardiovascular and Lymphatic Systems

Disease	Causative Agent	Most Common Mode of Transmission	Clinical Symptoms	Treatment and Prevention
Malaria	*Plasmodium* sp.	Vector (mosquito)	Chills, fever, vomiting headache	Quinine derivatives, artemisinin, vaccines in development
Chagas' disease	*Trypanosoma cruzi*	Vector (triatomine bug)	Swelling around the eye or on one side of the face, fatigue, fever, enlarged liver or spleen, and sometimes rash, loss of appetite, vomiting and diarrhea	Benznidazole or nifurtimox
Toxoplasmosis	*Toxoplasma gondii*	Vehicle	Flu-like symptoms, swollen lymph glands, or muscle aches and pains	No specific treatment for otherwise healthy people, medication is available for pregnant women
Leishmaniasis	*Leishmania* sp.	Vector (sand fly)	Cutaneous: volcano-like sores in the skin; Visceral: fever, chills, weight loss, followed by enlargement of liver and spleen	Cutaneous leishmaniasis is treated with topical applications containing the metal antimony, the oral drug miltefosine has recently been used for visceral leishmaniasis

REVIEW EXERCISES FOR CHAPTER 28

1. Describe the signs of severe sepsis and explain how it leads to septic shock.

2. How can infection lead to endocarditis?

3. Compare and contrast tularemia and brucellosis.

4. Compare and contrast plague and Lyme disease.

5. Compare and contrast yellow fever and dengue fever.

6. Compare and contrast Ebola with *Hantavirus* pulmonary syndrome.

SELF-TEST

1. Which of the following is **not** transmitted to humans by the bite of a mosquito?

   A. Malaria
   B. Yellow fever
   C. Toxoplasmosis
   D. Dengue fever
   E. All of the above are correct.

2. Which of the following is spread through contact with mouse droppings (urine, feces, saliva)?

   A. Ebola
   B. *Hantavirus* pulmonary syndrome
   C. Malaria
   D. Puerperal sepsis
   E. Lyme disease

3. Which of the following is caused by a spirochete?

   A. Lyme disease
   B. Leishmaniasis
   C. Plague
   D. Gas gangrene
   E. Infectious mononucleosis

4. Which of the following cannot be transmitted by the bite of a tick?

   A. Rocky Mountain spotted fever
   B. Lyme disease
   C. Brucellosis
   D. Tularemia
   E. All of the above are correct.

5. Which of the following is **not** caused by a protozoan?

    A. Malaria
    B. Chagas' disease
    C. Toxoplasmosis
    D. Leishmaniasis
    E. All of the above are correct.

6. Which of the following is transmitted by the bite of the human body louse?

    A. Plague
    B. Leishmaniasis
    C. Chagas' disease
    D. Epidemic typhus
    E. Rocky Mountain spotted fever

7. Epstein–Barr virus causes which of the following?

    A. Yellow fever
    B. Dengue fever
    C. Ebola
    D. *Hantavirus* pulmonary syndrome
    E. Infectious mononucleosis

8. Which of the following is characterized by a distinctive bulls-eye rash?

    A. Leishmaniasis
    B. Lyme disease
    C. Chagas' disease
    D. Rocky Mountain spotted fever
    E. Epidemic typhus

9. Which of the following causes acute bacterial endocarditis?

    A. *Streptococcus pyogenes*
    B. Enterococci
    C. *Staphylococcus aureus*
    D. Both A and C are correct.
    E. A, B, and C are correct.

10. True or False. Only gram-negative bacteria cause septic shock.

Answers

Review Exercises

1. Severe sepsis occurs when bacteria that are multiplying in the blood cause vasodilation and increased blood vessel permeability. As fluid leaves the circulatory system and enters the tissues, blood pressure drops, and organs may begin to fail. If blood pressure cannot be controlled by the addition of fluids, septic shock results and is often followed by death.

2. Bacteria that are circulating in the blood may become trapped in the heart of people who have abnormal valves. The bacteria multiply, causing inflammation of the inner layer of the heart, or endocardium. This inflammation, referred to as subacute bacterial endocarditis, can result in weakness, fever, and a heart murmur. *Staphylococcus aureus* can also cause a rapid destruction of heart valves, called acute bacterial endocarditis, which is usually fatal if not treated promptly.

3. Both tularemia and brucellosis have animal reservoirs. The reservoir for tularemia is small wild mammals, especially rabbits. The reservoirs for brucellosis include dogs and other domestic animals. For both diseases, transmission to humans can occur through vehicles of contaminated water or air. Tularemia can also be spread by tick bites. Early symptoms for both diseases can be similar and include fever, chills, and general malaise. The causative agents for the two diseases are different. Tularemia is caused by *Francisella tularensis*. Brucellosis is caused by several species of *Brucella*.

4. Plague and Lyme disease are both transmitted to humans by vectors. The vector for plague is the rat flea. The vector for Lyme disease is a tick. The reservoirs for both diseases are nonhuman animals: The plague bacillus multiplies inside of rodents and the Lyme disease spirochete multiplies inside of deer and mice. Symptoms for the two diseases are quite distinct. The plague is often characterized by swelling of the lymph nodes, forming buboes, as well as fever, thick urine, and putrid breath. Lyme disease is characterized by the formation of a bulls-eye rash around the original tick bite, followed by fever, headache, and muscle pain.

5. Yellow fever and dengue fever are caused by viruses that are members of the genus *Flavivirus*. These viruses spread to humans by the bite of mosquitoes. Symptoms for both diseases include fever and muscle ache. Also, more serious symptoms may occur for both diseases. The yellow fever virus may cause abdominal pain, vomiting, and bleeding from the mouth, nose, or eyes. Liver damage occurs, resulting in jaundice. Kidney failure may also occur. The dengue fever virus may cause dengue hemorrhagic fever, which results in high fever, liver enlargement, and possible death due to circulatory failure and shock. Mortality from the severe forms of both diseases is high.

6. Ebola and *Hantavirus* pulmonary syndrome are both serious, often fatal viral infections. Ebola is caused by the Ebola virus, which is transmitted to humans by contact with bodily secretions from an infected person. *Hantavirus* pulmonary syndrome is caused by *Hantavirus*, which is transmitted to humans by contact with mouse droppings. The reservoir for Ebola virus is not fully understood. The reservoir for *Hantavirus* is rodents. Death from Ebola virus results from systemic hemorrhage. Death from *Hantavirus* results from respiratory failure.

Self-Test

1. C	4. C	7. E	10. F
2. B	5. E	8. B	
3. A	6. D	9. C	

29

Infectious Diseases of the Nervous System

WHAT YOU WILL LEARN

This chapter introduces the infectious diseases of the nervous system. As you study this chapter, you will:

- review the structure of the nervous system;
- learn about the barriers to infection that protect the central nervous system;
- compare bacterial, viral, and protozoan diseases of the nervous system.

SECTIONS IN THIS CHAPTER

- The Nervous System
- Bacterial Diseases of the Nervous System
- Viral Diseases of the Nervous System
- Protozoan Diseases of the Nervous System

The nervous system controls all of the body's functions. Thus, pathogens that infect and damage the nervous system can have devastating effects, including loss of hearing, paralysis, permanent brain damage, and death. In order for a pathogen to infect the nervous system, it must be able to get through the defenses that normally protect the nervous system. This can happen as a result of damage from an accident or medical procedure, or as a result of inflammation caused by the infection.

The Nervous System

The nervous system receives signals from the environment and coordinates all the responses and functions of the body. Together, the **brain** and the **spinal cord** make up the **central nervous system** (Figure 29.1), which interprets incoming signals and sends out messages to control responses. The nerves that extend from the central nervous system to all parts of the body make up the **peripheral nervous system**. These nerves transmit signals from the environment to the central nervous system and carry signals from the central nervous system to all of the organs.

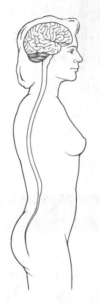

FIGURE 29.1. The central nervous system.

> **REMEMBER**
> The brain and spinal cord are protected by the skull, vertebrae, and a group of three membranes called meninges.

The brain and the spinal cord are well protected from damage and infection. In addition to being protected by the skull and the vertebrae, they are surrounded by a group of three membranes called **meninges** (Figure 29.2). The meninges consist of three layers: The outermost layer is the **dura mater**, the middle layer is the **arachnoid mater**, and the inner layer is the **pia mater**. The space between the arachnoid mater and the pia mater is called the **subarachnoid space**, and it is filled with a clear fluid called **cerebrospinal fluid**. The capillaries that allow access of materials from the blood to the cerebrospinal fluid are much more selective than other capillaries in the body. Because the substances that are allowed into the cerebrospinal fluid are so carefully controlled, this is referred to as the **blood–brain barrier**.

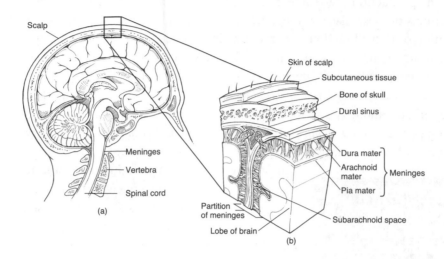

FIGURE 29.2. The meninges and cerebrospinal fluid.

Although the nervous system is very well protected, some microbes can penetrate these barriers. This most commonly happens when inflammation increases the permeability of the blood–brain barrier, allowing microbes to enter from the circulatory or lymphatic systems. Injuries to the skull or spine can also enable access, as can invasive procedures such as a spinal tap (lumbar puncture), in which a needle is used to withdraw a sample of cerebrospinal fluid. Microbes that cause infections of the nervous system may also have certain characteristics that make it easier for them to cross the blood–brain barrier.

> **REMEMBER**
> Inflammation can increase the permeability of the blood–brain barrier, allowing microbes to enter the nervous system.

As with most infections, infections of the nervous system often result in inflammation of the affected tissue. When inflammation occurs in the brain, it is called **encephalitis**. When inflammation occurs in the membranes that cover the brain, it is called **meningitis**. Meningitis and encephalitis can result from a variety of infections by both viruses and bacteria. If the spinal cord becomes inflamed, it is called **myelitis**. If both the brain and spinal cord are affected, it is called **encephalomyelitis**. Inflammation of the brain, meninges, and spinal cord are dangerous because they can result in damage to the nerve cells and may cause bleeding in the brain.

Bacterial Diseases of the Nervous System

Bacterial diseases of the nervous system result from either infection of the nervous system by bacteria that spread to the nervous system from another location in the body or by the production of toxin by bacteria that are located in another part of the body. The effects of these diseases range from inflammation of the brain and spinal cord to inhibition of nerve function. If nerve function is affected, death may result.

Bacterial meningitis is not common, but it is very dangerous. Several species of bacteria that initially cause upper respiratory infections may multiply in the blood and cross the blood–brain barrier. The bacteria multiply in the cerebrospinal fluid and cause inflammation. The three species of bacteria that cause most cases of bacterial meningitis are *Streptococcus pneumoniae, Haemophilus influenzae,* and *Neisseria meningitidis*. Symptoms of meningitis include fever, headache, stiff neck, and confusion. In extreme cases, meningitis can result in brain damage, stroke, seizures, or even death. Antibiotics can be used to treat bacterial meningitis, but it is essential that they be administered very early in the infection to minimize the damage. Several bacterial diseases of the nervous system are summarized in Table 29.1.

> **REMEMBER**
> Meningitis refers to inflammation of the meninges. Signs and symptoms of meningitis include fever, headache, stiff neck, and confusion.

TABLE 29.1. Selected Bacterial Diseases of the Nervous System

Disease	Causative Agent	Most Common Mode of Transmission	Clinical Symptoms	Treatment
Pneumococcal meningitis	*Streptococcus pneumoniae*	Direct contact	Sudden fever, severe headache, stiff neck	Cephalosporins; vaccine is available
Haemophilus meningitis	*Haemophilus influenzae* type b (Hib)	Direct contact	Sudden fever, severe headache, stiff neck	Ceftriaxone and cefotaxime, vaccine is available
Meningococcal meningitis	*Neisseria meningitidis*	Direct contact	Sudden fever, severe headache, stiff neck	Cephalosporins, vaccine is available
Meningitis	*Listeria monocytogenes*	Vehicle (foodborne)	Headache, stiff neck, confusion	Cephalosporins
Tetanus	*Clostridium tetani*	Parenteral	Lockjaw, stiffness in neck and abdomen, difficulty swallowing, fever, elevated blood pressure, muscle spasms	Immune globulin, removal of tissue; vaccine is available

TABLE 29.1. (continued)

Disease	Causative Agent	Most Common Mode of Transmission	Clinical Symptoms	Treatment
Botulism	*Clostridium botulinum*	Vehicle, parenteral	Infants: lethargy, poor feeding, weak cry, constipation; Adults: double vision, difficulty swallowing, dry mouth, muscle weakness	Supportive care, antitoxin
Hansen's disease (leprosy)	*Mycobacterium leprae*	Probably contact with respiratory droplets	Loss of sensation in areas of the skin, formation of nodules on skin, tissue damage	Multidrug combination including dapsone, rifampin, and clofazimine

Viral Diseases of the Nervous System

Several viruses are capable of infecting the nervous system and causing serious disease. **Viral meningitis**, also called **aseptic meningitis**, is the most common form of meningitis in the United States. Viral meningitis is usually a mild, nonlethal disease. It is frequently caused by *enteroviruses*, viruses that enter the body through the mouth and travel via the blood to the brain and surrounding tissues. **Encephalitis** can also result from viral infections, either by direct infection of the brain and spinal cord or as an indirect consequence of complications from a viral infection. Viral encephalitis may be caused by enteroviruses, herpes simplex virus types 1 and 2 (see Chapters 24 and 27), rabies virus, or by arboviruses. Arboviruses are viruses that are transmitted from infected animals to humans through the bite of infected arthropods such as ticks, mosquitos, or other blood-sucking insects. The characteristics of several important viruses that affect the nervous system are summarized in Table 29.2 and then discussed in more detail.

> **REMEMBER**
> Encephalitis, or inflammation of the brain, can be caused by several arboviruses that are spread by the bite of arthropods.

TABLE 29.2. Selected Viral Diseases of the Nervous System

Disease	Causative Agent	Most Common Mode of Transmission	Clinical Symptoms	Treatment and Prevention
Poliomyelitis	*Poliovirus*	Fecal–oral	In most cases, flu-like symptoms; In severe cases, loss of superficial reflexes, severe muscle aches and spasms, flaccid paralysis	No specific treatment; vaccine is available
Rabies	*Lyssavirus*	Parenteral (animal bite)	Flu-like symptoms progressing to anxiety, confusion, hallucination	Immune globulin and antirabies vaccine injections after an animal bite
Arboviral encephalitis	*Bunyavirus, Togavirus, Flavivirus*	Vector (arthropod)	Mild, flu-like symptoms, occasionally progressing to encephalitis	No specific treatment

Protozoan Diseases of the Nervous System

Although they are not common, protozoan infections of the nervous system can cause severe damage and frequently result in death.

TRYPANOSOMA

African trypanosomiasis, or **sleeping sickness**, is caused by flagellated protozoans in the genus *Trypanosoma*. *Trypanosoma brucei gambiense* causes the disease **West African sleeping sickness**, while *Trypanosoma brucei rhodesiense* causes the disease **East African sleeping sickness**. In both cases, the protozoans are transmitted to humans through the bite of the tsetse fly, which only exists in Africa. Although only about 40,000 cases of African trypanosomiasis are reported each year, the CDC estimates that about 100,000 cases actually occur.

> **REMEMBER**
> African sleeping sickness results from infection of the nervous system by a trypanosome called Trypanosoma brucei. Without treatment, this disease is fatal.

Very rarely, a case is reported in the United States; usually it is someone who has recently traveled to Africa.

The bite of the tsetse fly is often painful and may develop into a red sore called a **chancre**. The common symptoms of sleeping sickness include fever, severe headaches, extreme fatigue, irritability, swollen lymph nodes, and aching muscles. After the infection invades the central nervous system, neurological symptoms appear such as confusion, personality changes, slurred speech, seizures, and difficulty in walking and talking. Sleeping for long periods during the day but being wakeful at night is also common. For both types of trypanosomiasis, death will result if the person does not receive treatment. East African sleeping sickness is more severe than West African sleeping sickness and results in death more quickly. West African sleeping sickness is treatable even in the later stages with a drug called melarsoprol. Unfortunately, this drug is less effective against East African sleeping sickness. Efforts to prevent transmission of the disease by controlling the tsetse fly have had success in some parts of Africa. Vaccine development is also being attempted but is proving difficult because the pathogen can change its surface antigens during infection, thus evading the immune system.

NAEGLERIA

Naegleria fowleri is an ameba that is common in soil and freshwater and may be found in swimming pools that do not have enough chlorine. The amoeba enters the nose when a person is swimming or diving, crosses the blood–brain barrier, and infects the central nervous system, causing **primary amebic meningoencephalitis (PAM)**. Initial symptoms are fever, nausea, headache, vomiting, and a stiff neck. As damage to brain tissue occurs, the symptoms progress to confusion, loss of balance, seizures, and hallucinations. Although infection with *Naegleria fowleri* is very rare, when it does occur, it is almost always fatal. This is probably attributable to the fact that the disease progresses rapidly; thus, it is difficult to intervene effectively with drugs.

REVIEW EXERCISES FOR CHAPTER 29

1. Compare and contrast meningitis with encephalitis.

2. What are the signs and symptoms of polio? Is polio a severe disease?

3. Compare and contrast West African sleeping sickness with East African sleeping sickness.

4. Why is *Naegleria fowlerii* infection of concern if it is very rare in the human population?

SELF-TEST

1. Which of the following diseases is not currently preventable by a vaccine?

   A. Pneumococcal meningitis
   B. *Haemophilus* meningitis
   C. Tetanus
   D. African trypanosomiasis
   E. Rabies

2. Which of the following diseases is caused by a virus?

   A. Meningococcal meningitis
   B. Pneumococcal meningitis
   C. Poliomyelitis
   D. Botulism
   E. Listeriosis

3. Which of the following is transmitted to humans by a vector?

   A. Botulism
   B. Listeriosis
   C. Rabies
   D. Pneumococcal meningitis
   E. Arboviral encephalitis

4. Which of the following is most commonly transmitted to humans by the fecal–oral route of transmission?
   A. Poliomyelitis
   B. Botulism
   C. Pneumococcal meningitis
   D. African trypanosomiasis
   E. Primary amebic meningoencephalitis

5. What is the causative agent of primary amebic meningoencephalitis?
   A. *Streptococcus pneumoniae*
   B. *Trypanosoma bruci gambiense*
   C. *Neisseria meningitdis*
   D. *Naegleria fowlerii*
   E. *Flavivirus*

6. What is the causative agent of meningococcal meningitis?
   A. *Streptococcus pneumoniae*
   B. *Haemophilus influenzae* type B
   C. *Neisseria meningitidis*
   D. *Listeria monocytogenes*
   E. *Clostridium tetani*

Answers

Review Exercises

1. When inflammation occurs in the brain, it is called encephalitis. When inflammation occurs in the membranes that cover the brain, it is called meningitis.

2. Most people infected with poliovirus show no symptoms but are able to pass the virus to others. About 5 percent of people have generalized symptoms such as a sore throat, fever, and nausea, but the nervous system is not invaded, and the infection usually resolves itself quickly. In 1 to 2 percent of *Poliovirus* infections, meningitis can result, but this is also usually resolved quickly. However, poliomyelitis is considered a severe disease because it can be life-threatening. In these severe cases, poliovirus infects the nerves of the spine or the brain and causes a flaccid paralysis that may lead to a permanent inability to use limbs or to walk. Among people who develop paralytic polio, 2 to 5 percent of children and 15 to 30 percent of adults die.

3. Both West African sleeping sickness and East African sleeping sickness are transmitted by the bite of the tsetse fly. Both are caused by a protozoan in the genus *Trypanosoma*. West African sleeping sickness is caused by *Trypanosoma brucei gambiense*, while East African sleeping sickness is caused by *Trypanosoma brucei rhodesiense*. East African sleeping sickness progresses more rapidly and does not respond as well to drug therapy as the West African disease does.

4. Although *Naegleria fowlerii* infection is rare, when it does occur, it is almost always fatal.

Self-Test

1. D	4. A
2. C	5. D
3. E	6. C

INDEX

A

Acid fast stain, 49–50
Acidic, 21
Acidophiles, 119–120
Acids, 21
Acne, 484, 486
Acquired immunity, 378
Actinobacteria, 215, 230–231
Adenosine triphosphate, 88
Adhesins, 403
Aerobic respiration, 91
African trypanosomiasis, 536–537
Agglutination, 389, 430–431
Agricultural microbiology, 9
Agrobacterium, 219–220
Alcohols, 451
Aldehydes, 452
Algae, 248
Alkalophiles, 120
Alpha hemolysis, 229
Alphaproteobacteria, 212–213, 217
Alveolates, 252–254
Amebiasis, 505
Amebic dysentery, 505
Ames test, 156–157
Amino acids, 24, 101, 142–143
Ammonification, 291
Amoebae, 250–251
Anabolic pathways, 106–107
Anabolism, 86
Anaerobic respiration, 288
Animal viruses, 274–278
Animalia, 199
Anthrax, 525
Antibiotics, 462, 468, 472–476
Antibodies, 344, 381, 388–391
Anticodon, 141
Antigen–antibody reactions, 430–434
Antigenic determinant, 381
Antigenic drift, 345
Antigenic shift, 346
Antigen-presenting cell, 383–384
Antigens, 381–386
Antimicrobial drugs, 459–476
Antiviral drugs, 469–471
Apicomplexans, 254
Archaea, 237–242
Ascomycetes, 261–262
Aseptic packaging, 314
Atomic mass, 16
Atomic number, 16
Atoms, 14
ATP, 88

Autoclave, 445–446
Auxotrophs, 156

B

B cells, 386–388
Bacillus sp., 228, 403
Bacteria, 62–63, 116–126, 203–206, 212–215, 403–404
Bacterial growth curve, 117
Bacterial vaginosis, 513
Bacteriochlorophyll, 102
Bacteriology, 9
Bacteriorhodopsin, 241
Bases, 21
Basidiomycetes, 262–263
Bdellovibrio, 225
Beta hemolysis, 229
Betaproteobacteria, 213, 220–221
Biguanides, 450
Bilayer, 30
Binary fission, 116
Biochemical oxygen demand, 301
Biofilms, 294, 404
Biogeochemical cycles, 288–292
Bioreactors, 319–320
Bioremediation, 304
Bonds, 17–20
Botulism, 535
Brown algae, 255–256
Brucellosis, 525
Budding, 218–219, 275–276
Buffers, 21, 184

C

Calvin–Benson cycle, 101, 104–106
cAMP, 148–149
Campylobacter, 227
Candida albicans, 515
Capsomeres, 271
Capsule stain, 49–50
Carbohydrates, 22, 24
Carbon, 16
Carbon cycle, 288–289
Carbon fixation, 104–105
Carboxydotrophic bacteria, 220
Carcinogens, 156–157
Cardiovascular system
    anatomy of, 520–522
    diseases of, 523–527
Caries, 502
Carrier proteins, 59
Catabolism, 99–101
Catabolite repression, 148

cDNA, 182–183
Cell, 4, 89–90
Cell wall, 62–64
Cellular respiration, 91–97
Cellulose, 25, 75
Centers for Disease Control, 343
Cephalosporins, 465
Chagas' disease, 527
Channel proteins, 59
Chemical reactions, 20–21
Chemiosmosis, 102
Chemoheterotrophs, 287
Chemolithoautotrophy, 106, 239, 286–287
Chickenpox, 487
Chlamydia, 513
Chlamydiae, 215, 232
Chloroplast, 72
Cholesterol, 30
Chromista, 255–256
Ciliates, 253
Clostridium sp., 227, 312
Coastal waters, 298
Codons, 139
Commensalism, 330
Common cold, 496
Complement, 365–369, 405
Complex carbohydrates, 25
Compound microscope, 40
Conjugation, 159–160
Conjunctivitis, 487
Constitutive genes, 146
Co-repressor, 150
Corynebacterium, 230
Coulter counters, 125
Covalent bonds, 17–18
Coxiella, 223
Crenarchaeota, 238–240
Creutzfeldt-Jacob disease, 279–280
Cryptosporidiosis, 505
Culturing, 123–124
Cyanobacteria, 212, 216
Cyclosporidiosis, 505
Cytokines, 364
Cytomegalovirus, 514
Cytoplasm, 67
Cytoskeleton, 74–75

D

Deamination, 291
Dehydration synthesis, 22–23, 25
Deltaproteobacteria, 214, 224–226
Denature, 24

Dengue fever, 526
Denitrification, 291
Diarrhea, 502–503
Diatoms, 255
Diffusion, 60
Digestive system
    anatomy of, 500–501
    diseases of, 501–505
Dilution factor, 125
Dinoflagellates, 253
Diphtheria, 494
Diplomonads, 251
Disaccharides, 25
Disease, 4, 325–334. *See also*
    Infections; *specific disease*
Disinfection, 443
Disk diffusion method, 448–449, 471
Dissimilation plasmids, 162
DNA, 27–28, 67, 132–137, 277
DNA polymerase, 135–137
DNA probes, 183
DNA sequencing, 175, 185–187
DNA vaccines, 422
Domain, 200

E

Ebola hemorrhagic fever, 526
Ehrlich, Paul, 7
Electron(s), 14
Electron microscopy, 45–46
Electron shells, 17
Electron transport chain, 95–97
Electronegativity, 96
Elements, 15–16
Encephalitis, 533, 535–536
Encephalomyelitis, 533
Endocarditis, 523–524
Endocytosis, 61, 275
Endomembrane system, 70–72
Endospore stain, 49–50
Endospores, 68–69
Endotoxins, 410
Enteric bacteria, 223–224
Enterococcus sp., 229
Enzyme-linked immunosorbent assays, 433–434
Enzymes, 86–87
Epidemics, 350–351
Epidemiology, 341–343
Epsilonproteobacteria, 214, 227
Equilibrium, 22
Erysipelas, 484
Escherichia coli, 512
Estrogen, 30

Ethylene oxide, 452
Euglenids, 252
Eukaryotic cells, 8, 56–57, 69–75
Euryarchaeota, 240–242
Exocytosis, 61
Exons, 182
Exotoxins, 406–409
Extracellular matrix, 75
Extreme acidophiles, 242
Extreme halophiles, 241
Extremozymes, 321
Eye diseases, 487

F

Facilitated diffusion, 60
Fats, 28–29
Fatty acids, 29, 32
Fermentation, 97–99, 310–311
Fever, 364–365
Fimbriae, 66
Firmicutes, 214–215, 227–230
Flagella, 65–66, 74
Flagellar stain, 49–50
Fleming, Alexander, 7
Fluorescent-antibody tests,
    432–433
Food poisoning, 502
Food preservation, 312–314
Food production, 315–317
Foodborne disease, 311–312
Foraminifera, 253
Freshwater, 295–297
Functional groups, 22
Fungal infections, 515
Fungi, 199, 248, 259–263
Fusion, 275

G

Gamma hemolysis, 229
Gammaproteobacteria, 213–214,
    222–224
Gas gangrene, 525
Gastroenteritis, 503, 504
Gel electrophoresis, 184–185
Gene regulation, 144–150
Gene transfer, 157–162
Generation time, 117
Genetic engineering, 187–189
Genetically modified organisms,
    174
Genital warts, 514
Genomic libraries, 178–179
Genomics, 190
Giardiasis, 505
Glucose, 25, 59
Glucose effect, 148
Glycocalyx, 64–65
Glycolysis, 91–93, 97
Gonorrhea, 513
Gram stain, 47–50, 63
Gram-negative/positive bacteria,
    63–64

Green algae, 257
Group A streptococci, 485
Group translocation, 61

H

Haemophilus, 224
Halogens, 450–451
Halophiles, 120–121, 240–241
Hantavirus, 349–350, 526
Haptens, 382
Heat sterilization, 446
Helicase, 134–135
Helicobacter, 227
Helper T cells, 383, 386
Hemolysins, 229
Hepatitis, 504
Herd immunity, 350–351,
    423–424
Herpes, 486, 514
H1N1, 348
H5N1, 347–348
Hooke, Robert, 3, 7
Host defenses, 405–406
Hydrogen atom, 17–18
Hydrogen bond, 18
Hydrogen-oxidizing bacteria,
    220
Hydrolysis, 22–23
Hydrophilic, 30, 58
Hydrophobic, 30, 58
Hydrothermal vents, 298
Hyperthermophiles, 238–240
Hypertonic, 60
Hypotonic, 60

I

Immune response, 391–392
Immune system, 379–381
Immunity, 378, 406, 419
Immunizations, 422–428
Immunoglobulins, 388–391
Immunology, 9
Impetigo, 484
Inclusion bodies, 67
Inducible genes, 146
Induction, 146
Industrial microbiology, 9,
    318–321
Infections, 330–334, 340–341,
    402. *See also* Disease
Infectious mononucleosis,
    526
Inflammation, 362–364
Influenza, 344–348, 496
Interferons, 369–370, 471
Interleukins, 385
Introns, 182
Ion, 19
Ionic bond, 19
Iron-oxidizing bacteria, 221
Isotonic, 60
Isotopes, 16

J

Jenner, Edward, 7, 418–419

K

Kinetoplastids, 251
Kingdom system of
    classification, 199
Koch, Robert, 7, 327–328
Kreb's cycle, 93–94, 97, 100

L

Lac operon, 147
Lactic acid fermentation, 98
Lactobacillus, 228–229
Latex agglutination test, 203
Legionella, 222
Leishmaniasis, 527
Leprosy, 535
Light microscopy, 43–45, 46–49
Linnaeus, Carolus, 198
Lipases, 100
Lipids, 22, 28–29, 99–100
Lipopolysaccharides, 64
Lister, Joseph, 7
Listeria sp., 230
Lyme disease, 525
Lymphatic system
    anatomy of, 520–522
    diseases of, 523–527
Lysogenic cycle, 273–274
Lysosome, 70
Lytic cycle, 272–273

M

M protein, 405
Macrolides, 467–468
Macromolecules, 22–25, 28–32
Macronutrients, 122
Magnetosomes, 67
Major histocompatibility
    complex, 383
Malaria, 527
Marine environments, 297
Mass number, 16
Measles, 487
Medical microbiology, 9
Membrane attack complex, 367
Meningitis, 533–534
Mesophiles, 118–119
Metabolic pathways, 87–88
Metabolism, 86–107
Metchnikoff, Ilya Ilich, 7
Methanogens, 240, 242, 289
Metric units, 41
Microbes
    classification of, 195–206
    control of, 442–453
    description of, 1
    discovery of, 3–4
    ecosystems of, 286–288
    in food production, 315–317
    functions of, 3

habits of, 295–298
    metabolism, 86–107
    normal, 2
    pathogenicity of, 401–410
    types of, 8
    water quality and, 299–300
Microbial ecology, 9
Microbiology, 8–9
Microbiota, 329–330
Microenvironments, 293–294
Microfilaments, 75
Micronutrients, 122
Microscope, 3–4, 40
Microscopy, 41–42, 45–46
Mitochondria, 72
Molecular biology, 174–190
Molecules, 17, 22–32
Monoclonal antibodies, 428–429
Monomers, 22–23
Monosaccharides, 24
Mumps, 504
Mutagens, 151–152
Mutations, 150–157
Mutualism, 330
Mycobacterium, 230
Mycology, 9
Mycoplasma sp., 228

N

NAD/NADH + H⁺, 88–89
Naegleria, 537
Natural selection, 150
Necrotizing fasciitis, 484
Needham, John, 6
Negative selection, 155
Neisseria, 221
Nervous system
    anatomy of, 532–533
    diseases of, 534–537
Neuraminidase inhibitors, 470
Neutrons, 14
Nicotinamide adenine
    dinucleotide, 89
Nitrification, 291
Nitrifying bacteria, 220
Nitrogen cycle, 290–291
Nitrogen fixation, 3
Nitrogen-fixing bacteria, 219,
    290–291
Nosocomial infections, 334
Nucleic acid hybridization,
    204–205
Nucleic acid sequencing, 204
Nucleic acids, 22. *See also* DNA;
    RNA
Nucleotides, 27
Nucleus, 69

O

Oils, 28
Oomycetes, 255
Operons, 138, 146–147, 149

Opsonization, 367, 389
Organelles, 69, 72–74
Osmosis, 60
Osmotic pressure, 121
Otitis media, 494
Oxidative phosphorylation, 95–97
Oxygen, 121–122

P
Pandemics, 342
Parabasalids, 251
Parasitism, 330
Parasitology, 9
Passive immunity, 378
Pasteur, Louis, 6–7, 269
Pasteurization, 445
Pathogens, 190
Pathology, 328–329
Penicillins, 465–466, 475
Peptic ulcers, 503
Peptidoglycan, 48, 63
Periodic table of elements, 15
Periodontal disease, 502
Periplasm, 64
Peroxisomes, 71
Peroxygens, 452
pH, 21–22, 119–120
Phage typing, 204
Phagocytes, 64
Phagocytosis, 249, 360–362,
   405–406
Phenols, 449
Phospholipids, 30–31, 58
Photoheterotrophs, 287
Photophosphorylation, 101–105
Photosynthesis, 3, 72, 101–105,
   287
Phycology, 9
Phylogenetic trees, 197
Phylogeny, 196–197
Pili, 66
Plague, 525
Plantae, 199
Plasma membrane, 58–61
Plasmids, 67
Plasmolysis, 121
Pneumonia, 494–495
Point mutations, 152–153
Polar covalent bond, 18
Poliomyelitis, 536
Pollution, 300
Polyclonal antibodies, 428–429
Polymerase chain reaction, 175,
   179–181
Polymers, 22–23
Polypeptide chains, 24
Polysaccharides, 25–26, 64
Porins, 64
Primary amebic
   meningoencephalitis, 537

Prions, 279, 453
Probe, 183, 205–206
Products, 20
Prokaryotae, 199
Prokaryotic cells, 8, 56–57,
   61–69
Propionibacterium, 231
Protease inhibitors, 470
Proteins, 25, 74–75, 101,
   139–144
Proteobacteria, 212–214,
   216–227
Protista, 199, 246–260
Protons, 14
Protozoology, 9
Pseudomonas, 223, 453
Psychrophiles, 118–120
Purple nonsulfur bacteria, 217,
   222
Pyruvate oxidation, 93–94, 97

Q
Quinolones, 464, 468
Quorum sensing, 294

R
Rabies, 536
Radiation, 314–315, 447
Reactants, 20
Recombinant DNA, 174–175,
   176–177, 187–190
Red algae, 256–257
Redi, Francesco, 5
Redox potential, 96
Replica plating, 155–156
Repressible genes, 146
Repression, 146, 149–150
Reproductive system
   anatomy of, 511–512
   diseases of, 513–515
Resistance factors, 160
Resolution, 41
Resolving power, 41
Respiratory syncytial virus,
   496
Respiratory system
   anatomy of, 492–493
   diseases of, 494–496
Restriction enzymes, 175–176
Reverse transcriptase, 182
Ribosomal RNA, 199
Ribosomes, 67, 69
Ribozyme, 144
Rickettsias, 217–218
Rifamycins, 469
RNA, 27, 137–139, 199
Rocky Mountain spotted fever,
   525
Rough endoplasmic reticulum, 70
Rubella, 486

S
Salmonellosis, 502
Saturated fats, 29
Scarlet fever, 484
Self-tolerance, 393
Semmelweis, Ignaz, 7
Sepsis, 523
Serial dilution, 125
Serial endosymbiotic theory, 73
Serological techniques, 203
Sewage treatment, 301–303
Sheathed bacteria, 221
Shigellosis, 502
Shingles, 487
Simple diffusion, 60
Simple microscope, 40
Skin
   anatomy of, 358, 482
   diseases and lesions of,
      483–487
Sleeping sickness, 536–537
Slime molds, 257–259
Smear, 47
Smooth endoplasmic reticulum,
   70
Sodium, 19
Soil, 295
Spallanzani, Lazzaro, 6
Species, 201–202
Specimens, 46–49
Spectrophotometer, 126
Spirillum, 221
Spirochaetes, 215, 233
Spontaneous generation, 5–6
Sporogenesis, 68
Stains, 47–50
Staphylococcus sp., 228, 405,
   485, 523–524
Sterols, 30–31
Strains, 202
Strep throat, 494
Streptococcus sp., 229, 485,
   523–524
Streptomyces sp., 231
Substrate-level phosphorylation,
   93
Substrates, 86–87
Sulfate-reducing bacteria,
   225–226
Sulfonamides, 463
Sulfur cycle, 291–292
Sulfur-oxidizing bacteria, 221
Sulfur-reducing bacteria,
   225–226
Superantigens, 409
Symbiosis, 330
Syphilis, 513

T
T cells, 383–386

Taq polymerase, 185–187
Taxonomy, 196
Tetanus, 534
Tetracyclines, 468
Therapeutic index, 462
Thermophiles, 118–120, 242
Thylakoids, 73
Toxic shock syndrome, 513
Toxins, 406–410
Toxoplasmosis, 527
Trachoma, 487
Transcription, 137–138
Transduction, 160
Transfer RNA, 139, 141
Transformation, 158–159
Translation, 139–144
Transmission of infections,
   332–333
Transport proteins, 59
Trichomoniasis, 515
Trypanosoma, 536–537
Tryptophan, 150
Tuberculosis, 495
Tularemia, 525
Typhoid fever, 502
Typhus, 525

U
Unsaturated fats, 29
Urinary system
   anatomy of, 510–511
   diseases of, 512
Urinary tract infections, 512

V
Vaccination, 418–419, 424–428
Vaccines, 419–422, 424–428
Vacuoles, 71
van Leeuwenhoek, Anton, 4, 7,
   327
Vectors, 178
Vibrio, 223, 341
Virions, 270
Viroids, 279
Viruses, 8, 267–280, 453
Vitamin D, 30

W
Warts, 487, 514
Wetlands, 300
White blood cells, 359–360
Whooping cough, 495

Y
Yellow fever, 526

Z
Zoogloea, 221
Zygomycetes, 260–261